The State and Agriculture in Africa

THE STATE AND AGRICULTURE IN AFRICA

EDITED BY
NACEUR BOURENANE
THANDIKA MKANDAWIRE

CODESRIA BOOK SERIES

The State and Agriculture in Africa

First published in September 1987 by
CODESRIA BOOK SERIES
364A Regents Park Road
London N3 2LJ

Copyright © CODESRIA

CODESRIA is the Council for the Development of Economic and Social Research
in Africa head-quartered in Senegal. It is an independent organisation whose principal
objectives are facilitating research, promoting research-based publishing and creating
multiple fora geared towards the exchange of views and information among African
scholars. Its correspondence address is B.P. 3304, Dakar, Senegal.

ISBN 1 870784 00 6 (Paperback)
    1 870784 01 4 (Hardback)

Cover designed by Simon Aquah
Typeset and designed by Grassroots, London N3 2LJ
Printed by Whitstable Litho Printers Ltd, Whitstable, Kent
Production organised and co-ordinated by Grassroots Book Production (01-346 8276)

ACKNOWLEDGEMENTS

This volume is part of the continuing work by CODESRIA on African agriculture. It consists of selected and revised papers from the work of a CODESRIA Multinational Working Group on the State and Agriculture in Africa which were presented at a conference held in Addis Ababa in 1984.

CODESRIA would like to express its gratitude to the Swedish Agency for Research and Economic Cooperation (SAREC) for financial support during the preparatory phase of this project and to the Danish Agency for International Development (DANIDA) for support of the research activities and publication of the research results.

CODESRIA also wishes to thank the Institute for Development Studies of the Addis Ababa University, the host institute for the conference, and the Economic Commission for Africa for extending to us its conference facilities and for provision of valuable background material to the participants.

Thandika Mkandawire
Executive Secretary

CONTRIBUTORS

Tahani Abdelhakim
Is senior researcher at the Institut Agronomique Méditerranéen in Montpellier, France.

Slimane Bedrani
Is Director of the Centre de Recherches en Economie Appliquée pour le Développement and Professor at the Institut des Sciences Economiques of Algiers University.

Abdel Bouami
Is professor of economics at the Faculty of Law and Economics of the University of Casablanca, Morocco. His main research interest is in food and agricultural development.

Naceur Bourenane
Was formerly senior researcher at the Centre de Recherches en Economie Appliquée pour le Développement, Algiers University. His main research interests are on rural development, agro-industry, and the problems of social reproduction.

Kary Dembele
Is presently professor of sociology at the Ecole Normale Supérieure de Bamako, Mali. Formerly he was Deputy Director of the Institut de Productivité et de Gestion Prévisionnelle (IPGP) and head of its Department for Human Resources Management.

Emmanuel Hansen
Was formerly senior lecturer in political science at the University of Ghana, Legon. He has published widely in the areas of political thought, the state and the military. He has taught in universities in the United States and Britain. His current interests are in the area of the state and food agriculture in Africa. He is currently Visiting Research Fellow at the Institute of Commonwealth Studies, University of London.

Patricia McFadden
Is a lecturer at the Institute of Development Studies, University of Dar-es-Salaam. Her interests are mainly in gender issues, liberation struggles and agri-business.

Guy C.Z. Mhone
Was lecturer in the Department of Economics, University of Zimbabwe at the time of writing his contribution. He taught at the State University of New York (Oneanta)

Lucian Msambichaka
Is an associate research professor and Director of the Economic Research Bureau of the University of Dar-es-Salaam. He is a co-author of *Agricultural Development in Tanzania; Policy Evolution, Performance and Evaluation. The first two decades of Independence.* His main research concern is in food policy and agricultural development.

Thandika Mkandawire
Is Executive Secretary of CODESRIA. His research interests include state policies, technology and development as well as food and agricultural development.

Mohamed Raki
Is professor of rural economy at the Institut Agronomique et Vétérinaire Hassan II, Rabat, Morocco. His research concern is in food and agricultural development.

Emmanuel C.W. Shula
Is research fellow at the Rural Development Studies Bureau, University of Zambia. His interests cover development administration and all aspects of underdevelopment and policy performance. His research work has involved the rural subsistence sector, rural cooperatives, integrated regional projects, women projects and out-of-school youth projects. Currently he is coordinating a growth and and linkage study in the Eastern Region of Zambia for IFPRI, NFNC and RDSB.

Thomas Shopo
Has been a senior research fellow at the Zimbabwe Institue of Development Studies since 1983. He is a former lecturer of economic history at the University of Zimbabwe, and has also worked in the Research Unit of the Parliament of Zimbabwe.

Taladidia Thiombiano
Is professor of economics at Ouagadougou University and Director of both the Centre d'Etudes de Documentation, de Recherches Economiques et Sociales (CEDRES) and the Ecole Supérieure des Sciences Economiques. His research interests include the financing of development, quantitative techniques in economics as well as agriculture and agro-industry.

Sunday T. Titilola
Is a research fellow in agricultural economics at the Nigerian Institute of Social and Economic Research (NISER), Ibadan, Nigeria. He has authored several articles on Nigeria and Africa's food "crisis". His research interests range from food and rural development policies to macro-economic policies, food production and rural development.

Raji Virashawmy
Is senior lecturer in economics at the University of Mauritius, author of several articles on the Mauritian and African economies. His research interests are in cultural values and economic development.

Adrian Wood
Is a senior lecturer in the Department of Geographical Sciences at Huddersfield Polytechnic where he specialises in development issues. He taught geography at the University of Zambia from 1975 to 1982 and was a senior research fellow at that University's Rural Development Studies Bureau from 1983 to 1985.

CONTENTS

THE STATE AND AGRICULTURE IN AFRICA: INTRODUCTORY REMARKS*

Thandika Mkandawire

INTRODUCTION

It is now more than two decades since much of Africa became independent. The immediate post-independence period was characterised by high expectations about improvement in the living conditions of the least privileged of the African population (principally the peasant agricultural producers), the reduction of dependence in the management of national economies, and the consolidation of nationhood in nations plagued by various forms of heterogeneity and conflict—social, ethnic, religious, racial etc. Reality was to confound most of these expectations as the performance of African economies and political systems failed to live up to them. In close to half of the African economies, growth of the Gross Domestic Product has been negative during the last two decades. Even in those cases where positive rates of growth have been achieved the growth has either been negated in per capita terms by high rates of population growth or has excluded large sections of the population.

Of greater salience is the fact that for a continent in which more than 70 percent of the labour force finds its livelihood in agriculture, Africa has performed poorly in agriculture in general and specifically in food production. Africa is the only continent which has experienced a decline in per capita food production during the last two decades. Self-sufficient in food production at independence, Africa is today a net importer. This agrarian crisis has been rudely brought to world attention by recent horror films of famine victims.

It should, however, be emphasised that this general image of agrarian stagnation is misleading since it conceals profound processes of change taking place in African agriculture. It also glosses over vast

* The author would like to thank N. Bourenane, A.S. Bujra, Sam Moyo, Tom Shopo and Zene Tadesse for comments on earlier drafts of this paper. The usual caveat about responsibility for errors holds.

differences in performances among classes of producers, among commodities, among countries and among regions within the same country. More significantly, the image conceals historical processes of social differentiation induced by and reinforcing the differential access to resources that are provided by state and international agencies to African agriculture.

Most of the activities and relief programmes spawned by the famine have tended to treat the current crisis as episodic, reflecting ecological disasters. Although these disasters have played a far from insignificant role in the generation of the present crisis, social factors have been decisive even in those cases where ecological disasters would seem, at first sight, to have been the sole cause. In recognition of this, CODESRIA has organised four research networks to deal with the role of the four major actors in African agriculture, namely, the state, aid donors, transnational capital and the peasant household. The present collection of articles is confined largely to the role of the state.

This division of the theme has its pitfalls. First, it is unduly "statecentric" and ascribes to the state a degree of autonomy that is not congruent with its relationship with the actors in the national and international economy. Nevertheless, while foreign actors set out the global parameters within which the state acts, whether the state is populist or not, agrarian based or not, depends to a large extent on internal factors—the configuration of social forces within the country, the resource endowment of the country, the nature of the struggle for independence etc.

TOWARDS A TYPOLOGY OF AFRICAN ECONOMIES

The collection brings together case studies based on a common framework for examining the nature of the agrarian crisis. Designed so as not to straitjacket the debate, the framework was intended merely to impart a minimum of coherence and comparability to the collective enterprise without glossing over or in any way obscuring the irreducible specificity of each national experience. To achieve this, it was important to maintain contacts among the researchers in the group through the coordinators of the group, correspondence, exchange of material, workshops and conferences. Preliminary versions of these papers were presented at a conference of researchers and policymakers held in Addis Ababa in 1984. As will be clear, the collection embraces a wide range of countries differing in complexity, accessibility of information and research infrastructure. Undoubtedly some of the differences are reflected in the unevenness of the collection. However, the collection does show the fruitfulness of continent-wide studies even when the studies are informed by different intellectual traditions and theoretical perspectives.

Seventeen countries were identified for case studies taking into account different ecological conditions, types of decolonisation, social and cultural aspects of the countries, the nature of the states and the strategies or economic policies pursued by them in different countries. The choice also took into account the revenue base of the countries considered. The states discussed in the volume are classified as either "merchant" or "rentier" states, depending on their revenue base i.e. if, as in the case of the former they depend on surpluses extracted from agriculture through various taxes and levies or if, as in the case of the latter, they obtain their revenue in the form of rent from the country's mining activities. In the former category, there are Ghana, Burkina Faso, Morocco, Malawi, Mali, Mauritius, Tanzania, Swaziland and perhaps Egypt. In the latter we have Algeria, Nigeria and Zambia. Zimbabwe is an intermediate case where mineral and agricultural exports play equally significant roles as sources of foreign exchange earnings, if not state revenues. It will be clear from the studies in this book that the "merchant" state, heavily dependent on agricultural surpluses, bears a different relationship to agriculture from the mineral-rich "rentier" state.

As is clear from these studies, the state has played a wide range of roles in African economies. In cases where the main thrust has been the destruction or marginalisation of peasant production, the state has either acted as the mediator of capitalist penetration or directly entered production. Where the desire has been to carry out agrarian reform through greater peasant production, the state has either been interventionist in rather paternalistic and commandist ways or has sought to mobilise peasants toward self-determination and participation (cooperatives, self-management etc.).[1] Which of these roles the state ultimately plays depends on a wide range of factors—the revenue base of the state, forms of colonisation and decolonisation, configuration of local class forces, the extent of foreign capital penetration of the economy, the degree and form of peasant mobilisation during and after the struggle for independence, the global economic conjuncture etc.

To take into account these different roles, we have had to introduce another typological characteristic, viz. whether the state policy is "capitalist" or "populist/nationalist/etatist." Here again there are some ambiguities. First, a number of countries have changed strategies in the course of time. Second, the classification blurs some rather significant differences among countries. Thus placing Algeria and Zambia together might seem like a procrustean use of concepts. However, we have only used extensive state ownership of means of production as the main criterion. Obviously the historical reasons for such ownership have important implications on the overall political economy of the country. Algeria's national liberation struggle and its popular base led to the adoption of a populist and fiercely nationalist position. Zambia, with a different process of decolonisation, began with a liberal, pro-foreign investment policy. When no such foreign investment was forthcoming and when, indeed, repatriation of profits was the order of the day, the government was compelled to assume a more active role in the economy to stem the outflow of surplus and to push up the rate of investment. The "leftist" twist to this basically nationalist programme was provided by the relatively well-organised working class in the country's mining industry.

INTRODUCTION. 1		
	Oil & Mineral "Rentier" States	"Merchant" States
Capitalist	Nigeria (1973-) Zimbabwe	Malawi, Mauritius Morocco, Swaziland Egypt (1970-) Zimbabwe Nigeria (1960-72) Ghana (1966-) Sahel Burkina Faso Mali (1969-)
Nationalist/etatist/ Populist	Algeria Zambia	Tanzania (1965-) Mali (1960-68) Ghana (1961-66) Egypt (1951-70)

Combination of these characteristics yields the matrix in Table Introduction 1. Such classificatory schemes are always problematic and, as we have indicated above, are full of ambiguities. They do, however, provide useful heuristic and organisational tools and if they are taken in this spirit, and not as rigorous conceptual frameworks, they serve a useful purpose.

THE "MERCHANT" STATES

Ghana
Ghana is a classical "primary export-oriented" economy come to

grief. The successful and rapid introduction of a cash crop (in this case cocoa) and the generation of enormous profits by merchant capital and the colonial state without any prior inflow of capital into the sector gave credence to the view that trade provided African countries with a "vent for surplus" which could be extracted quite cheaply without disrupting other economic activities. Indeed, apologists of colonial exploitation elaborated models in which the indigenous population were unburdened of unwanted "leisure" through trade. However, as several studies (including Hansen's chapter in this book) show, this was merely a convenient myth to rationalise colonial exploitation. The fact of the matter is that the production of cocoa did involve redeployment of productive resources (labour and land) toward export crops away from other crops such as food crops. Hansen recounts Ghana's struggles, at least at the level of rhetoric, to alter this pattern of resource allocation and stimulate food production. Every attempt since Nkrumah to increase food production has come to nought. Common to all these attempts is the assignment of a passive role to the peasantry. Nkrumah sought to modernise agriculture through state farms. Busia was to do this by encouraging large-scale commercial farmers. The Acheampong regime, with its "Operation Feed Yourself", was to do it by exhortation and encouragement of large-scale commercial farmers. Of even greater salience was the state's attempt to involve transnational corporations in food production in order to supplement its "Operation Feed Yourself" programme. In the event, the programme failed to attract TNC investment in food production despite seemingly attractive investment conditions. Limann was to seek to further entrench large-scale capitalist farming. In the early 1980s, the "populist" Jerry Rawlings regime was apparently continuing with the Acheampong model.

It is important to note that ever since the overthrow of Nkrumah, the state has tried to nurture a rural capitalist class out of bureaucrats, military commanders etc. Hansen considers such an attempt illusory. First, a number of these social groups have only half-heartedly entered agriculture and have continued to maintain their other positions in the economy so that most of them are absentee farmers. Second, corruption has been so rampant that funds provided for agriculture on favourable terms have been diverted to speculative activities by these elements.

Hansen is clearly in favour of a strategy in which the majority of agricultural producers—the peasants—are central at the level of formulation and implementation. However, if Ghana must have a successful reorganisation along capitalist lines, then those capitalists must be from the farming community itself:

To think that one could build up a rural capitalist class from the ranks of the petty bourgeoisie presently located in the bureaucracy, the army or the distribution sectors of the economy is a socio-economic illusion. This can only be done from the ranks of the farming community but so long as they do not have any influence or control over the state agencies responsible for making available the capital for such purpose it is difficult to see it happening.

The conditions under which the emergence of an agrarian capitalist class is possible remain a moot point in the African agrarian question. However, there are, at least according to the World Bank, some "success stories" of capitalist agrarian transformation in Africa—two of the most commonly cited examples (Malawi and, more recently, Zimbabwe) are discussed here.

Malawi

The case of Malawi provides a sharp contrast to both Ghana's pre-and post-independence experiences.

Unlike Ghana which witnessed a peasant export "boom" in the colonial era, Malawi never experienced such a phase. This was due to a deliberate colonial policy to maintain Malawi as a "labour reserve" for the South African and Rhodesian plantations and mines. This "labour reserve" role required that peasants should not have any sources of income outside wage labour in the white plantations and mines. The nationalist movement was bent on reversing this role by creating gainful employment within Malawi itself.

After trying to pursue a smallholder-based agricultural strategy during the country' first five years of independence, the Malawi government moved toward a trimodal system consisting of large-scale capitalist farmers, small "kulak" farmers known as "Achikumbe" and peasant household farmers. Mhone demonstrates that the development strategy, quite outspokenly capitalist, was based on specific and logical relationships between these three sectors. The dynamic sectors were to be the large-scale commercial farmers heavily financed by surpluses extracted from peasant farmers through nonequivalent forms of exchange orchestrated by various parastatals, especially the Agricultural Development and Marketing Board (ADMARC). The low returns to peasant labour would also perpetuate the "labour reserve" function of the small peasant household, only this time the labour would be used by capitalists within Malawi. Overseeing all this process is a highly personalised authoritarian regime whose entire leadership is heavily involved in agriculture.

Malawi, often touted as "a success story" of capitalist agriculture, indicates some of the socio-economic preconditions for the emergence of a link between the state and an agrarian capitalist class.

Swaziland

McFadden's study brings to the collection an economy which probably has, relative to anywhere else in Africa, the highest involvement of agri-business in it agriculture. Indeed, Swaziland's agriculture epitomises "bimodalism" taken to its extreme. While communal tenure (Swazi Nation Land) covers 56 percent of total land area it contributes only 12 percent of total crop production. Individual tenure farms, on the other hand account for 43 percent of Gross Domestic Product and three quarters of total exports. Hence any discussion of state policy and agriculture must touch upon upon the relationship between the state and agri-business.

Like Malawi, Swaziland was a "labour reserve" economy in which peasant agriculture was ignored or blocked. Colonial economic policy toward "native" agriculture was designed largely to force the population into wage labour in the mines and "white" farms and to allow for the reproduction of labour outside the capitalist system itself. With respect to the latter, female labour, in the absence of males, was central to the reproduction of labour for the capitalist sector.

McFadden argues that state policy in the post-independence era has accepted this premise with the only change being intensification of agri-business penetration of agriculture. In addition, proletarianisation is increasingly within Swaziland itself and not externalised to South African mines. The introduction of "outgrowers" schemes attached to TNC-dominated "nucleus" farms has introduced a new dimension in capital's subsumption of the labour process in Swaziland.

The state's main role has involved provision of land, and ensuring availability of a docile labour force through repressive labour laws. The remarkable thing about Swaziland is how the "traditional" political system has been harnessed to the state apparatus to discipline labour. "Misconduct" in the capitalist sector is penalised by traditional authorities who impose sanctions on workers within the "traditional" sector, e.g. the "right" to land can be nullified by the king for causing trouble to capital. McFadden stresses the importance of the alliance between the traditional monarchy, the neocolonial state and captial in defining agrarian policy in Swaziland.

Tanzania

Few countries' pronouncements on rural transformation have attracted as much attention as those of Tanzania. As a consequence the country has received substantial material and financial support for its development programme. Judgment on what has been the outcome of the experience varies widely. However, with respect to one aspect of this experiment—increased agricultural production through modernisation—there is near-unanimous agreement that Tanzania has failed. State officials have tended to blame this failure on unfavourable climatic conditions. Msambichaka argues that the failure "lies solely in the perverse agricultural policies which have been pursued throughout the decades".

First, policies pursued have continued to exhibit the same commodity bias as those of the colonialists. In terms of virtually every input—land, labour, extension services, infrastructure, credit—state policies have favoured export crops. And yet the performance of this favoured sector has been poor. The culprit here, Msambichaka argues, is state pricing policies which have been more designed to extract surplus from agriculture than to ensure its expansion.

In response to adverse state policies, peasants are increasingly withdrawing from the official markets as they rely on parallel markets in which they receive higher prices.

Msambichaka argues against the assumption that food production can be assured by primitive methods of production. He pleads for the introduction of new seeds, gradual mechanisation and increasing reliance on small-and large-scale irrigation for water.

Msambichaka's chapter and the "self-evident" character of his proposals raises the question as to why a state putatively dedicated to rural transformation could pursue policies that ultimately led to the deterioration or both farm and nonfarm incomes.

Mauritius

With Virashawmy's chapter on Mauritius we shift to a country which further underlines the importance of the specificities of agrarian policies in Africa. First, we have the insular character of the country. Second, we have the fact that the country is peopled by different racial and ethnic groups none of which were "natives" to the island. Third, we have plantation agriculture as the dominant mode of organisation. Fourth, local agrarian capitalist interests financially, managerially and technically dominate the production of sugar, the major plantation crop. Fifth, the presence of transnational capital is largely confined to the marketing of sugar. Sixth, as a result of the configuration of class forces, Mauritius has a basically social-democratic structure. And finally, there is the island's relatively poor resources base, making the country extremely dependent on one crop despite the progress made in diversifying the export base through a shift toward "non-traditional" exports.

One major characteristic of plantation economies is the imposition by the state of measures that tend to stifle the emergence of an indigenous merchant or industrial class outside the plantocracy itself. The state, to safeguard the interests of the plantocracy, will refuse to protect a nascent, import-substitution industrialisation for the simple reason that these costly industries will turn the terms of trade against agriculture and may compete with agriculture for scarce resources— labour and capital.

However, over time, the exigencies of accumulation and the uneven development within the plantation sector itself will bring in other social forces and factions that might opt for other accumulation trajectories. The historical experience of Mauritius bears out this pattern. The plantocracy has had to divest itself of some of the state power and share it with other social groups. As a result of this, agricultural policy has had to take into account the interests of these new groups. Some import-substitution industrialisation, including food production, has been embarked upon. However, with the fall of world food prices, the populist coalition in favour of import-substitution industrialisation has had to succumb to IMF and World Bank pressures for "structural adjustment" in favour of export orientation and tourism.

Morocco

A former French protectorate, Morocco never experienced the same kind of settler colonial exploitation as its neighbour Algeria. The hierarchical social structure of traditional Morocco was maintained albeit at the service of the colonial power. In 1954 only 12 percent of land was controlled by colonial settlers (as compared to 35 percent in Algeria). Consequently, the achievement of independence was translated into an attempt to modernise agriculture, while safeguarding the colonial social heritage and without questioning the relations of production and the dominant social relations.

After initial hesitation as to what agricultural policy it would adopt, the state decided to promote and develop agricultural entrepreneurs who would be capable of stimulating modernisation of the rural sector and ensuring the development of an essentially export-oriented agriculture.

Through measures such as credits, prices and irrigation equipment the state sought to encourage private investment in irrigated and irrigatable areas. This was facilitated by the high prices for phosphates and later by international borrowing. The result was a spectacular growth in the production of agricultural exports. However, this was achieved at the expense of peasant production. Located in non-irrigated areas, they could not benefit from public investments in irrigation. In addition, while export agriculture grew rapidly, the same was not true of traditional food crops (especially cereals). To meet domestic demand and stave off social explosions, the state has had recourse to the international market for imports of cereals, incurring enormous debts in the process. Today the state wishes to adopt corrective measures. The question that immediately arises is whether the state has the means to carry out these reforms in the light of the unfavourable conjuncture— declining prices of phosphates, problems of export markets for agricultural products and threats of protectionism from the European Community. The entry of Spain, Portugal and Greece will further increase pressures for protection for precisely those commodities that Morocco exports to Europe.

Egypt

In our classification of countries, Egypt appears under two categories characterising changes in the nature of the state and state policy over a period of 30 years. Abdelhakim reflects this periodisation.

Egypt provides us with a historically distinct situation in agriculture. Its agriculture is almost entirely dependent on irrigation and concentrated on cotton. The country's history differs from that of the other countries treated in the volume in the sense that even prior to the dethronement of the monarchy and the termination of the British mandate, Egypt had a full-fledged landed aristocracy and a national bourgeoisie.

One outstanding feature of the 1952 revolution was the land reform proclamation that was intended to break the hold of the landed aristocracy, if not in the economy as a whole, at least in agriculture. Before the revolution, 2,000 landowners owning 200 feddans or more and constituting 0.1 percent of those owning land, owned 20 percent of the land. Immediately after the implementation of Act 178 of 1952 which set the limit of land owned by an individual at 200 feddans, the land owned by this group dropped to 5.9 percent. The 1961 law lowered the limit to 100 while the 1969 law was to lower this further to 50 feddans. As for smallholders (less than 5 feddans), before the land reform they constituted 94 percent of landowners but owned only 35 percent of the land. Following the land reform, their share went up to 46 percent reaching 57.1 percent in 1969 only to drop to 49.7 by 1974. The result of all this was a reduction of the concentration of land but with increased fragmentation of land at the other end. Most significantly, however, was that the "radical reform" laws of 1952, 1961 and 1969 tended to benefit the "rich peasantry". Following the Nasser regime, the 50 feddans ceiling was removed and there was a re-emergence of large estates whose share of land jumped from 6.1 percent in 1969 to 16.3 percent in 1974.

During Nasser's time, prior to the export of oil, the state's dependency for revenue on surpluses in agriculture made it an unqualified "merchant state". The major mechanism for extracting surplus was the monopsonistic control of markets for a wide range of products. The new populist regime of Nasser carried out a policy of active state intervention not confined only to mediating the new property and land use relationships in agriculture but including the extension of the material productive base (notably through the Aswan Dam) and a process of integrated industrialisation. In the latter phases, beginning with the Sadat regime, state economic policy was dominated by wholesale reversion to private control.

In addition, ceilings imposed on the size of landholdings during the Nasser regime have been removed and there has been a dramatic return to large landholdings in agriculture.

Why the reversal of the land reform programme? Abdelhakim suggests the nature of the land reform programme did not augur well for a lasting peasant-based agriculture which requires an alliance between the state and the peasantry. Such an alliance must have the support of the peasantry while closing the doors to other possible rival agricultural classes constituted by large commercial farmers. Such a strategy requires some kind of social upheaval that sires a state committed to excluding formations other than the peasantry from any leading role in direct agricultural production. Abdelhakim argues that such was not the case in Egypt where, while the aristocracy may have been eliminated by the land reform, the "rich peasants" emerged as the major gainers from the land reform programme. In addition, greater reliance on the market, especially after the Nasser regime, contributed to the reversal of the gains made by the peasantry in the post-revolution land reform phase. Egypt's experience confirms that in a market economy unless there is concerted political will for the continutation of a peasant-based strategy, it can be reversed quite quickly and the dynamism and the gains of output growth may be hijacked by social classes other than those for whom the reforms were intended. The case of Egypt also points to some of the pitfalls of the "peasant-based strategies" that are being pushed in Africa without taking into consideration the processes of differentiation that market forces unleash and the political basis of a lasting state-peasant alliance.

Zimbabwe
Zimbabwe straddles the resource-base typology we posited above, being both a major agricultural and a mining export economy. Furthermore, Zimbabwe is the only country in which the focus is on the preindependence era. The choice of the period was intended. Increasingly,

the agriculture performance of the UDI period is being cited as a "success story" of capitalist agriculture. Indeed the bimodal character of that country's agriculture is seen more as a model and a justification for the co-existence of large-scale farmers and peasant farmers.[2]

Although the avowed intention of the ruling party is land redistribution, the "success" of the inherited agricultural system has made the government wary of killing the "goose that lays the golden egg".

Shopo's study of the state and agriculture and UDI not only underscores the racist nature of the regime but also highlights the centrality of the state in a wide range of activities pertaining to agriculture. The state provided cheap credit; it searched for markets through sanctions-busting; it ensured the availability of cheap labour through direct repression and maintenance of "labour reserves" (in the form of "tribal trust land"); it protected white commercial farmers from "overproduction" from the peasant producers by segmenting the market; it provided relatively good infrastructure to the land occupied by commercial farmers.

Partly as a result of these state measures, agriculture became lucrative so that the "familial" farming patterns of settler agriculture were gradually being replaced by agri-business.

Shopo's account points to the dilemmas of the post-independence regime. Since its source of political legitimacy differs from that of the state that nurtured the inherited agriculture patterns, the maintenance of the existing system poses severe political dilemmas for the regime. Should the system be merely "Africanised" by encouraging the emergence of a black landowning capitalist class or should the reform process be speeded up? Can the privileges thus far enjoyed by commercial farmers be generalised so as to benefit all farmers, as some of the advocates of "bimodalism" would like us to believe?

In addition there is the inherited relationship between industry and agriculture. Throughout the UDI period the terms of trade favoured commercial agriculture and indeed the share of agriculture in gross profits increased. One effect of this was the low level of surplus transferred from agriculture to industry, a fact that might partially explain the rather low levels of investments in industry for much of the UDI period. Can the present government reverse the "rural bias" of the past allocation of the economic surplus to finance its industrialisation programmes without scaring commercial farmers—a rather pampered lobby in Zimbabwe policy-making?

THE SAHEL

Burkina Faso and Mali, ranked among the "least developed countries" of the world and geographically disadvantaged by being landlocked, are countries in which ecological factors have contributed substantially to the agrarian crisis. However, even in these drought-prone countries, the studies in this book by Dembele and Thiombiano clearly demonstrate that the role of the state, aid and transnational capital in shaping agrarian transformation have been decisive.

Mali

Since independence, state policy toward the economy in general and agriculture in particular has witnessed two major phases. The first phase was that of "Malian socialism" in which the state played a much more interventionist role than in the second phase. During this phase the emphasis was on bureaucratic and authoritarian control over peasant surpluses through a maze of party structures. Among the various aspects of the process, there was compulsory labour on public farms on days usually reserved for working on individual plots, punctuated by the periodic embezzlement by the local bureaucracy of earnings from peasant labour. Here the government tried to mobilise the tattered remnants of the traditional power structure to strengthen its control over agricultural production.

As could be expected, the outcome of all this was low productivity on "collective farms", passive resistance by peasants (through withdrawal from official markets and greater reliance on parallel markets, and migration to the urban areas by the younger population seeking to escape the stranglehold of the state and party). Eventually, the ruling party lost its legitimacy and in 1968 the government was overthrown and a new phase started.

The new phase involved greater liberalisation as "collective farms" were abolished and USRDA structures were dismantled. For a while, there were noticeable increases in output and improvements in the balance of payments. In 1972 a new Five-Year Plan was launched. During this plan 93.6 percent of the funds came from outside in the form of aid and loans. Two interesting results emerged. First, growth in industrial crops was higher than that in food crops. Second, among the latter, "wage foods" grew faster than "peasant foods".

Dembele argues that as a result of policies pursued under these two plans, there has been greater social differentiation in the countryside with the emergence of "pilot or model peasants", bringing along with them new behaviour patterns and deterioration of traditional customs and skills (such as weaving and blacksmithing).

In the 1981-85 Plan the government claimed it was adopting "a grassroot integrated participatory endogenous development" also called "grassroot self-centred" development. Dembele argues that this approach was basically paternalistic and contemptuous of the peasantry since much of the planning in the various "grassroot" projects was to be done by bureaucrats and technocrats.

Finally, Dembele stresses the importance of policies of water management to deal with Sahelian phenomena.

Burkina Faso

For the implementation of its economic policies, Burkina Faso is almost entirely dependent on foreign loans and aid. The role assigned to Burkina Faso by the colonial masters is reminiscent of that of the "labour reserve economies" of Southern Africa. For a long time the country, especially the Mossi region, has functioned to some extent as a labour reserve for the Ivory Coast. This meant that no viable peasant-dominated export crop production emerged as in Senegal or Ghana. It also meant that no national bourgeoisie emerged. More significantly, it meant that at independence the state was controlled by an alliance of clerks, school teachers and a sprinkling of petty traders and hawkers who pursued a basically "business-as-usual" policy at least up to the 1972-73 drought. As a matter of fact, for a long time state participation in international economic life was limited to serving as a management facility at the disposal of sundry agri-business companies and international agencies. In addition to other problems, then, this country came to be cast in the role of a guinea pig for social experiments. If there was a common underlying theme to agriculture it was to encourage the production of crops that would in some way be taxed to provide revenue for the state.

Both the state's own preoccupation with taxable commodities and aid donors' biases and interests led to an emphasis on export crops, especially cotton.

Thiombiano's chapter touches upon a series of issues that are of great relevance to the whole issue of the agrarian crisis in Africa. First, as we indicated earlier, the image of stagnation and even decline in African agriculture conceals important processes in African agriculture. Thus, for instance, while cereal production stagnated in Burkina Faso, cotton and sugar production thrived. Increases in output were not only based on extensive cultivation but were the result of rather dramatic increases in productivity. Cotton productivity increased from 111kg per hectare in 1961 to 979kg per hectare in 1982. Significantly, export crops produced entirely by peasants did poorly, a point suggesting that the export character of a commodity may not be as significant as the social context of its production: what class dominates and what the social relations of production of that crop are.

A second point raised by Thiombiano's chapter is the issue of parallel markets which undermine any statistical efforts to set the record straight. In the case of Burkina Faso, the state has fixed producer prices for both cotton and cereals. However, while the state enjoys monopsonistic powers in the case of cotton, it has allowed other merchants into the buying and marketing of cereals. The immediate effect is that for cereals there is a discrepancy between official producer prices, official consumer prices and actual market prices of as much as 500 percent. As a result there is no correlation between official prices and cereal production figures released by the state. Indeed, we simply do not know the dimensions of the food crisis in Africa. It has been noted, quite correctly, that if one took into account official figures of food production and food imports and matched those against population growth in Africa, the continued existence of large parts of Africa would simply be inexplicable.

THE "RENTIER" STATES

Let us now turn to three "rentier" states. Two of these are oil-exporting countries (Algeria and Nigeria) while one is a mineral-exporting country (Zambia). In the three cases, for much of the period covered in the studies, the revenue imperative was not decisive in terms of agriculture policy. Indeed, most of the agriculture programmes were implicitly or explicitly "welfarist"—based on the transfer of some of the mineral export revenue toward the transformation of agriculture or provision of social services to the rural population. In all three cases other sectors of the economy other than the major export sector have suffered from something like the "Dutch disease" in which the export boom has thwarted other exports by either competing with them for factors of production or by so revaluing the exchange rate as to make other products uncompetitive on the world market. Consequently, the high rates of industrialisation facilitated by the mineral rents have not transformed the structure of exports away from the export of primary products.

However, these similarities based on mineral wealth conceal crucial differences engendered by different historical paths toward independence and the configuration of social forces that have coloured state policies.

Zambia

Zambia is mineral rich and its exports include not only its near-legendary copper but also cobalt, lead and zinc. The country is also self-sufficient in coal.

Policy towards agriculture has mirrored the vicissitudes of the country's minerals in the world market and the state's reliance on mineral rents to meet its fiscal needs. Wood and Shula provide an historical background which, when juxtaposed to Shopo's account of

Rhodesia, underlines the complexity and diversity of the colonial heritage even for countries as closely twinned as the two ex-Rhodesias. While in the case of Zimbabwe settler-farmer interests were well articulated in state policy (see Shopo's chapter in this volume), in Zambia the settler farmers had to contend with two other power interests—namely those of the mining industry and (especially after World War II) the colonial government's "development and welfare" perspectives on peasant production. The mining interests were keen on obtaining cheap wage goods for its labour force and cheap supplies of labour. This placed the mines in a rather contradictory position. Being interested in cheap supplies of labour, they were not particularly keen on the commercialisation of peasant production. This, however, meant that they would have to rely either on imports or on a highly organised white settler farming community. The settler farmers wanted, in their turn, to be protected from imports and peasant production and strove for a monopolistic position in the supply of wage goods to the mines. The colonial government, especially after World War II, had a vision of rural "development and welfare" that envisaged the birth of "progressive farmers" among the African population. The picture was not made simpler with the creation of the Federation Rhodesia and Nyasaland and the increasing subservience of Northern Rhodesia's interests to those of Southern Rhodesia.

The government of independent Zambia was faced with this unresolved set of contradictions. The state was however cushioned by the substantial surpluses then available to it from the copper industry and was in no apparent hurry to resolve these contradictions. Instead, it embarked upon a flurry of agricultural experiments some of which were patently contradictory. State farms, cooperatives, "progressive farmers" schemes, "integrated rural development", large-scale capitalist farms and even "kibbutz"-type arrangements—all were tried. None of these led to substantial increases in production and some were abandoned in mid-course but the costs could easily be absorbed by a state awash with surpluses from copper. It is also important to note that underlying much of the policies towards agriculture was a rather "welfarist" and not "productionist" perspective. The state did seek to transfer some of the copper wealth to the rural sector and was therefore less concerned with the production implications of its financial allocations to agriculture.

Then came the crunch in 1974. Copper prices collapsed on the world market while the prices of virtually all Zambian imports went up. Matters were not made easier by Zambia's deep commitment to the liberation struggle.

With the demise of Zambia's major source of foreign exchange, other sources had to be sought. First, recourse was to private bank loans. When these became increasingly difficult to obtain, the second recourse was the IMF. Zambia had to swallow the standard IMF package—devaluation, removal of subsidies, privatisation, contraints on state expenditure etc. As far as agriculture is concerned, the "structural adjustment" programmes forced the state to accept full-fledged capitalist agriculture. It also diminished the role of statutory boards such as Namboard in the agricultural sector. Most significantly, it forced the state to abandon or downplay the equity considerations that had coloured agrarian policy in the post-independence era.

Algeria

As a result of oil revenues, agriculture plays a very minor role as a source of income to the state.

Algeria, like Zimbabwe, was a settler colony in which the indigenous people were deprived of the best of the arable lands. As in the case of Zimbabwe, settler domination could only be ended by an armed struggle. However, unlike Zimbabwe, settler agriculture collapsed as a result of the massive departure of the "colons" and the virtual collapse of the colonial state. The bitter and prolonged struggle for national sovereignty gave the national movement a populist and, initially, a radical character that was greatly to influence the character of the state and its policies.

This led, in the post-independence period, to the creation of a relatively strong base of agricultural production under state control. The productive base was constructed from farmland abandoned by former settlers and a small number of other large landowners. Produce from these farms—mainly wine and citrus fruits—was originally intended for the French metropolitan market; with independence that outlet was no longer certain.

Partly as a response to popular demands and wartime mobilisation of the population, the post-independence state decided to run the state farms on a self-management basis. Bedrani and Bourenane argue that steps were, however, quickly taken to deflect the embryonic process from its initial goal of worker management; in place of worker control, a permanent multi-faceted bureaucracy has risen bent on controlling every phase of the agricultural process: input supplies, production, marketing.

Alongside the state-controlled sector, there thrives a private sector of considerable size, based on small individual plots. The majority of the agricultural population is involved in this sector, which covers two thirds of the arable land. However, the quality of farming in this sector is relatively low, its potential correspondingly limited. Production here is mainly for subsistence, the surplus going to the urban markets.

Largely as a result of oil revenue, the levels of investment achieved in the 1970s in the nonagricultural sectors created employment opportunities offering levels of income far superior to levels of income obtainable from land. This process was accompanied by the state's social and commercial policy which tended to make the structure of demand in the rural areas for agricultural and nonagricultural consumption goods similar to that of the urban areas. The combined effect of these was the abandonment of less profitable agriculture activities in favour of those with a high return. Production of vegetables, fruits etc. for the domestic market became more important than the international market. Here we have the recurrence of the "Dutch disease".

Due to lack of means among small peasant producers, much of this production for the domestic market was confined to those with capital. It also tended to be speculative and short-term in character.

This had two effects: "devalorisation" of working on land, and a significant reduction in the share of revenue from agriculture in total peasant incomes.

The fall in oil prices has exposed the vulnerability of this kind of agricultural development.

Nigeria

Nigeria is another oil-exporting, OPEC member state included in this collection. This is about the only similarity between Algeria and Nigeria in terms of resources (both natural and human), political system, development strategy and social structure. Nigeria, with close to 100 million people is the most populous country in Africa. From a predominantly agricultural "merchant" state less than two decades ago, Nigeria is today a major oil exporter with oil revenue accounting for 96 percent of the country's exports in 1980, as compared to only 25.9 percent in 1965.

Nigeria has experienced many changes in government through elections and military coup d'etats. However, throughout all this period the capitalist path of development has never been seriously tampered with. Most of the conflicts have either been fractional (i.e. about which fractional interests of capital are predominant) or about the "indigenisation" of capitalism. This is not to deny the presence of other social forces that have articulated different social projects and preferences, but merely to underline the fact that no radical social experiment such as those of Nkrumah and Nyerere has so far been attempted.

Titilola's study is an account of the numerous policies that have been promulgated by the state to come to grips with the Nigerian agrarian crisis symbolised by the decline in the production of Nigeria's erstwhile major foreign exchange earners and a huge food import bill. Virtually all these policies have failed to meet the stated objectives despite vast amounts of money allocated to, if not spent, on agriculture. Because of the oil revenue, the "revenue" imperative domination, of the agricultural pricing policies of most African governments was removed and by 1975 the marketing boards that had existed ever since the colonial days were abolished. Consequently, agricultural policy has been basically focussed on increasing production rather than maximising the surplus extracted from the sector. In line with this taxes on export crops were removed or kept to the minimum. In addition, over the entire "oil boom" period food prices were allowed to rise faster than the general consumer index. Terms of trade improved for a wide range of agricultural commodities. Indeed a whole range of schemes were introduced to subsidise agricultural production.

Although at the rhetorical level state policy was anchored to the "basic needs" strategy, actual practice has involved an attempt to establish capitalist relations of production in Nigerian agriculture. Land tenure has been readjusted to permit commoditisation of land; credit has favoured large-scale commercial or "progressive farmers"; the World Bank (with all the "class biases" of its approach) has been given a major role in agriculture. In addition, at times, import substitution policies for agriculture have been adopted to protect agriculture.

Titilola points out that although there is greater social differentiation in the rural areas suggesting the emergence of new social relations, in terms of output, the results of these attempts to introduce capitalist production have generally not been impressive. Some of the reasons given include labour shortages due to massive rural-urban migration spawned by the industrialisation underwritten by oil. In addition, although prices have favoured agriculture, supply responses to price incentives have been low as a result of low levels of technical capacity, poor deliveries of necessary inputs, lack of farm-tested research results and poor marketing infrastructure and erratic changes in government food import policies. The last point points to a crucial weakness of the "get-prices-right" orthodoxy that pervades much of the debate on the African agrarian crisis.

BY WAY OF CONCLUSION

A comparative analysis of these case studies makes it possible to identify a number of issues that have an important bearing on theoretical perspectives and debates on the role of the state in the process of agrarian transformation and resource allocation in Africa.

The first of these relates to the direction of state policies or the nature of the "biases" informing state policies. Failure of African agriculture has been attributed to various kinds of "biases"— "commodity bias" (cash versus subsistence crops, exports versus food crops), "spatial bias" (rural versus urban areas), "sectorial bias" (industry versus agriculture, formal versus informal sectors), "market bias" (external versus domestic markets"), "scale bias" (small-scale versus large-scale), "gender bias", "technological bias" (capital-intensive versus labour-intensive, appropriate versus inappropriate technologies), "class bias" (capitalist versus working classes, feudal class versus peasants), "fractional bias" (state or bureaucratic bourgeoisie versus agrarian or industrial bourgeoisie). Explicitly or implicitly much of the writing on African agriculture subscribes to one or several of these not always mutually compatible "bias" explanations of the crisis. However, although a reasonable case can be made for each of these biases, it seems essential, if one is to avoid eclecticism, that some ranking be imposed among them without necessarily falling into the trap of reductionism. Conflation of these "biases" can lead to inconsistency in the analysis of the dynamics of policy formation in Africa.

Most of the studies here lean toward socially defined biases—class or fractional biases. However, in virtually all of them there is reference to "commodity bias" either as a consequence of these class or fractional "biases" or parallel to them. Crouch and Janvry[3] in a study of Caribbean agriculture have introduced a classification of crops that goes a long way to reconcile these two biases. They distinguish crops by their characteristics on the supply and demand side. On the supply side we have (a) crops that are imported, (b) crops that are produced under capitalist conditions and (c) crops that are produced under peasant conditions. On the demand side we have (a) peasant foods, (b) wage foods, (c) industrial and luxury crops and (d) export crops. Their conclusions on the Caribbean are pertinent here and seem to confirm various cases in this collection. First, imported foods have offset stagnant domestic production and have been low priced. Wage foods (rice and sugar) and industrial inputs and luxuries (cotton and beef) produced under capitalist conditions have been dynamic with decreasing prices. Peasant-produced wage foods (corn and beans) and peasant foods (manioc and sweet potatoes) have generally stagnated as evidenced by increasing domestic prices. Finally, export crops pro-duced both on capitalist farms and on peasant farms have generally undergone favourable although unstable growth and pricing. The classification introduced by Crouch and Janvry is useful because it brings out two essential characteristics of each crop—the social context of the production and consumption or use and its role in the process of accumulation. Although different dietary traditions and climatic conditions may lead to a different classification of food, this scheme should provide a basis for empirical studies on the nature of "biases" in Africa.

A second theoretical issue is the possibility of capitalist transformation of Africa agriculture. The "Dependence School" literature generally tends to deny the possibility of such a transformation on the basis of arguments that have been generally used about the possibility of industrialisation in the periphery.[4] The main argument is that at this historical juncture the emergence of capitalist farmers and a rural proletariat is neither the dominant process now nor is it necessary for global capitalist accumulation. Rather, the trend is to incorporate and maintain, albeit in a stunted form, precapitalist relations of production. Various studies here caution againt such a hasty conclusion. The determination of various regimes to establish capitalist relations of production and the support such regimes are receiving from aid agencies, state policies to attract transnational capital to agriculture, intensified research into the conditions of African agriculture, cheap availability of labour as a result of population pressure and the crisis in subsistence economies—all these are creating favourable conditions for capitalist agriculture. In retrospect, colonial attempts to introduce capitalism were halfhearted when compared to what a number of African governments are doing. Unencumbered by the fear of the emergence of an indigenous bourgeoisie or "native revolts" over land rights, African governments are changing land tenure systems at a pace that would have made many a colonialist cringe.

Not that we are about to see complete subsumption of the labour process in African agriculture by capital. Peasant agriculture has proved resilient even in the advanced countries of Europe.[5] In addition, as studies on Malawi, Swaziland and Zimbabwe clearly demonstrate, peasant agriculture serves an important role in the cheap reproduction of labour for the capitalist sector not in a functionalist sense but as a result of the unevenness of the accumulation process itself. Nevertheless, there are signs that we will see the extension of "bimodalism" in African agriculture. The wage-profit nexus will become more salient as capital subsumes the labour process in a number of commodities.

A third issue relates to the widespread move toward the privatisation of African agriculture, the dismantling of a whole range of parastatal organisations that have thus far dominated agriculture and the general withdrawal of the state from active involvement in agriculture. There are both internal and external pressures accounting for this. Internally, there is the emergence of social classes that are resisting the state's role in agriculture. In a number of cases these classes are placed in agriculture itself and have been nurtured by past state policies and now find state policies an encumbrance. In other cases, these are retiring state bureaucrats who now wish to use their earnings or contacts to venture into commercial farming unencumbered by the taxes and restrictions that they may in the past have actively imposed on the peasants. The presence of local classes interested in agriculture has made "privatisation" less politically onerous since it does not unequivocally mean foreign ownership as it clearly did at independence. This is a point often forgotten by those who intepret the current wave of "privatisation" as seemingly an outcome of a learning process which has ineluctably impressed upon the once-recalcitrant African policy-makers the evils of planning and virtues of the market.[6]

In addition we have the external pressures on the state. Towards the end of the 1970s the onslaught on the state role in the economy in general, and in agriculture in particular, was mounted. The major salvo, however, was to await the "Berg Report".[7] The report placed virtually all the blame for the agrarian crisis on the pervasive presence of the state in African economies. Coincidentally, this was at the time when most African governments were hurting severely from the second "oil crisis". Many of them were engaged in what is now known as "policy dialogues" (the current euphemism for the dressing-down and arm-twisting of policy-makers by foreign financiers).

As a result of these "policy dialogues" and the obvious failure of past policies, many African governments were succumbing to pressures for change. By the end of 1985, *African Economic Digest* could report "...the philosophy of opening up all sectors of the economy to the forces of competition, coupled with suitable incentives to producers to dismantling of public enterprises has come out loud and clear".[8] In similar vein, another magazine could report rather dramatic switches in policy in Tanzania where "commercial farmers who a year ago would scarcely imagine getting a go ahead to enter large-scale farming have now been assured of getting land leases and guarantees of 33, 66, 99 and even more years".[9] Zambia was actively attracting white farmers from Zimbabwe and encouraging indigenous capitalist farmers and agri-business. Zimbabwe's "transition to socialism" has been stalled as "pragmatism" slowed down land reform and agri-business was being attracted to the country. These are not merely anecdotal accounts of events in selected states but examples of a widespread phenomenon in Africa.

Curiously, though, all this is taking place when evidence from African and non-African countries cited by "free market" ideologues as "success stories" clearly points to the pervasive intervention of the state beyond the simple manipulation of parametric variables (infrastructure, research etc.). The significant question then is not whether or not the state should play a leading role in the transformation of African agriculture. Rather, it is under what conditions does the state assume such a key role in capitalist transformation and not become a fetter on capitalist accumulation without providing alternative forms of accumulation, as has happened in a number of African countries? How do the workings of the state apparatus mesh with the workings of the market?

A fourth point is the question of agriculture and industrialisation. Virtually all the governments studied here have stated that industrialisation, including that of agriculture itself, is their principal objective. For the "rentier" countries much of the industrialisation has been financed by mineral export revenues. However, for most of the other countries, it is clear that in the absence of external sources of investible surplus, surplus will have to come from the nonmineral sectors. Most specifically, agriculture will bear a heavy burden here. This immediately raises an interesting problem with the current wave of promotion of capitalist agriculture in Africa and calls for strengthening the political position of the agrarian capitalist class in policy formation.[10] Byres[11] has argued that in the classical transition to capitalism, accumulation had as a prime condition the "crucial configuration" that tended to favour capitalist accumulation. More specifically, he argues, the agrarian question was resolved in favour of the "urban bourgeoisie". Since the "urban bias" was important to capitalist expansion, will the current attempts to strengthen a rural bourgeoisie and turn state policies in their favour adversely affect the emergence of an industrial bourgeoisie as has been argued to be the case in Ivory Coast where a planter bourgeoisie wields power over the state?[12]

A fifth issue is the managerial capacity of the state. In virtually all the studies here we run up against the question of the technical-administrative capacity of the state to implement its avowed policies. In addition to problems of the internal capacity of the state, there is resistance to state policy by other social groups. In Africa these efforts to challenge traditional attitudes of the state to rural society are quite pronounced, their manifestations multiform: the partial or wholesale refusal to plant certain crops when earnings therefrom drop; the partial or total boycott of official marketing circuits for crops whose prices are fixed by the state or which are under state monopsony; the refusal to follow scrupulously the technical implementational instructions issued by official extension agencies. We clearly see that the functionalist assumption that because the state is necessary to capitalist accumulation it will have the "political will" or capacity to intervene effectively is not valid.

A sixth issue of the agrarian scene in Africa is the low level of technological mastery of production in Africa. If as, Walter Rodney noted, Africans went into colonialism with a hoe and emerged after colonialism with a hoe, two decades of so after independence the picture remains distressingly the same for vast numbers of the African peoples. However, side by side with this image of stagnation are impressive technical breakthroughs in agricultural technologies elsewhere. Since agricultural innovations are location-specific, transformation of the technical level of African agriculture will require local research. Here local initiative will have to contend with the transnationalised interests of agri-business with its monopoly of new biotechnologies, fertilisers, machinery etc.

The seventh issue is the fundamental question of the state and the peasantry. Africa remains a continent in which the peasants still enjoy relative autonomy over the process of production. This is largely due to relatively easy availability of land, although in some parts of Africa the picture is changing rapidly. The failure to eliminate peasant production is seen by some as the source of the failure of agrarian transformation in Africa. The leading exponent of this view is Hyden, who states:

> Development is inconceivable without a more effective subordination of the peasantry to the demands of the ruling classes. The peasants simply must be made more dependent on the other social classes if there is going to be social progress that benefits the society at large.[13]

Whatever the validity of this assertion, the states in the countries included in these studies behave as if they accepted the argument. As a result, we witness the amplification of ongoing processes of social differentiation in the rural areas, of land alienation and concentration through various schemes of "privatisation", of the increasing subsumption of the labour process by capital and of a more repressive presence of the state apparatus in the rural areas. Sometimes the processes lead to the emergence of islands of relative prosperity in a sea of increasing marginalisation.

Finally, this collection should be seen for what it is: part of an ongoing process of reflection by African scholars to understand the major social phenomena of their respective countries and to contribute to the transformation of the societies of their continent. Much more work needs to be done.

NOTES

1. For the classification of state roles see McCall, M. and Skutch, M., "Strategies and Contradictions in Tanzania's Rural Development: Which Path for the Peasants?, in Lea, D. and Chaudri, D.P. (eds.), *Rural Development and the State*, Methuen and Co., London, 1983.
2. The argument given is that large-scale farmers can serve as the backbone of the farmer lobby in pursuing reform. Indeed Eicher chides those opposed to large-scale farmers for failing to appreciate the positive role that such farmers play in policy formation in favour of agriculture. See Eicher, Carl, *Transforming African Agriculture*. This argument has been forcefully advanced elsewhere by Bates and Hart. See Bates, H., *Markets and States in Tropical Africa; The Political Basis of Agricultural Policies*, University of California, Berkeley; Hart, Keith, *The Political Economy of Agriculture in West Africa*.
3. Crouch, Louis and Janvry, Alain, "The Class Basis of Agricultural Growth", *Food Policy*, February 1980.
4. Founou, Bernard, "Limites des Alternatives Capitalistes d'Etat ou privées à la Crise Agricole Africaine", *Africa Development*, Vol. X, No. 3, 1985.
5. Goodman, David and Redclift, Michael, *From Peasant to Proletarian: Capitalist Development and Agrarian Transitions*, Basil Blackwell, Oxford, 1981.
6. For the incredulous it should suffice to cite only one example. The OECD blames past failures in Africa on "the penchant of first generation of impatient political leaders to try statist approaches to the organisation and control of economic activity from production to input supply to marketing and pricing". To the OECD's pleasure all this is changing now: "...pragmatism is beginning to gain a foothold in several governments long noted for their ideological approach to development policy-making." OECD, *Development Cooperation*, OECD, Paris, 1984, p. 12.
7. World Bank, *Accelerated Development in Sub-Saharan Africa: An Agenda for Action*, Washington, D.C., 1981.
8. *African Economic Digest*, 21 December 1985.
9. *African Business*, January 1986, p. 48.
10. See Bates, Eicher and Hart, *op. cit.*
11. Byres, T.J., "Agrarian Transition and the Agrarian Question", in Harriss, John (ed.), *Rural Development: Theories of Peasant Economy and Agrarian Change*, Hutchinson University Library, 1982.
12. Bonnie Campbell argues that the Ivorian bourgeoisie has sought to block the emergence of such an indigenous industrial burgeoisie. Bonnie Campbell, "Inside the Miracle: Cotton Production in the Ivory Coast", in Barker, Jonathan, *The Politics of Agriculture in Tropical Africa: Transnational, National and Local Perspectives*, Sage Publications, Beverly Hills, 1984.
13. Hyden, Goran, *Beyond Ujamaa in Tanzania: Underdevelopment and an Uncaptured Peasantry*, University of California, Berkeley, 1980.

1. NATIONAL FOOD POLICIES AND ORGANISATIONS IN GHANA

Emmanuel Hansen

INTRODUCTION

The 1972-74 food crisis brought Africa's food problems to the attention of the world in a dramatic way. For many people outside Africa that was the first time they came face to face with the fact that Africa's food agriculture is in serious trouble. But for many it was seen as a natural disaster. Now a decade later we are facing another major food crisis. Over ten African countries are facing a serious drought problem. At a meeting of the FAO's 13th Regional Conference for Africa held in Harare in July 1984, the Director-General, Mr Edouard Souma, painted an even more bleak picture of Africa's food prospects. He warned of impending disaster if production rates keep falling behind population growth rates.[1] The international community responded to the first crisis with aid and the same international community is being called upon to respond to this crisis also. The dramatic nature of the problem and the response to it as an emergency in a way hides a critical aspect of the food problem in Africa. For the food crisis did not suddenly descend upon us like a bolt from the sky. It has been building up for some time. Second, the emergency nature of the problem tends to give the impression that it is only in the Sahel region where serious drought conditions exist at the moment that the problem is. This also tends to reinforce the view in the minds of many that our food problems are the result of natural conditions, like drought, floods etc.

It has to be noted, however, that the Sahelian drought of 1972-74 and the current one are only dramatic manifestations of the problem. Many African countries have been suffering from food problems of one sort or the other for almost two decades, and now the combination of adverse weather conditions, poor organisation of social production, the current world crisis, poor economic policies, have all contributed to bring what was once only a periodic and temporary shortage during the "lean season" into an almost permanent crisis.

PROFILE OF GHANA'S AGRICULTURAL ECONOMY

Ghana provides a very interesting, though tragic, case for study. For as a country it does not seem to have ecological and social conditions which should make it vulnerable to such food problems, which it has been experiencing for more than a decade now.[2]

Ghana is not in the arid or semi-arid zone or in the Sahel regions like Niger, Burkina Faso, Mali, Ethiopia, Mauritania, parts of Senegal or Somalia: nor has it large tracts of land which cannot be used because they constitute desert or because of swampy conditions. Although now and then people with the ambition of going into large-scale production inveigh against the frustrating and complex land tenure system it is not an obstacle to production. There is really no land problem as such in Ghana. There are hardly any problems of land reclamation in the country. Although soil fertility is not rated very high, it is good enough to support crop production and animal raising without much difficulty. There is relatively dependable rainfall, not very evenly distributed in the Northern and Upper Regions, but well enough to sustain life and a reasonable level of crop production. The agricultural cycle has not been interrupted by such natural disasters as floods, rampaging locusts or other insect infestations or long periods of drought. Although in 1975-76 drought in the Northern and Upper Regions did not help matters it cannot be used to explain almost a perennial food problem. Furthermore, there have been no civil disturbances like wars such as we have had in Mozambique, Zimbabwe, Uganda and Chad to interfere with agricultural production. It would seem from the above therefore that there are no ecological, climatic or physical reasons why Ghana should have such endemic food problems.

Let us look at the social character of the agricultural economy. Ghana's agricultural economy is based on the "peasant model". Over 90 percent of food and agricultural production is in the hands of small peasants with average holdings of not more than two hectares. Ghana has no settler agriculture like colonial Algeria, Kenya, Tunisia or Zimbabwe. Nor has it plantation agriculture as in the case of Mauritius, Cameroon, the Ivory Coast or Liberia to interfere with food production. Multinational involvement in agriculture is a relatively recent phenomenon in Ghana and its impact at present is very negligible. The significance of this is that historically there has been no land alienation as was the case in Kenya, Algeria or Zimbabwe. This meant that there was no ruthless displacement of peasants, although some form of peasant displacement is now beginning to take place especially in the Northern and Upper Regions where capitalist rice farming is beginning to take hold.[3] At present this has hardly any impact on the agricultural economy of the country as a whole. There has been very little rural proletarianisation. Although the country has for a long time been known for its extractive industry—gold mining—this is a capital-intensive industry, and its labour requirements are not such as to disrupt agricultural production. Ghana is not like Zambia, which derives its export receipts mainly from the copper mining industry. Besides, the gold industry produces only about 8 percent of export receipts, and it does not determine the local social formation. It would seem therefore that the peasant has control over production, though not, especially in the case of export crops like cocoa, and over marketing. What is significant about this is that the peasant enjoys relative autonomy, and it is claimed that under such conditions he can attend to the demands of export crop production and food production at the same time.

In addition Ghana has been credited with having a good economic and social infrastructure: a network of connecting roads which, if somewhat fallen into disrepair in recent years, by the standards of developing countries is still good; a large pool of trained manpower which in recent years has been migrating outside the country, and a relatively good infrastructure for health and education, although in recent years there has been a marked decline in the quality of these services. It would seem from what we have said so far that there are no physical, ecological, climatic or social reasons why Ghana should have such serious food problems. Yet for the second time running since 1982 Ghana has had to appeal to the international community for emergency food aid.[4] What, one is tempted to ask, is the problem with Ghana's food agriculture?

To answer this question it would be necessary to review the whole structure of food production and distribution in the country and see how Ghana's food policies have evolved and to what extent such food policies respond effectively to the food needs of the country. We shall come to this in a moment.

FOOD AND DEVELOPMENT

The relationship between food and development has been made many times and there should be no need to reiterate the point but since it seems our leaders have not heard or, if they have heard, do not seem to pay much attention to it, we shall continue to say it. Ghana, like all African countries is committed to programmes of development and modernisation; so at least the leaders say. At the concrete level it implies sustained economic growth above the level of the rate of population growth so that a surplus could be accumulated which could be used to improve the living standards and quality of human life in general. We should be able to provide food, shelter, clothing, health and education for our people. We should be able to provide security and guarantee them freedom from arbitrary arrest and humiliation. Our people should be able to enjoy a reasonable degree of freedom from anxiety for themselves and their families. These days policy-makers both foreign and domestic are busy making frantic efforts to provide what are called "basic needs", in effect the critical minimum for the population to be able to reproduce itself and to provide a modicum of elementary standards of decency. Commendable as these efforts are, we must recognise quite clearly that to live in the biological sense of the world alone is not enough. After all, animals manage to do this. We should strive to live like humans which implies not only the satisfaction of material needs but also the attainment of a certain measure of psychic well-being. We should strive to achieve a sense of existential fulfilment. A minimum condition for achieving such a state of affairs is the provision of adequate and high-protein food to feed the growing population without which no meaningful development can take place. And it is precisely in this sphere that our achievements leave a lot to be desired. Herein lies the tragic character of our post-colonial development. Long ago Arthur Lewis recognised this when he said:

> The most certain way to promote industrialization (develop-ment) in the Gold Coast is to lay the foundation it requires by taking vigorous measures to raise the food production per person engaged in agriculture.[5]

It is 30 years since independence, long enough to bring a young adult into the world. What have we done to provide this minimum condition? Before we try to answer this question let us take a look at the current food situation.

CURRENT FOOD SITUATION

For almost two decades now Ghana has been struggling with food problems. Now the situation has reached crisis proportions. It used to be said that Ghana's food problem was nothing more than a reflec-tion of change in the patterns of consumption particularly of the urban elite. Although there has been a change in food consumption patterns such arguments as explanatory factors of the food problem are no longer tenable. Food is the most common topic of conversation (and not only among housewives) in Ghana. Each time people meet they try to exchange valuable information on where to get what: meat, fish, beans, edible oil, sugar, yams, bread etc. Queuing for food and other domestically consumed items has become so internalised as part of social behaviour that visitors both African and overseas have marvelled at how Ghanaians could queue so patiently and for so long for such basic needs of man. The fact of the matter is that they have no choice. So widespread has the food problem been and so much attention both public and private does it occupy that for some time the mere arrival of a ship with a few thousand tons of wheat, flour, tinned fish etc... was enough to attract front-page coverage in the media. Now it is not the arrival of the ships that attracts front-page coverage; it it merely the promise of assistance. Such has been the state of desperation.

Ghana's food problem manifests itself at two levels: first, there is a shortage of staple foods like yams, cassava, maize, rice etc. Second, there is an even more serious shortage of high-protein foods like cereals, pulses, meat, fish and eggs. For a large number of people meat, fish and eggs are almost out of the question. They are either unavailable or, for the mass of the people, unaffordable. A further point is that the food problem is not a periodic or a temporary short-fall as in the days of old when there was usually a stable food situation, periodically broken by drought or some natural disasters ; nor is it confined to the "lean season", the period of two months or so before the next harvest, when surpluses are almost completely exhausted; nor is it confined to specific geographical locations where ecological conditions make them more vulnerable. Although there are local and regional variations in the severity of the problem one important characteristic of the current food crisis is its national character. The current food problem is *national* not local. There have been food shortages in both rural and urban areas.

The fact of the matter is that with a population growth rate of between 2.7-3.0 percent per annum food production has not been able to match the annual growth rate of population let alone outstrip it. Now, to make matters worse, it would seem that there has been an absolute fall in the rate of food production. The *Five-Year Development Plan 1975-1980* had this to say with regard to the production of staple food crops:

> Except in the case of a few of them, production of most of these crops (cereals, starchy staples, pulses, vegetables, fruits) in the past years especially between 1970-74 had not increased significantly in response to domestic demand for these basic staples.[6]

In more recent years it has been estimated that taking 1975 prices as constant the agricultural sector as a whole has declined by 7.5 percent between 1976 and 1981. Of this general decline food has been the most seriously hit.[7] Perhaps Ghana's food problem will be better appreciated if it is realised that even before the problem reached crisis proportions, food imports constituted almost 20 percent of all imports, and this, for a country where 35 percent (1974) of GDP is from agriculture, is an indication that there is something really wrong. Joseph Bognar, the Hungarian economist, commenting on food imports in 1962 wrote:

> Food imports amounted to 19.18% of all imports in Ghana in 1960. It is obvious that such continued high levels of food imports could not be tolerated; assuming an unchanged structure, imports in 1970 would amount to £350 million as compared with that of £129.6m. in 1960. Such a volume of imports cannot be balanced by exports... It is necessary to act very quickly to increase the agricultural production by 1965 by between 40 and 50% in order to avoid increased food imports.[8]

Not only have we not been able to maintain the levels of production at that time but, as I have indicated, we have even suffered a drop in absolute levels. In the past such shortfalls have been financed through imports, themselves financed through exports of other agricultural products, like cocoa and forest products which collectively make up 75 percent of export receipts. But now, with falling agricultural production, crippling debt service and the high cost of fuel, Ghana's capacity to import food has been reduced to almost zero. One can understand the full nature of the crisis if it is realised that at the moment fuel imports alone consume 50 percent of export earnings. The seriousness of the whole thing is brought home clearly if it is realised that Ghana does not even depend very much on fuel for industrial purposes as it uses hydro-electric power. The fuel then is mostly for vehicular transport, and if Ghana now consumes 50 percent of export earnings on vehicular transport at a period when she is suffering from such crippling transportation problems due to inability to buy spare parts and keep vehicles on the road, one would wonder what would happen when full transport capacity is restored.

Nothing demonstrates the current food crisis in the country more than a look at the price system. Since the 1950s the prices of foodstuffs have been going up. The retail price index for local food has risen faster than any other item of consumable expenditure.[9] It has been estimated that between March 1963 and September 1965 the prices of locally produced food increased throughout the country by 82.3 percent. Over the same period the prices of imported food went up by 30 percent.[10] Kodwo Ewusi has also estimated that foodstuffs have been responsible for the overall price rise at urban, rural and national levels. He contends that price indices indicate that local food became the most expensive commodity in rural and urban areas between 1964 and 1969. He further claims that 90 percent of the increase in the overall price levels can be traced to the rise in food prices.[11] The price situation in recent times has been even more severe. If we take 1963 as a baseline (1963 = 100), by 1970 the total national consumer price index was 188.5; for imported food it was 150.1 and for locally produced food it was 210.0. By 1977 the figures were 759.9 for imported good, 1,033.9 for locally produced food and 1,382.5 for all commodities. Again, taking 1977 as a baseline (1977=100), by 1980 food was 392.5, but by April 1983 it had reached 2,658.9 for food and 2,039.6 for all commodities.[12] It has further been estimated that between 1979 and 1982 the wholesale prices of maize and millet increased by over 500 percent, while the price of yams increased by 1,500 percent.[13] Table 1.1 below makes the situation clear.

However, it is when one relates food prices to earnings that one grasps the full impact of the situation. At the current price of labour at ₵ 40 per day, for those lucky enough to find any wage employment, it would take the worker more than a week to buy an American tin of rice (3kgs), more than a day to buy 3 kgs of maize, ten days to buy a tuber of yam, over ten days to buy a bottle of edible oil, more than a day to buy an American tin of garri (cassava grains). One finger of plantain costs more than half a day's wages and one egg costs a little over one-third of a day's wage. Considering the fact that many workers do not even get the minimum wage, the situation is probably worse than the above figures suggest. The plain fact is that most people count themselves lucky to have one square meal a day. The government secondly allocated ₵ 30 per day for the feeding of one student at university. The university authorities with access to food at "controlled prices" have estimated it at ₵ 90 per day. Assuming the government figure is realistic this has serious implications for the current living standards of the workers. If ₵ 30 is considered adequate for the feeding of one university student per day, then at the current cost of labour at ₵ 40 per day, the government is asking the worker to provide for all his other needs—maintenance of his family, education, health, transport and housing etc.—at the cost of ₵ 10 per day! Recently the Secretary for Finance and Economic Planning was quoted as saying that in the last decade the wages of the worker had declined by 80 percent in real terms.[14]

TABLE 1.1

Market prices of selected food items in Accra

Item	Unit	Average prices (₵)		
		January 1982	January 1983	April 1983
Maize	Bag (50 kg)	500	1,000	4,000
Maize	American tin (3 kg)	30	80	200
Kenkey	Ball	1	4	10
Cassava	Tuber	10	15	25
Garri	American tin	15	65	190
Yam	Tuber	45	80	160
Rice	Bag (50 kg)	600	900	1,200
Rice	American tin	80	200	200
Plantain	Bunch (100)	160	250	1,000
Plantain	One	4	5	20
Tomatoes	3	4	10	40
Meat (beef)	1lb	20	30	60
Bushmeat	(grasscutter)	250	450	600
Chicken	2lb	45	75	180
Eggs	Crate (30)	80	90	180
Eggs	One	3	4	8
Sugar	American tin	70	150	300
Sugar	Cigarette tin	5	15	25
Bread	Loaf	4	disappeared	disappeared

Note: ₵ 2.75 = $1.00.
Source: J.A. Dadson, "Ghana, Food and the Nation", *West Africa*, 11 July 1983.

Quite apart from the prices which have made many goods unaffordable for the mass of the population, the sheer time and effort spent in the frustrating quest has had a considerable impact on productivity and morale, particularly in the public sector. Meanwhile, this undesirable situation has created opportunities for certain classes of persons to engage in corruption on an enormous scale. Such opportunities for widespread corruption are not confined to the privileged classes who control the distribution of goods, but spread to relatively lowly elements in such strategic establishments. Thus, as the manager becomes an important element in the distribution of goods, his secretary who controls access to him also assumes an importance out of proportion to his or her actual role. In declining importance the clerk who writes out the invoice, the receptionist who directs the visitor to the appropriate officer, the man at the gate, the security man without whose good offices one cannot even enter such premises, all assume overwhelming importance as each person occupies a crucial position in the chain which leads to the final decision to acquire scarce goods. It is this symbiotic and interacting relationship between the managerial class or the petty bourgeoisie who control the distribution process and the junior personnel which creates the compact network on which the corruption feeds and acts at both vertical and horizontal levels. This interacting and symbiotic relationship acts as a mediating influence on the otherwise serious undercurrent of class conflict which has been building up particularly in the urban areas.

This elaborate system, *Kalabule*, as it is called in local parlance, has completely thrown out of gear the distribution system in the country. The hostility which is sometimes expressed towards market women is the result of the humiliation and frustration on the part of the mass of the workers, who, unable to locate the real source of the problem or affect any change in the system, vent their anger and frustration on the market women since it is at this level that they confront the cruel logic of the market mechanism. Specific measures designed by succeeding governments, even including harsh punishments like public floggings and the razing down of public markets (acts of desperation more than rational policy responses), have not succeeded in eliminating *Kalabule* as a system.

It has to be said that critical though the present situation is, there is no famine in Ghana. There is no starvation but there is a serious food shortage which has affected consumption and nutrition levels and is now beginning to affect public health. There has been a reported increase in the incidence of skin diseases. Although it is often claimed that this is due to the use of certain types of locally manufactured soap (itself the result of the crisis in the agricultural economy), one cannot discount the effects of poor feeding. The Principal Medical Officer in charge of the Child Welfare Clinic (Kwashiorkor) in Kumasi recently called attention to the alarming rate of death at the hospital.[15]. Kwashiorkor, it has to be noted, is a nutrition-related disease. In 1983, 140 prisoners were released before their time was due as a result of an outbreak of cholera and skin diseases. It was believed that among the possible causes of the outbreak were malnutrition and underfeeding[16] Considering that children constitute the most important capital for development in any underdeveloped country the long-term effects of the deleterious nature of the food problem on children can only be imagined.

One of the effects of the food crisis, coupled with the general economic deterioration in the country, has been the large-scale migration out of the country of professionals, academics and technicians. That was how it started. Now anyone with the means to migrates.

Skilled, semi-skilled or unskilled attempt to do so. From the days when being a Ghanaian was in itself a recommendation, Ghanaians have now become unwelcome visitors in the West African sub-region. In spite of the claims of the local news media, labour migration on such a scale is more a comment on the contemporary situation in Ghana than a reflection of the behavioural peculiarities of the individuals who migrate.

There is no disagreement among observers, social scientists, politicians and policy-makers about the nature of Ghana's food crisis or the current economic problems in general. Political leaders and administration freely comment on the general agricultural decline and inveigh periodically against the *Kalabule* system, thus admitting its existence. Flt Lt Rawlings, noted for his commendable frankness, had this to say in a message marking the 27th anniversary of Ghana's independence:

> After 27 years of political sovereignty very few of us at any level of authority or in any sector of activity or in any type of endeavour, find ourselves in a position truthfully and honestly to be satisfied with the conditions of the nation, our regions, our cities, our towns and villages or indeed in the conditions of our own personal individual lives. The hard and bitter truth is that in several years of national effort, the governments and people of our country have failed to work together to implement policies and programmes necessary for the achievement of reasonable standards in our living conditions. Basic needs for food, health, transport, shelter and recreation remain largely unfulfilled and out of reach of the majority of our people. There is evidence of mounting despair, deepening apathy and growing withdrawal and rejection of national responsibility on the part of ordinary people.[17]

Thus there is no disagreement about the current state of affairs.

The disagreement is about what has led to such a state of affairs and what can be done about it. The usual reasons given by our politicians and policy-makers—bush fires, drought, and the sudden influx of Ghanaians deported from Nigeria—important though they are, only explain temporary shortages but not the endemic kind which Ghana has been experiencing for more than a decade now. And in the case of the returnees one could even argue that in the long run it was not without some benefit to Ghana. In the first place, the orderly, calm and prompt manner in which they were evacuated by the government, already stretched by crippling shortages of all kinds, won the admiration of the international community. It convinced the outside world that in spite of the general run-down of the economy there was still administrative and managerial talent in the country. In the second place, the callous and inhuman treatment meted out to the returnees in Nigeria and the wide publicity given to their plight by the international media aroused the public conscience in the West and made aid donors and other voluntary organisations respond to Ghana's aid needs with a greater degree of sympathy and urgency than they would have done under normal conditions. The incident helped to centre attention on Ghana which she was able to use effectively. She even succeeded in getting donor countries to accept her argument that relief aid should not be for the exclusive use of the returnees since they were already participating in the collective consumption of the country. Hence such explanations as to the causes of Ghana's food problems should be rejected and more fundamental factors should be sought. This brings us to the discussion on food policies in Ghana.

FOOD POLICIES

Since World War II, two main ideas have emerged which have shaped food and agricultural policy in the country. One view is that what is often seen as a food problem at the level of the market is not the consequence of production problems but market and distribution problems. One of the most forceful exponents of this position is La Anyane, formerly professor of agriculture at the University of Ghana, Legon. He writes:

> Some of us may be surprised to learn that we currently have in Ghana enough food to feed the people of this country *adequately* but that we are unable to realise this abundance of food that exists because of our inefficient marketing organization.[18]

In 1965 a statement attributed to the Minister of Agriculture had expressed a similar view.

> There is adequate food in the country. The problem has been the failure of food agencies to get them to the market. I intend to set up a separate body within the Ministry of Agriculture to take up the distribution of food...[19]

For these people not only is there no problem with production, there is also no problem with the form of production. Whatever bottlenecks exist lie largely within the sphere of marketing and distribution. Of course production and marketing are in a way closely related because if the producer is unable to market his produce in one season the chances are that he will cut back in the next season so that what starts as a marketing problem becomes a production problem. However, the view that there is adequate productive capacity within the current form of production to feed the nation and that whatever problems arise with food supply can be attributed squarely to marketing seems to have been at the base of official policy towards food agriculture from the time of colonial rule till the end of the Second World War, when it was challenged by another view. What is interesting is that although such assertions are no longer made with such confidence, food policy is still being made with such assumptions in mind. Old ideas die hard. At the end of the Second World War a view emerged first among colonial officials which was later to receive full expression in the post-colonial period. It states:

> It is now becoming apparent that the peasant had methods used in cultivating and preparing crops which will have to give way to mechanisation if the country is to continue to progress rapidly.[20]

This admitted for the first time that the form of production which had characterised the agricultural economy in the past would have to change if the country was to continue to progress. For those attracted to such a type of thinking the answer lay in transforming the basic form of agricultural production in the country through the establishment of large-scale mechanised farms. A more recent expression of this position was a statement attributed to the Commissioner for Works in March 1976 in which he urged the nation to switch from traditional to modern methods of farming:

> The small farms must give way to plantation type of farm. Seasonal crops designed to depend on the rainy seasons must give way to all-year cropping assured by irrigation.[21]

The dominance of Keynesian economics in the post-war period inclined colonial officials and their African successors towards the view favouring the dominance of the state in seeking this agricultural transformation. It was this position which easily lent itself to the attraction of the CPP as it fitted into its own view of socialism. It was this which led to the development of large-scale mechanised agriculture during the Nkrumah period. Tony Killick is therefore right in his observation that Nkrumah's economic policies were in the mainstream development orthodoxy of the period.[22] These two ideas have provided the basic tenets on which food and agricultural policies have been formulated in Ghana from colonial times till the present.

Let us now review the history of the agricultural economy of Ghana and see to what extent these two ideas have shaped policy with regards to food and agriculture.

FOOD AND AGRICULTURAL POLICY IN THE COLONIAL PERIOD

Colonial interest in Ghanaian agriculture arose out of demands to produce raw materials to feed metropolitan industries. In Africa two distinct methods were employed to achieve this, depending on the history and geographical and ecological factors of the given area and the needs of the Empire at the time. Raw material production for export was undertaken either through peasant agriculture as was the case in most areas of West Africa, with the exception of Liberia and the Ivory Coast, or was done by settler farmers on the basis of commercial agriculture plantations run by white settlers or large multinational companies as was the case in East, Central and Southern Africa.

In cases where peasant agriculture was preferred there was often no large-scale alienation of land or harsh labour codes to induce labour to work either on the European farms or on plantations. In many cases where the latter method was preferred, the weight of the colonial state was invoked to support settler farmers or the plantations by the use of legislation forbidding any competition from potential native growers by designating certain crops as "European crops". Where no settler farming or plantation was undertaken the peasants were "encouraged" to grow certain crops which had export outlets.

In Ghana, due to ecological factors (the place was unhealthy for white settlement), the long history of African resistance to land alienation which is crucial for settler or plantation agriculture, the failure of early experiments and attempts at plantation agriculture for export crops, the resistance to plantation agriculture from planter interests in the Caribbean, and the phenomenal success of peasant agriculture, peasant agriculture was preferred as the primary mode of producing export crops to feed British industries.

Initial colonial interest was in palm oil until by the first decade of the century when it was overtaken by cocoa, the production of which rapidly set the country on the pattern of a mono-crop economy. Dependence on peasant agriculture meant that the country was saved from massive land alienation, and harsh labour legislation and initially all the social consequences associated with the development of a landless peasantry. It also meant that the British authorities, in order to get the peasants to respond enthusiastically to the production of the crop which came to be regarded as the economic life blood of the country, had to get control of and influence over those who had control over the peasants, and in the political structures and authority patterns of Ghanaian society at the time, these were the Chiefs. Hence the need to cultivate the Chief, enlist his support and install him as a junior partner in the colonial administration. In the course of time these practical considerations developed into an elaborate philosophy of rule called "Indirect Rule" by which colonial officials sought to give philosophical underpinnings to the system of their administration.

One consequence of putting production into the hands of peasants was that it made it possible for some of the more enterprising to accumulate some capital which was either used to expand production or invested in real estate or trade. In the course of time a differentiation of the rural community became clearly discernible on the basis of the position which a group occupied in the structure of colonial agricultural production.[23]

In time four main sectors of the rural community could be identified. A small group of "rich" peasants who had a number of holdings and engaged wage labour and were in some way tied to the authority patterns of the traditional political structures emerged. Some of these were also partly in agriculture and partly in trade as "trader farmers", or were engaged in usurious activities on the side. There was a second group of what might roughly be equated to Mao's middle peasants. These also had a number of holdings though their holdings were much smaller. They relied on wage labour but only to a limited extent. They grew mostly cash crops. Like the rich peasants they hardly grew any marketable food crops except in bad times. The poor peasants were those who had very few or small holdings, usually one or two and were hardly in a position to engage hired labour. They relied mostly on family labour and grew mostly food crops both for domestic consumption and for the market. Although they might grow a few export crops they relied predominantly on food crops. In a way they were the opposite of the "rich" peasants in terms of crop differentiation. They often had no power in rural society as they did not stand in any close relationship to the rural power structure. Most growers were and still are to be found among this group. A fourth group has also emerged, the agricultural labourers or what could be called for want of a better term, a rural proletariat. This has developed largely on the basis of export crop production and initially they were migrants from Burkina Faso or Togo but the situation has changed in recent times. Now they could be migrants from other parts of the country. It could be seen from this that even at the very beginning food crop production did not attract the most enterprising of the African peasants.

The native peasant producer has been praised for his industry in building up the cocoa economy and Geoffrey Kay and Stephen Hymer have gone to the extent of arguing that he built the industry in spite of definite obstacles put in his way by the Department of Agriculture.[24] Although this praise is well deserved one must not forget the role of the colonial state in fostering, promoting and encouraging peasant interest in export crop production. Early in the century the government opened a Botanical Station at Aburi which later formed the nucleus of the Agriculture Department. From its nursery seedlings were distributed to farmers.[25] It has to be noted that the formation of the Department of Agriculture itself was tied to the promotion of export crops. The Department of Agriculture organised model farms from which peasants learned improved farming techniques. The Department also provided extension work and passed on the results of its research to farmers. Furthermore, cocoa and other export crop growers benefited and were encouraged by the provision of a ready and guaranteed market, though not at a guaranteed price. This was to come later. Similar support, though not to the same level, was given to other export crops like coconuts, rubber, kola, palm oil etc. Today, in spite of a severe food crisis, food crops do not get the same type of state support as exports like cocoa, either in production, marketing, research or the provision of credit. It is perhaps sufficient to point out that we are yet to have yam, cassava, millet or maize research stations. It does not of course mean that no research work is carried out on these crops. The food crop grower was left on his own as if production could be increased by its own momentum. On this point it is worth noting La Anyane's comments:

> At the some time as the Government encouraged diversification it can be said that it contributed to making cocoa a sole crop of the future. Governor Hudgson organised the marketing of cocoa on such a comfortable level that the cocoa farmer had no incentive for growing other crops.... before the arrival of the firms on the cocoa marketing scene Government had taken initiative in fostering the marketing system. Hodgson's scheme was not only for cocoa, but also for coffee.[26]

Nor has the situation changed much for the food crop grower. As late as 1968 the government was lamenting:

> Lack of capital to invest in food production has been one of the principal causes of small-scale farming in the country.

Food farmers have little to invest from their resources and need credits in cash or kind to secure the hired and mechanised services to produce the production of inputs.[27]

Where any encouragement was given for the production of food crops, as was the case with rice production in the Western Region in the 1920s with the establishment of the Esiama Rice Mills, it was because such food crops have potential as export crops or because it was seen as a measure of import substitution more than in response to meet the demands of domestic food supply. The external orientation of the economy was amply reflected in the policy of food production. The interesting thing is that there has been no marked departure from this policy.

Quite apart from the fact that food production for the local market was not particularly encouraged or pushed, colonial policy on export crop production adversely affected local food production. Early in the century colonial records were full of warnings as regards impending food shortages as peasants began to destroy food farms to make way for the growth of export crops. The response of the colonial state to this was to exhort peasants to grow more food to alleviate possible food shortages as if that by itself was enough to stem the tide. What is more surprising is that the colonial state did not seem to see the contradiction between this and urging peasants to grow more export crops, all the more so when export crops had a greater monetary value and guaranteed market.

This policy of structuring the entire agricultural economy of the country to suit the external demands for export crops and leaving food production to propel itself by its own momentum was rigourously pursued throughout the colonial period.

It was only with the coming of World War II that the policy was changed and anything near to a food policy was enunciated. The change was in response to the exigencies of the war. A statement from the Department of Agriculture makes the position clear:

The influence of the war has been paramount in forming the policy of the Department. The main factors controlling this policy have been: the need to reduce imports generally, especially foodstuffs and to save shipping space, draw less on supplies vital to the United Kingdom and conserve foreign exchange... Consequently a growing proportion of the Department's efforts has been thrown into the campaign to increase the production of indigenous foodstuffs...[28]

There was the need to increase domestic production of food to reduce imports and to satisfy the food needs of the large European population which had grown up as a result of the war: the Royal Air Force, the Royal Navy, the merchant marine and the army and Pan-American Airways. There was also the need to increase production to feed black troops the recruitment of whom had also reduced the labour supply in agriculture. Here the state intervened in food at two levels: at the level of production and at the level of the market. At the level of production the state undertook to grow certain vegetables, particularly the ones favoured by the European community—green peppers, carrots, lettuce, cabbages, cauliflowers and radishes. Up till now these vegetables are still called "European vegetables". With respect to African crops there was no direct intervention in production. The state merely encouraged the peasants to grow more food. At the level of marketing the state decided to buy up what could not be sold on the market at an agreed price. It also appointed a marketing officer who inspected local food prices and kept these under constant review. This policy achieved impressive results. It was the first time that the country had a purposeful plan to grow food for the domestic population. But as we have already indicated, the agricultural economy of the country had always operated in response to the needs of the external economy and at that time the needs of the external economy demanded the restructuring of food production to meet domestic needs. It is therefore not surprising that soon after the war the policy was reversed to the pre-war state. The only thing which made the colonial government adopt some of the policies of the war period was a serious food shortage which hit the country soon after the war, caused largely by maize rust. But this time the intervention was only at the level of the market but not at the level of production. But what has to be noted is that this policy of intervention in domestic food supply was seen by the colonial state, as it was during the war, as an emergency measure, and not as a basis on which to form a national food policy.

FOOD POLICY UNDER THE FIRST REPUBLIC, 1957-66

It was during the time of the First Republic that the state initiated a policy of active intervention in food and agriculture in a purposeful way. It was part of the Nkrumah government policy of state intervention in the economy generally. What was strikingly novel about this (apart from the brief period of the war years) was the plan to intervene at the level of production. The mechanism for this was the establishment of state production units. Four main working units finally came into operation. These were the State Farms Corporation, the Workers' Brigade, the United Ghana Farmers' Council and the Young Farmers' League. The plan was to transform agricultural production from small-scale peasant production units to large-scale mechanised farms relying on heavy inputs like fertilisers and treated and improved seed varieties. Unlike the policy during the colonial period when it was thought that traditional peasant agriculture was capable of providing the food and agricultural needs of the country adequately, the CPP's policy of attempting to transform agriculture through large-scale mechanised farming was based on the opposite view. Nkrumah in particular typified the new position. In an address to a Pan-African Conference of farmers he said the major task "is the creation of a complete revolution in agriculture on our continent—a total break with primitive methods and organizations...".[29] The transformation was not to come about through a change *within* the peasant mode of production. The peasant mode of production was to be supplanted by large-scale mechanised farming organised on a collectivist basis by the state. However, in one sense the new policy shared a common outlook with the old policy: both tended to neglect food agriculture. An examination of the production records of the State Farms Corporation shows that food did not figure as a priority in its plans. Only 40 percent of the acreage was devoted to what could be called food crops and of these only rice and maize received any serious attention. In 1963, in response to the deteriorating food condition in the country the State Farms were instructed to turn their attention to food crops, only to be instructed again almost immediately to revert to industrial crops to feed local industries. Hence, in spite of stated declarations attention was still focused on export and industrial crops. In actual implementation of plans there had been no departure in any meaningful way from colonial food policy.

FOOD POLICY UNDER THE SECOND REPUBLIC

After the overthrow of the Nkrumah government by a military coup in 1966 the succeeding National Liberation Council (NLC) and Progress Party (PP) governments embarked on a policy of disengagement of the state from the economy in general and from agriculture in particular. The PP manifesto states:

> The PROGRESS PARTY affirms its support for and confidence in private enterprise. Our Government will constantly search for and devise other schemes to promote, assist, reward and encourage private risk taking.[30]

Although the PP was unable to go as far as it would have liked to in the pursuit of privatisation of the public economy because of the obvious political costs, it dismantled the state enterprises. The Young Farmers' League and the United Ghana Farmers' Council were abolished. The Workers' Brigade was changed into the Food Production Corporation and its operation reduced. Although the state did not reverse the policy of seeking to transform the peasant-based agriculture it was through large-scale commercial farming with private entrepreneurs that it was now to be achieved. The state was to assist the petty bourgeoisie in the bureaucracy, in commerce and industry and in the professions to go into agriculture. The state was to provide incentives in the form of provision of agricultural credit and inputs like fertilisers, high-yielding seeds and weed-killers. It was an attempt to change the form of agriculture by changing the structure of the productive class in agriculture. It was in the Northern and Upper Regions that this system was to make itself felt most, particularly in the area of rice farming. Hence the bourgeoisification of rice farming, which reached its apogee under the government of the National Redemption Council (NRC) and th Supreme Military Council (SMC) under General Acheampong, had its beginnings in the Second Republic. The interesting thing is that in both cases—the Nkrumah policy of seeking agricultural transformation through state enterprises and the Busia policy of seeking the same end through private enterprise—the peasant who produced the bulk of the country's food needs was by-passed. This policy, therefore, like the previous one, did not make much impact on food production. In addition it would seem that Dr Busia, recognising that food production was unable to keep pace with the rate of population growth tended to show more interest in reducing the rate of population growth than in increasing food production. This, at any rate was how his frantic efforts to reduce the birth rate appeared to the general public.

FOOD POLICY UNDER THE NRC/SMC:
OPERATION FEED YOURSELF

After the efforts of the First Republic the single most ambitious programme to respond to Ghana's food problem was "Operation Feed Yourself" (OFY). This was launched with great fanfare in February 1972 after the fall of the Second Republic. It was a crash programme aimed at increasing food production and increasing national self-reliance. Initially it enjoyed considerable popular support as students from the country's universities and youth organisations responded to the call of the new military leaders to help increase food production by willingly giving their labour to harvest vegetables, sugar cane and rice on state farms.

What was significant about the OFY was that it set targets to be achieved although, as it turned out, they were unrealistically optimistic. It also for the first time devoted attention to the production of staple crops for the domestic market—crops like maize, cassava, plantain, yams, tomatoes etc. And, perhaps most importantly, it set up an institutional framework for the implementation of its plans.

The OFY programme was organised in three major phases. Phase one was to be devoted to increasing food production for domestic consumption, phase two was to concentrate on the production of raw materials for local industries and phase three to diversify and increase production for export as a source of earning foreign exchange.

The country was divided into several agricultural zones and in each zone the crop which could be most profitably grown was encouraged.

The state made food a political issue and hinged its legitimacy on it. It sought to use mobilisation methods to increase production; it ran agricultural programmes on television and other mass media and encouraged educational institutions, prisons, the Armed Forces and private commercial firms both foreign and local to establish production units. The general citizenry were similarly encouraged to take up backyard gardening to supplement the family food by producing food crops and rearing animals like poultry, pigs and rabbits.

Furthermore, as we have said earlier, a definite institutional framework was created to plan and implement the OFY programme. The Commissioner for Agriculture was the head of the OFY programme. Under him was a body called the National Operations Committee which was the instrumental mechanism for implementing the OFY programme. The Committee grouped together representatives from various government departments and agencies connected with food production and distribution. Important units dealing in one way or the other with food which were represented on the Committee were the Department of Agricultural Economy, the Department of Fisheries, the Seed Multiplication Unit, the Crop Production Unit and the Agricultural Mechanisation Unit. At the Regional level there were similar institutional bodies dealing with the OFY programme. Here the programme was headed by the Regional Commissioner. Under him was the Regional Agricultural Committee which was responsible for carrying out policies and plans formulated by the National Operations Committee.

In addition to the bureaucratic structures, there were also state agencies for production. These were made up of such organisations as the State Farms Corporation, Food Production Corporation, Settlement Division, Food Distribution Corporation, Grains and Legumes Development Board, State Fishing Corporation and the Pomadze Poultry Enterprises. Furthermore, Regional Development Corporations as centres of economic development were set up in every region. They were to plan programmes and stimulate economic activities in their respective regions. They were also expected to organise agricultural programmes for the production of food and industrial raw materials to feed local industries. Of these the most important from the point of view of investment, scope of participation and range of activities was the Upper Regional Agricultural Development Project (URADEP). It was an "integrated" agricultural development project funded by the World Bank, the United Kingdom and the Government of Ghana. It was meant to be an integrated rural development project to increase food and agricultural production, raise rural incomes, provide amenities and raise the general standard of living and the quality of life. Depending on its success a similar project was to be set up in other areas and in the Volta Region in particular. The project was to be carried out over a three-year period, 1976-81.

The project was badly conceived and badly managed. The peasants—the main targets of the scheme—were left out of all aspects of it. There was no place for peasant participation either in its conception, planning or implementation. It was therefore not surprising that this, plus a combination of bad management, inter-departmental wrangling which hampered operations, and corruption, led to the failure of the project. Mohamed Chambas writes:

> It is clear that the aims of the project have not been met on some of its most important dimensions. Given the poverty of the rural peasants in this region, how many of them can meet the requirements of the commercially organised FRCs? There is now a clear danger that as the project fails to have impact on the productivity of the mass of peasantry, food shortages will strike more frequently, causing the more able members of the rural community to migrate South as usual, perpetuating the status of the Upper Region as a reservoir of cheap labour for the cocoa and other economic and domestic sectors.[31]

Chambas concludes that the main beneficiaries of the project have been the donor countries, the international suppliers of capital, the local and national petty bourgeoisie in the agricultural bureaucracy, business, commerce and the service sector, and the landlords and rich peasants. For the majority of the peasants in food and subsistence agriculture, URADEP brought no visible benefits to relieve them of the drudgery of work or improve rural incomes or provide social amenities. But what is most surprising is that with such a record of poor performance a second project, the Volta Regional Agricultural Development Project (VORADEP) was initiated along the same lines.

The government also established the Ghana National Reconstruction Corps to involve the youth in agriculture and economic activities in the country. It was to concentrate mainly on food agriculture in the pursuit of the programme of food self-sufficiency. In its operations it was basically a youth settlement farm for large-scale farming in the food-growing areas. Like the other projects of its type it did not achieve much through mismanagement, overmanning, misconception of projects and corruption. It finally became an instrument for recruiting youth to support Acheampong's "Union Government" scheme.

As part of the programme of increasing local self-reliance in food production the government also encouraged the formation and promotion of *Nnoboa*—a traditional group farm concept where many farmers banded together to assist each other in turn, particularly during planting and harvesting. Its actual impact on food production was however negligible.

Important as it was to formulate definite policies for the production of local staples and set up organisations to execute them, there were certain problems with each set-up. In the first place, as can be seen above, the set-up was rigidly bureaucratic. The bureaucracy assumed a predominant position in the formation and implementation of policies and the bureaucratic structure did not make provision to accomodate inputs into the policy-making organs from peasants or peasant organisations at both national and regional levels. The peasant was regarded as in colonial days as a passive agent whose only role was to obey orders.

In addition to these structures and production units certain programmes were put into operation designed to ease up the bottlenecks and obstacles in food production, and create further incentives for food growers. Import duty on agricultural machinery was removed and fertilisers, improved seeds and other farm implements were sold to peasants at subsidised prices. But this in itself brought certain problems. In the first place although heavy agricultural machinery was easily available for those who had the capital for it, simple farm implements like the hoe and cutlass were hardly available, and a flourishing "black market" in cutlasses ensued. So bad was the situation that in certain areas only Regional Commissioners were empowered to distribute cutlasses to peasants. This did not help matters. It only increased the power of the bureaucracy. This was particularly serious considering the fact that the hoe and the cutlass still constitute for the majority of the peasants and particularly food producers the basic instruments of production. Furthermore, although the peasant was to benefit from such extension services hardly if ever did the peasant who in fact bore the major burden of food production benefit from such schemes. The only beneficiaries were the big commercial farmers whose class ties with the petty bourgeois in control of the state enabled them to command such resources. With the exception of rice and maize these were in most cases not even food crop growers.

Another area in which the state sought to encourage food production was the provision of credit. Schemes were provided by which credit was granted on generous terms to senior civil servants and army officers who wanted to go into agriculture. In addition the state encouraged commercial banks and lending institutions to adopt more flexible lending policies and provide credit to aspiring agricultural entrepreneurs. Of all the financial institutions giving credit for agricultural purposes the Agricultural Development Bank was the most liberal. But even here the small food crop grower did not fare any better. Until recently, before a farmer could qualify for credit from the Agricultural Development Bank (ADB), he was expected to have cleared at least six acres of cultivable land and have an acceptable security. If we realise that the average holding in the country is not more than two hectares and that for food crop growers it might even be smaller, these two requirements put any institutional credit outside the reach of the majority of the peasants. The Bank itself might have realised the problem for in 1969 it introduced the Credit Commodity Scheme or the Group Loans Schemes by which a number of growers, usually of food crops such as maize, rice, cassava but also sometimes cotton and ginger, got together to constitute a group on the basis of a common crop. The minimum number for the group was to be ten members. It had an elected leader and an executive who guaranteed the loan and were held responsible for repayment. It is estimated that between 1969 and 1976 the ADB disbursed a total of about ₵ 52 million in loans to 241,009 farmers cultivating crops such as maize, cassava, yam, rice, tomato etc.[32] However, this was just a drop in the ocean compared to the credit needs of the small-scale food producers: most of the loans continued to go to large-scale producers. One reason which has often been given for the poor response of financial institutions to the demands for agricultural credit from small-scale producers, particularly food growers, is the fear of default. A study of the ADB Credit Commodity Scheme, however, revealed that between 1971 and 1977 the recovery rate among small-scale producers was between 23 and 62 percent with a mean of 43 percent while among the large-scale producers, over the same time period, the recovery rate was only 23 percent.[33]

In 1976 the Ghana Commercial Bank also initiated a Group Lending Scheme similar to that of the ADB's Credit Commodity Scheme. Here farmers were encouraged to form a Farmers Association and encouraged to grow food crops. Members of the association could continue to cultivate their own individual farms but in addition they had to cultivate a group farm on a co-operative basis. In return the bank advanced credit for agricultural machinery and other inputs. Just as in the case of the ADB Scheme the group leaders had to guarantee the loan and all loans must be repaid before the next loan would be granted. Membership was limited to a maximum of 50 and a minimum of five. Each member was expected to cultivate a minimum of four and a maximum of 25 acres.

Another source of institutional credit was the rural banking system. It was started in 1976 in recognition of the fact that most of the existing schemes did not respond adequately to the credit needs of the small-scale producer. Hence at its inception the focus was placed on the small-scale producers, most of whom are food producers. One would therefore see in this scheme an attempt to capitalise on food growers. The bank's definition of a small-scale producer, however, turns out to be anyone with a holding of not more than 100 acres. Considering that only 1.8 percent of agricultural producers in the country have holdings of more than 50 acres, in real terms it means that practically every farmer is qualified under the scheme to get financial assistance from the bank. In its conception it was hoped that the rural bank would eliminate the rural private money-lender whose sharp business practices had put many farmers into debt, and become an instrument for rural development. By 1978 five rural banks had been set up. By December 1983 the figure had reached 83. The idea of the rural bank has grown very rapidly in recent years and bank officials have spared no pains to express satisfaction at the progress made so far. In 1978 it was able to make a ₵ 1.2 million investment in government bonds. Whereas bank officials were enthused about this, it prompted the criticism that the bank which was specifically set up to develop the rural areas had become a conduit through which funds were drained from the rural areas to the urban areas.[34]

Commendable as some of these efforts are it is quite clear from the account we have given above that access to institutional credit was still heavily weighted in favour of the large-scale producer, and that for the majority of small-scale producers who produce more than 90 percent of the country's food crops the rural private money-lender still provides the only readily available source of credit in the rural community.

In addition to the above the state also intervened at the level of marketing and distribution. A number of state agencies were established to assist in the marketing and distribution of food. The best known of these were the Cattle Development Board later to be called Meat Marketing Board, the Food Distribution Corporation (FDC) and the State Fishing Corporation (SFC) (although the genesis of this dated from much earlier). The SFC is a production unit, although it has numerous distribution outlets. They were to buy bulk food either from the rural areas or overseas and retail it at state-designated prices. This was expected to help stabilise prices and alleviate the severe food problem especially during the "lean season". The assumption underlying the strategy was that inefficient marketing and distribution constituted one of the main obstacles to adequate food supply in the urban areas. However, the effectiveness of these agencies was hampered by three main factors. One was the lack of an adequate number of suitable transport vehicles. A second was the lack of adequate storage facilities and the third was bad pricing policies. The FDC had far fewer vehicles than it needed to do a halfway decent job. In addition, most of its vehicles were huge articulators which were useful on the main roads but could not ply the feeder roads which led to the farms. The main problem with food transportation was from farm to main road transport. For in many areas food had to be carried many miles by head porterage which limited the amount of food which could be so transported. The result of this was that large quantities of food each year rotted on the farms. With regard to this particular problem the vehicles of the FDC, few as they were, were not particularly useful. At one time during the military regime of General Acheampong there was even talk of using army helicopters to evacuate food and other agricultural products: not even the USA can afford such a luxury. The agencies' less than adequate storage facilities meant that they suffered from a high spoilage rate. This was particularly so with the FDC. Sometimes it had to put food on the market when it could not be sold to cover even costs so as to avoid spoilage. But perhaps the worst problem of these agencies which completely undermined their effectiveness, was their pricing policies. The assumption was that these agencies were to sell at prices which the poor wage-earner could afford. But the prices were so infrequently reviewed in line with the cost of living that there existed at one particular time a high disparity between the agencies' prices and the commercial market prices. A particularly notorious example of this was the Meat Marketing Board which continued to sell meat at about ₵ 4.20 (four cedis and twenty pesewas) per kg when the going commercial price was about ₵ 15 (fifteen cedis) per pound. It was the same with the State Fishing Corporation. The result of this was that there was such pressure on these agencies that only the very highly placed members of the petty bourgeoisie with interlocking ties to such agencies were able to satisfy their food needs from such bodies. One had to be rich and powerful enough to be able to buy food at "control prices". The poor wage-earners whom the scheme was to benefit ended up subsidising food prices for the rich and powerful. Furthermore, the huge price differential encouraged speculative activities on the part of the managerial class in control of such agencies as they diverted stock to private vendors for huge profits. Thus, these agencies, the Meat Marketing Board, the Food Production Corporation and the State Fishing Corporation all became mechanisms for personal appropriation and enrichment on the part of the petty bourgeoisie and their hangers-on in control of the state apparatus.

In spite of its weaknesses and shortcomings the OFY programme has been the most ambitious and purposeful attempt so far to respond to the domestic food and agricultural needs of the country, both in programmatic and institutional terms. In one aspect it marked a departure from previous food and agricultural policies. This was with respect to the Special Agricultural Scheme.

It was introduced in 1974 as a complement to OFY. It was a scheme by which the state sought to involve transnational corporation (TNCs) in food and agricultural production. This was something which even the colonial state had been anxious to avoid for obvious economic and political reasons. However, with the emergence in government circles of the view that agricultural success in the country lay in transforming peasant agriculture into large-scale mechanised farming, the role of foreign capital in the form of the transnational corporation became an important issue, especially considering the inability of the state to fulfil this role and the lack of a strong enterprising and productive bourgeois class in the agricultural sector.[35] As Ghana's rate of production showed no visible signs of outstripping its population growth in the early 1970s policy-makers began to look with admiration at the Ivory Coast and the East African countries, particularly Kenya, which they saw as a model of agricultural success. To them the secret of their success lay in the role of foreign investment in agriculture. Besides, reasons were not lacking as to the progressive role of the TNCs. They were repositories of skills, technical know-how and capital. Their large-scale operations enable them to take advantage of economies of scale. They are linked to international capital through which they have extensive marketing outlets; and in Ghana in particular they had huge stocks of idle capital in the form of profits and dividends waiting for repatriation. What better plan could be devised than a scheme to mobilise such resources for the mutual benefit of foreign capital and the nation? This was the thinking behind the scheme. There was hardly any consideration of the political and economic consequences of putting such strategic industry into the hands of foreign capital.

As we have already pointed out, the scheme was to get transnational corporations to go into agriculture, producing food crops for the local market, raw materials for local industries and export crops. The specific crops included cereals like rice and maize, industrial crops like cotton, palm oil, coconut and kenaf and crops with export potential like avocado, pineapples, banana and ginger. Also included were livestock and poultry. The projects were to be financed from the companies' accumulated profits and dividends awaiting repatriation.

State participation in the scheme included equity shares, land acquisition and site selection and the provision of social infrastructure like roads, electricity supply etc. In addition the state also provided, among other things, guaranteed minimum prices, exemption from duty and levy on machinery, and a five-year tax holiday. In spite of these attractive conditions the response was not particularly enthusiastic. By 1979 only eight companies had accepted to participate in the scheme. Table 1.2 shows the participating companies.

The scheme has been in operation for a decade now. What impact has it had on food and agricultural production in the country? Quite apart from the fact that the scheme has not had any significant impact on food and agricultural production in the country it has had some deleterious effects on Ghana's agricultural economy and the economic situation in general. Kwesi Jonah has, in an exellent study of the politics of indigenisation in Ghana, demonstrated how the operation of the scheme has resulted in the de-indigenisation of Ghanaian capital and the displacement of some 3,000 peasant producers from the Western Region. Surely this does not augur well for food production.[36] The first two years of OFY did show some modest success. Cultivated acreage of maize went up from 997,000 acres in 1970 to 1,050,000 acres in 1974; and production went up from 378,000 to 478,000 long tons within the same period. The record for rice registered a similar success story: production went up from 48,000 to 72,000 long tons between 1970 and 1974.[37] For the first time in over two decades Ghana did not import any rice in 1974. Government leaders were enthusiastic and proudly announced that the country had attained self-sufficiency in rice and maize. The announcement, however, turned out to be premature. As in the case of the Asian countries like India which embarked on the "Green Revolution" the impressive increases in rice and maize were obtained at the expense of more traditional crops like yams, millet and guinea corn.[38] This was particularly the case in the Northern and Upper Regions where rice farming on a large scale was initiated. After 1975 production started to decline and in the following year the situation, due to drought in the north-east section of the country, was so bad that Ghana had

TABLE 1.2
Special Agricultural Scheme: Participating Companies, 1976

Name of Company	Shareholders	Type of project	Project Cost (₵)	Amount Approved for transfer December 1983 (₵)
1. Benso Oil Palm Plantation Ltd.	UAC Int. 60%, Govt. of Ghana 40%	Oil palm plantation, processing of palm fruit	27,200,000	15,437,580
2. SCOA Farms Ltd.	SCOA (Ghana) Ltd. 60%, Ghanaian private 40%	Fruits and vegetables for export, cereals for local market, ornamental plants	3,650,000	935,000
3. Demco Ltd.	John Holt (Liverpool) 20%, NIB 40%, ADB 40%	Agriculture plant pool	1,000,000	–
4. Nasia Rice Co.	Barclays Bank 33 ⅓%, NIB 33 ⅓%, ADB 33 ⅓%	Rice farming	490,000	–
5. Glamour Farms	Glamour Ltd. 60%, Ghanaian private 40%	Poultry complex	3,000,000	150,000
6. Danafco Farms	Boltsten Holdings Ltd. 60%, Ghanaian private 40%	Dairy farming	1,500,000	–
7. Ghana Livestock Ltd.	Barclays Bank Int 27 ½%, Standard Bank 15%, Ghana Commercial Bank 22 ½%, ADB 20%, Bank of Ghana 15%	Cattle Ranching	1,000,000	–
8. Ghana Industrial Farms Ltd.	CFAO (Ghana) Ltd. 30%, Ghana Commercial Bank 40%, Shell Int. 30%	Cereals and vegetables, poultry complex	5,750,000	Project abandoned

Source: Jonah, Kwesi, *The Politics of Economic Decolonisation: The case of Ghana's Indigenisation Policy* (Unpublished M.A. thesis, University of Ghana, Legon, 1980) p. 170; Adei, Stephen, "Transnational Corporations and Self-Sufficiency in Food Production in West Africa", paper presented to ISSER Conference on Food Self-Sufficiency in West Africa, University of Ghana, Legon, 1-3 May 1984, p.15.

to appeal for international food aid. It was a great setback for the government which had hitched its legitimacy to the provision of adequate food for domestic consumption.

Thus the programme which was meant to respond to the food needs of the country failed. One reason was that in spite of public declarations to the contrary the emphasis still continued to be on export and industrial crops. We have seen how in the areas of capitalisation, extension or research disproportionate attention was given to export or industrial crops. It is perhaps not too far-fetched to suggest that many public officials probably saw the whole OFY programme as an emergency measure, for barely two years after the inauguration of the programme the government declared it a success, announced that the country had achieved food self-sufficiency and launched the second phase "Operation Feed Your Industries".

Thus the concern for food took a back seat. The divided attention failed, and in the long run the country was unable to feed either her industries or her citizens. As for the members of the petty bourgeoisie located in the bureaucracy, the Armed Forces or the distributive sectors of the economy they saw the whole venture as an opportunity for personal enrichment. Some of those who benefited from the huge loans from the state failed through inefficiency and bad management, as they still kept their paid jobs and only made occasional visits to the farms. A successful first generation absentee farmer is a rare commodity. Others used their farms merely as agricultural fronts and continued to engage in those speculative activities in which they were well at home. Fanon's description of this class is as true today as it was when it was written:

> The national bourgeoisie of underdeveloped countries is not engaged in production, nor in invention, nor building, nor labour; it is completely canalised into activities of the intermediary type. Its innermost vocation seems to be to keep in the running and to be part of the racket. The psychology of the national bourgeoisie is that of the businessman, not that of a captain of industry.[39]

To think that one could build up a rural capitalist class from the ranks of the petty bourgeoisie presently located in the bureaucracy, the army or the distributive sectors of the economy is a socio-economic illusion. This can only be done from the ranks of the farming community but so long as they do not have any influence or control over the state agencies responsible for making available the capital for such purpose it is difficult to see it happening.

FOOD POLICIES UNDER THE THIRD REPUBLIC

The failure of the government's food and agricultural policy contributed to the general economic and political malaise in the country and created the conditions in which a faction within the military could justify its overthrow in 1979. This set off a chain of reactions which finally led to the restoration of civilian rule in 1979 under the leadership of Dr Hilla Limann. This was the Third Republic.

Under his leadership the government continued to show keen interest in Acheampong's policy of seeking to attract transnational corporations into agriculture. Added importance was given to this policy by the government's new Investment Code which widened the scope and area of TNCs' and foreign capital's involvement in food and agriculture. The main policy document in agriculture, *Action Programme for Agricultural Production, 1980-81*, repeated what previous policy-makers had already said: the need to produce not "merely adequate staple food-stuffs to satisfy our domestic needs but eventually surpluses for export".[40] It stated the government's intention of utilising "the private small-scale producers, the private commercial or corporate farms and the parastatal organizations".[41] It is difficult, however, to see any clearly thought out schemes and social institutions by which the objectives were to be attained. By any standard the programme was not an improvement on the OFY. In some ways it was even a step backward. After a couple of years it was clear that the policies were not making any impact on food and agriculture in the country. The decline continued and the general economic situation became worse. In December 1981 Dr Limann's government was overthrown and a new government, the Provisional National Defence Council (PNDC) with Flt Lt Jerry Rawlings as Chairman, was set up.

FOOD POLICY UNDER THE PNDC

Rawlings initiated his coup with a promise to seek a revolutionary transformation of Ghanaian society. With regard to food and agricultural policy, he proclaimed a "Green Revolution". But this turned out to be more an attempt to generate public enthusiasm for food and agricultural production than the beginning of new initiatives in food agriculture.

On the hand, some of the initial activities of the regime and its supporters were clearly harmful to agriculture. This was particularly so with regard to the ban on the importation of food and agricultural inputs, and the actions of certain over-enthusiastic supporters, particularly armed soldiers, in trying to force down the prices of food in the absence of increased output. Other reasons like the long drought and fires which destroyed farms did not help matters. In 1983 the food situation was particularly severe. Of late the regime has refrained from such activities and more positive steps are being taken to enhance food supply. Attempts have been made to reorganise some of the institutions connected with food production and distribution. The Ministry of Agriculture has been decentralised with the appointment of regional directors so that more decisions could be taken at the local level. There have been plans to restructure the Food Production Corporation and the Ghana National Reconstruction Corps. The Ministry of Agriculture is to take over their farms for use as service centres.

The policy of emphasising a few selected crops, in this case maize, cassava and rice, is a sensible one, but the fact still remains that there have been no new innovations or policy initiatives in the social organisation of production to achieve increased productivity other than appeals to producers. The idea that agricultural success depends on heavy capital investment mainly from TNCs and international finance organisations has not been abandoned and in spite of proclamations to the contrary and the overtly populist posture of the regime, the peasant farmer is still not at the centre of government policy towards food and agriculture. There have been no marked departures from previous policies which have proved to be disasters. As in the past, more effort seems to be expended on distribution than production. A National Mobilisation Committee has been formed but it seems to be more concerned with monitoring the distribution of food and assessing the food needs of specific areas than organising production. It is clear from what we have been saying that since independence we have not been able to work out a coherent and integrated national policy to respond adequately to the food needs of the country. Colonel Bernasko, a trained agriculturist and one time Commissioner for Agriculture, was right when he declared:

> Since Ghana achieved independence we have struggled over food problems... By and large it can be said that in spite of the stupendous efforts of the past we have not brought the food problem under control. We have had shortages of cassava and shortage of plantain. It appeared every successive Government had been embarrassed at one time or the other by the chronic food shortages.

The question then becomes: what is to be done?

First, we must abandon the tendency to treat the food problem as merely an emergency issue which needs a one-time solution. There is a tendency on the part of both the government and the public to relax once a particularly good year appears, thinking that the problem has been solved, only for it to reappear the following year and then to be followed by another frantic effort. There should be a continuous and sustained attack on the food problem. Second, food production should be raised to the level of an ideological commitment and a social organisation of production involving general mobilisation should be planned and carefully implemented. A comprehensive national food policy which will integrate and co-ordinate production, distribution and marketing, extension, storage and research into a single all-embracing policy carefully conceived and efficiently implemented should be worked out. Operation Feed Yourself had some good points which could be integrated into such a new policy. The crisis in the economy calls for radical solutions. We cannot wait till we get all the capital and the resources before we begin to plan the radical transformation of our agricultural economy. It is precisely because we are in crisis and we do not have resources that we need a radical solution. There is need to transform the agricultural economy to provide for the domestic needs of our people and also to feed our local industries, but this can only be done through appropriate policies and undertaken by a class which is currently in agriculture and not by a rentier class.

NOTES

1. *Guardian*, 24 July 1984, also see *People's Daily Graphic*, 5 July 1984.
2. For descriptions of Ghana's agricultural economy see: Dickson, K., *A Historical Geography of Ghana*, London, Cambridge University Press, 1971; Dickson, K. and Benneh, G., *A New Geography of Ghana*, London, Longman, 1970; Boateng, E.A., *Geography of Ghana*, London, Cambridge University Press, 1967; La Anyane, S., *Ghana Agriculture*, London, OUP, 1963; Willis, J.B., *Agriculture and Land Use in Ghana*, London, OUP, 1962.
3. See Shepherd, Andrew, *Capitalist Rice Farming in Northern Ghana*, unpublished Ph. D. Thesis, Cambridge University, 1979.
4. *West Africa*, 5 March 1984.
5. Lewis, Arthur, *Report on Industrialization in the Gold Coast*, Accra, 1953.
6. Republic of Ghana, *Five-Year Development Plan, 1975-1980*, Accra, 1977, p. 3.
7. Gyekye, L.O., "The Role of Group Lending Schemes in Food Production in Ghana: An appraisal", paper presented at the ISSER Conference on Food Self-Sufficiency in West Africa, University of Ghana, Legon, 1-3 May 1984, p. 1.
8. Quoted in Benneh, G., "Some Policy Issues in the Quest for Food Self-Sufficiency in Ghana", paper presented at the ISSER Conference on Food Self-Sufficiency in West Africa, University of Ghana, Legon, 1-3 May 1984, p. 5.

9. Nyanteng, V.K. and Van Apeldoorn, G.J., *The Farmer and the Marketing of Foodstuffs*, ISSER, University of Ghana, Legon, Technical Publication Series, No 19, p. 2, also see appendix 2.

10. Quoted in Jones, Trevor, *Ghana's First Republic, 1960-1966: The Pursuit of the Political Kingdom*, London, Methuen, 1976, p. 245. .

11. Ewusi, Kodwo, "Road Transport Facilities and their Effect on local Food Prices in the Sixties", *Economic Bulletin of Ghana*, No. 3, 1971.

12. *Quarterly Digest of Statistics* ‡Ghana£ June 1985, p. 45.

13. Benneh, op. cit.

14. *West Africa*, 26 March 1984, p. 655.

15. *West Africa*, 30 July 1984, p. 1556.

16. *West Africa*, 18 July 1983, p. 1691.

17. *People's Daily Graphic*, 8 March 1984.

18. "The Role of Marketing in Present Day Ghanaian Agriculture", USAID/Ghana 8th Annual Agricultural Conference, Accra, 1976.

19. *Ghanaian Times*, 13 July 1965. Quoted in Killick, Tony, *Development Economics in Action*, London, Heinemann, 1978, p. 188.

20. Gold Coast, Department of Agriculture, *Annual Report*, 1947-48, p. 1.

21. *Africa Contemporary Record, 1976.*

22. Killick, op. cit.

23. For the early beginning and development of cocoa agriculture see Hill, Polly, *Migrant Cocoa Farmers of Southern Ghana*, London, Cambridge University Press, 1963.

24. Kay, Geoffrey, *The Political Economy of Ghana*, London, Cambridge University Press, 1972; Hymer, Stephen, *The Political Economy of the Gold Coast and Ghana*, New Haven, Yale University, Occasional Paper No 73, 1969.

25. For details of this see La Anyane, *Ghana Agriculture*.

26. Ibid., p. 73.

27. Ghana, Ministry of Agriculture, *Annual Report*, 1968, p. 8.

28. Gold Coast, Department of Agriculture, *Annual Report*, 1941-42.

29. Quoted in Killick, op. cit.

30. *Progress Party Manifesto*, 2 August 1969, p. 9.

31. Chambas, Mohamed Ibn, *The Politics of Agricultural and Rural Development in the Upper Region of Ghana, Implications of Technocratic Ideology and Non-Participatory Development*, unpublished Ph. D. Thesis, Cornell University, 1980.

32. Kodjo, Solomon N., *Financing the Small Scale Farmer in Ghana*, unpublished M.A. Thesis, University of Ghana, Legon 1979, p. 112.

33. Ibid., p. 113.

34. Ibid., p. 125.

35. For the role of TNCs in African agriculture see Dinham, Barbara and Hines, Colin, *Agribusiness in Africa*, London, Earth Resources Research Publication, 1982.

36. Jonah, Kwesi, *The Politics of Economic Decolonisation: The Case of Ghana's Indigenisation Policy*, unpublished M.A. Thesis, University of Ghana, Legon, 1980.

37. Ghana, *Five-Year Development Plan, 1975-1980*, p. 3.

38. For the experience of India see Griffin, Keith, *The Political Economy of Agrarian Change*, London, Macmillan, 1974.

39. Fanon, Frantz, *The Wretched of the Earth*, Penguin, 1967, pp. 119-20.

40. Ghana, *Action Programme for Agricultural Production 1980-1981*, Accra, p. 3.

41. Ibid., p. 13.

2. AGRICULTURAL AND FOOD POLICY IN MALAWI: A REVIEW

Guy C.Z. Mhone

Maize is by far the most important crop as it is the staple food for most people. It is the policy of the Government to ensure that Malawi is self sufficient in foodstuffs even in a poor cropping season. In addition, *it is now possible to* export maize and there is thus a need to produce a large surplus—each year. This increased need for maize should be the result of *higher yields per acre*, and no encouragement should be given to growing of maize on land at present used to grow other crops.

Higher maize yields per acre will themselves increase a farmer's income and will also leave him with the time and land needed to produce other cash crops that offer him a higher return per acre. It is most important that this policy is understood by all concerned as any tendency to increase the proportion of cultivated land planted with maize would be directly contrary to the nation's long-term objectives.

> *Guide to Agricultural Production in Malawi 1979/1980*,
> Ministry of Agriculture and Natural Resources,
> Malawi Government, p. 31 (author's emphasis)

INTRODUCTION

The above quotation is as succinct and direct a statement of policy as one can expect from the Malawi government. It is a statement of food policy, in particular that on maize, directed at peasant farmers, who are the main producers of staple foods in Malawi. For anyone familiar with the Malawi political situation the dictatorial and paternalistic undertones of the statement are unmistakable. As indicated in the statement, food policy is not an end in itself; it is an element of a larger strategy of achieving long-term national objectives.

Malawi's food policy is one of meeting the minimal basic needs of its population in what is ambiguously referred to as self-sufficiency without any stipulation of its qualitative aspects. Once self-sufficiency is assured food policy gives in to other objectives.

This paper is an attempt to place Malawi's food policy within the context of the government's long-term objectives. The paper relies heavily on the critical literature on Malawi by such writers as Mwakasungula, Mkandawire and Kydd and Christianssen.[1] The aim of the paper is to recast some of the views of these writers into a more general paradigm of the political economy of Malawi in an attempt to show that the Malawi government, despite it rhetoric, does not have a positive policy directed at improving the welfare of the masses, and that, in fact, the logic of the society's socio-economic and political structure demands the exploitation of the masses.

In other words, the debate on whether Malawi's so-called "economic miracle" has benefited the masses is secondary to the question of whether Malawi's economic policies are intended to benefit the masses.[2] In this paper we argue that Malawi's economic policy and in particular agricultural and food policy is designed to prop up a fascist regime and only incidentally and peripherally are the interests of the masses positively taken into account within a developmental perspective, the populist and paternalistic rhetoric by Dr Banda notwithstanding.

That Malawi's post-independence economic performance has been reasonable if not good compared to other similarly endowed sub-Saharan countries is generally accepted.[3] It is true however that while the World Bank has been particularly impressed by this growth the IMF has been lukewarm if not sceptical about the country's relative performance, which the IMF contends has been average.

In any case for the period 1967 to 1979 no one can credibly argue that the economy was static or that it regressed. Official statistics indicate that an average annual growth rate of 7.4 percent in monetary GDP was achieved between 1967 and 1979. At least two issues arise from this performance. First, it has been of interest among observers and academics alike as to whether this performance has been due to Dr Banda's pro-Western and pro-capitalist stance implying a market-oriented economy. Second, observers and researchers have been interested in the welfare and equity implications of the growth performance.

The diverging views on the above two issues are adequately summarised by Kydd and Christianssen, Mkandawire, and Mwakasungula, and in fact each of these writers takes a clear position with respect to them. The above writers have however convincingly demonstrated the need to question any enthusiastic claims to be made about the "market-oriented" nature of the growth and its presumed trickle-down effects in improving the welfare of the masses. Despite differences in interpretation these writers complement each other in delineating the main features of policy and structural changes that occurred particularly in the agricultural sector during the growth period. Drawing on the work of these writers this paper will first attempt to outline the main features of agricultural and food policy since independence; second, implications of the policy will be discussed; and finally, the conclusion will locate the foregoing within a paradigm of the imperatives of the Malawi state.

MAIN FEATURES OF AGRICULTURAL POLICY AND DEVELOPMENT SINCE INDEPENDENCE

One of Banda's legacies will be his attempt to turn what historically had been a curse into a blessing. The lack of colonial interest in the development of Malawi was due to the fact that it had no minerals that were considered economically worth exploiting. It does, however, possess a diverse range of minerals which have so far been viewed as uneconomical. Malawi's agricultural potential was, however, appreciated during the colonial days even if it was not adequately exploited due to the country's lack of complementary and more enticing resources such as minerals, and due to its being landlocked. The neighbouring countries of Zimbabwe and Zambia let alone South Africa to the south offered much more attractive settler opportunities. Nevertheless the economy of colonial Malawi evolved on the basis of plantation agriculture, underdeveloped peasant farming and the export of labour to its better endowed neighbours to the west and south.

The agricultural nature of the economy and its role as a labour reserve economy were further reinforced by the peripheral role assigned Malawi during the Federation of Rhodesia and Nyasaland.

The reputation of Malawi as an "imperial slum" was shared by both colonialists and neighbouring countries, where Malawians worked, mostly in menial or undesirable occupations, and it left an indelible mark on nationalist politicians, including Dr Banda, who were determined to salvage Malawi from the imperial junk heap. The Malawi nationalist movement enhanced Malawi's reputation partly by acting as the vanguard in the break-up of the Federation and becoming the first independent country in Southern Africa. But even more, Dr Banda had vowed to turn Malawi's legacy of agricultural specialisation, albeit an underdeveloped one, into an asset that would become the key to development in a manner that would make its neighbours envious. Dr Banda has demonstrated his seriousness about this by personally holding the post of Minister of Agriculture and Natural Resources since independence.

Any ambiguity that may have existed as to Malawi's post-independence future path was dispelled immediately after the "cabinet crisis" of September 1964, a bare two months after independence and two years after self-government. After ousting the rebellious "ex-ministers" and firmly establishing himself in the driver's seat Dr Banda proclaimed his unequivocal commitment to capitalism, the West, and the international economic system. Further, despite the belated reservations on Malawi development strategy expressed by the World Bank, Dr Banda even based his economic strategy on classical economic prescriptions.[4] The basic economic strategy was one based on suppressing consumption and redistributing surpluses in favour of investment in order to guarantee a high growth rate. Agricultural policy was subordinated to this overall strategy.

In accepting classical development prescriptions Dr Banda had also committed Malawi to an export-led development strategy based on the country's major agricultural exports consisting of tobacco, tea, groundnuts and cotton primarily. In the first five years of independence export promotion was pursued without drastically tampering with agrarian relations and their structure. In fact, export promotion was pursued together with the desire to advance smallholder peasant production. The nationalist government, and in particular Dr Banda, having been a very outspoken critic of colonial agrarian malpractices such as semifeudal tenancy systems and forced labour, was compelled, even if in a token manner, to be seen visibly and symbolically to remove past inequitable practices.[5]

Thus the initial half-hearted support for smallholder communal agriculture was not unexpected. Further, during the first five years such a populist policy was necessary as an antidote to the ruthless suppression of opposition and the progressive reinforcement of the dictatorial one-man-rule that Dr Banda was unleashing during this period. It is not surprising that this populism was summarily discarded at the peak of Dr Banda's power and repression in the early 1970s.[6] Smallholder populism was promoted as long as Dr Banda's power had not been consolidated and as long as colonial memories were fresh. After having firmly consolidated his political power as sole and unchallenged ruler of Malawi in the political sphere, Dr Banda proceeded to reinforce this power with a material base in the early 1970s.

The smallholder promotion policy that Dr Banda had pursued immediately after independence was in part a legacy of the liberalism of a dying colonialism which had attempted to make amends by containing and even reducing the encroachment on customary land.

In fact this colonial policy had in part succeeded in reversing the trend toward the dominance of estate agriculture in cash crop farming.[7] However, after the late 1960s this trend was reversed back to the early colonial structure. Banda's trumpcard was his holding company Press Holdings in which he had more than a 99 percent share and which he deployed systematically and strategically so as personally to have a stake and control in the commanding heights of the Malawi economy particularly in finance, agriculture, retailing and distribution.

The government could restrain consumption by decreeing appropriate incomes and prices policies, but it needed to control the disposal of investible surpluses. This was accomplished through Press Holdings' purchase of majority equity in the two major banks while the Agricultural Development and Marketing Corporation (ADMARC) and the Malawi Development Corporation held minority shares. All three organisations were intertwined through interlocking directorates in which final authority and control resided in Dr Banda through Press Holdings' dominance. With this set-up Dr Banda was able to initiate the development of estate agriculture as the backbone of agricultural strategy in Malawi. The period 1969 to 1973, marking the height of Banda's repression, was characterised by what Kydd has referred to as the era of "Presidential fiefdom" whereby Dr Banda not only extended Press Holdings' control over tobacco estate farming and monopoly over flue-cured and burley tobacco, but began to endow his "entourage" or close top political aides and their relations with assets in the estate and industrial sphere.[8]

The second period, stretching from 1973 to 1977, was characterised by Dr Banda's emphasis on dispensing political patronage by encouraging political leaders and top civil servants to purchase estates on the basis of credit from ADMARC or the banks with his backing.[9] Investment in estate farming has for long remained the only viable option for the aspiring capitalists since the retail and distribution, and manufacturing sectors are monopolised by Press Holdings solely or with joint participation of foreign companies.

Initially, this dispensation of patronage through estate farming entailed the Africanisation of existing settler farms which had been abandoned or sold to ADMARC. In the course of the 1970s, however, the development gained such momentum that the privatisation and individualisation of communal trust land was embarked upon. This not only increased the number of estate farmers but it also increased the area of cultivable land devoted to estate farming. Dr Banda had often commented that communal ownership of land was not only a disincentive to agricultural development but also prevented financial institutions from lending money to prospective land developers.[10]

Accordingly, legislation had been passed in 1967 to facilitate the move toward the privatisation of land through the granting of leases over communal land upon the request of an individual with the backing of the appropriate political leaders and financial institutions and the consent of the local chief or headman.[11]

This growth in estate farming was facilitated by ADMARC's huge profits accumulated during the growth period. Since ADMARC dealt with the purchase and resale of peasant produce, and the sale of modern agricultural inputs to smallholder farms, the profits came from the forced taxation of peasant farmers.[12] Thus not only was potential and actual cultivated land progressively being taken away from the peasants but the latter were actually made to finance its development as well.

The privatisation of communal land was further reinforced by the reinforcement of the colonial "master farmer" approach to agricultural development in the form of the "Achikumbe" or "progressive or master farmer" scheme. This scheme provided agricultural information and inputs to a select number of what were considered enterprising and promising communal farmers, who would in the process be given leasehold over the land they occupied. The rationale was that such "demonstration" farmers could induce spread effects or positive externalities in having their neighbours emulating them. Farmers could qualify to become "Achikumbe" by sustaining a stipulated level of productivity for a number of years in a sequence. During the colonial period the system was roundly criticised by nationalists since it was inequitable, particularly in that such farmers were actually subsidised through the taxation of their poorer neighbours. Further, such farmers always identified with their colonial benefactors and were seen as a reactionary force.[13]

When Dr. Banda reactivated the "Achikumbe" programme he rationalised it in classic capitalist ideology and saw it as complementing estate farming and as a progressive force among the communal farmers. Essentially the "Achikumbe" were small-scale peasant farmers engaged in commercial agriculture. Such farming necessarily implied the enlargement of farms and the use of hired labour. It should be noted that in the late 1950s the then Director of Agriculture had pointed out that the programme could not possibly be expanded because of the high population density rates and the consequent shortage of cultivable land. In fact he suggested that "the ultimate solution to this agrarian problem [was] in part dependent on the growth of industry [as an] alternative occupation to agriculture and in the introduction of a sound system of land tenure".[14] This agrarian problem was undoubtedly more acute in the Malawi of the 1970s, but where the colonial administrator was unwilling to tread for fear of disrupting communal life, Dr Banda, armed with capitalist ideology, ran roughshod over communal lands by dispossessing the peasants and proletarianising a large number of them through the development of estate and "Achikumbe" farming.

By the late 1970s agrarian policy had reverted to its early colonial pattern of an essentially bimodal structure with estates on one side and smallholder farms on the other. However, the post-independence situation also represented a greater differentiation among the peasantry (the smallholder farmers) due to the fostering of the "Achikumbe". The "Achikumbe" represented a continuum of farming from those utilising only family labour to those using seasonal casual labour, and finally to those that employed wage labour but not on the same scale as the estates, which are close to being agribusinesses with absentee owners. The majority of the peasants outside of the "Achikumbe" group remained subsistence farmers who peripherally engaged in cash crop farming.

Government, together with aid agencies or foreign-borrowed money, subsidised commercial estate agriculture even further through substantial investments in economic infrastructure to facilitate the distribution and transportation of agricultural produce, inputs, research and information. While the most sophisticated research information was produced by the government for use by estate farmers and "Achikumbe", peasants were relegated to listening to agricultural radio programmes and reading agricultural pamphlets, or, as was presumed, to emulating their more fortunate "Achikumbe". In the course of events the World Bank was alarmed by the inattention given to peasant farmers and under the rationale of attempting to accomplish equity began experimental schemes in integrated rural development to benefit peasant farmers.[15] The schemes essentially entailed the coordination of concentrations of peasant farmers in a manner that allowed them to exploit infrastructure, inputs, information and extension manpower that had been provided through aid funds.

There has been a tendency to exaggerate the difference and presumed conflict between the strategy of the World Bank in showing concern for the peasant farmer through integrated rural development and that of the government in promoting large-scale farming.[16] Dr Banda had in the early 1970s approached the World Bank to finance some of his development schemes including the promotion of large-scale farming. The World Bank for a number of reasons was reluctant to do so particularly given the ambiguous and anomalous relationship that existed among the four bodies: government, Press Holdings, ADMARC, and the banks. Banda's private and official interests were all intertwined and inseparable. Further, the utilisation and dispensation of funds blurred the line between patronage and corruption and economic justification. It was the cold reception he got from the World Bank that made him resort to the exploitation of indigenous investible surpluses through ADMARC and the banks, which even engaged in external borrowing to finance some of Banda's schemes through Press Holdings. It is clear therefore that Banda had preempted the World Bank by going it alone in the development of estate farming. The World Bank therefore had no alternative but to find something to do with peasants since it was essentially redundant in the other sector (the estate sector), its altruism notwithstanding.[17]

The Malawi government has pursued a consistent policy with respect to the role agriculture has had to play in the development of Malawi. Briefly, it sees its task as one of ''raising the living standards of the rural people... by estate development in agriculture that would provide employment and foreign exchange''.

Now it is true that almost all Third World countries, particularly the non-oil non-mineral rich ones, are preoccupied with foreign exchange. But very few such countries have subordinated and oriented all domestic economic structures towards the maximisation of foreign exchange to the degree that Malawi has. The key to Malawi's agricultural and food policy is to be found in Dr Banda's concern with foreign exchange. It was mentioned earlier that both prior to and after independence Dr Banda was determined to turn what seemed to be a hopeless situation into a promising one. He fully recognised the importance of agriculture to Malawi in the international division of labour and specialisation, and, on the basis of his belief in capitalist ideology, he was determined to reinforce this by taking full advantage of what he believed gave the country the best comparative advantage.

Thus Malawi's agricultural policy is primarily one of producing cash crops that will fetch the highest prices in the international market subject to the maintenance of self-sufficiency in food production. However, since in Malawi food self-sufficiency has seemed to mean the maintenance of the barest minimal standards of consumption necessary to reproduce labour power this means that food policy is a secondary goal to cash crop farming.[18] Now, Dr Banda is a firm believer in capitalism as implied by the unequal and exploitative relations it entails for progressive accumulation but he is sceptical about markets.[19] Thus the proper balance of necessary cash crops and food to be produced is strictly controlled and regulated. In this respect again he was perpetuating an old colonial tradition.

Banda was quite aware that Malawi was a price-taker on the international market but while the vicissitudes of this market could not be controlled, like the colonial government, he was intent on making domestic production stable and predictable. This was achieved by a number of intermeshing controls. First, a proper balance had to be maintained between estate and smallholder farming through land policies, "Achikumbe" certification and the provision of agricultural support services including finance. Second, to this balance was added a segmentation of crop specialisation and markets such that each sector (estate, "Achikumbe" and communal) was allowed to produce only certain crops and was allowed to market them through prescribed channels. There could be crop substitution only among those that were high-yielding in foreign exchange within each sector.

Thus the estate sector has had the monopoly of growing tobacco (flue-cured and burley), tea, sugar and cotton which have a high cash value on the international market. The estates are also free to market their crops directly on to the international market thereby receiving the full value for their crops. Smallholder farmers have the monopoly over the production of maize and other subsistence crops all of which historically have had low foreign exchange value, and cash crops such as groundnuts, fire- and sun-cured tobacco and cotton of a particular variety.

Smallholder farmers are obligated to market their cash crops through ADMARC and may privately market subsistence food crops. Interestingly, in the past this sector has not had to be encouraged to produce maize since it had to do so for its own survival. The recent drastic increase in producer prices of maize are not, as some suppose, a result of the World Bank having convinced the Malawi government to accommodate itself to some of the Berg Report recommendations on the use of price incentives. Maize is now a foreign exchange earner, given persistent droughts in neighbouring countries. If food imports by neighbouring countries turn into a secular trend and are paid for in hard currency one would not be surprised to see a reassignment of maize as an estate crop provided the prices are high enough.

The above stipulations in production and marketing allow the government to achieve three goals which are, in order of importance: first, that foreign exchange is maximised to finance elite and urban imports and Dr Banda's symbolic projects; second, that through ADMARC the forced appropriation of investible surpluses is undertaken to ensure the availability of surpluses that will continuously regenerate Press Holdings and estate farming; third, that subsistence food production is assured in a manner that reproduces the workforce albeit at a low level of consumption.[20] It is in this context that the quote from the *Guide to Agricultural Production in Malawi* given at the beginning of this paper should be interpreted. The three goals cited above are the long-term agricultural policy objectives of the Malawi government. From the above it is obvious that the enhancement of the welfare of the majority of the Malawian people is not a matter of direct concern but one based on faith in the trickle-down effects from the consumption and expenditure behaviour of the few that are the direct beneficiaries of agricultural policy.[21]

THE IMPACT OF AGRICULTURAL AND FOOD POLICY

Employment

The main impact of agricultural and food policy in post-independence Malawi has been the quantitative increase in the proletarianisation of the peasantry in the 1970s consequent upon the development of estate farming. It was noted earlier that initially, estate development was carried out through the indigenisation of the ownership of existing estates, but that in the latter part of the 1970s estate development was undertaken at the expense of communal cultivable land.[22] At the time of independence it had been quite clear both from comments by previous colonial officials and by the new African government, and, in particular, Dr Banda, that Malawi had high population density rates, particularly in the Central and Southern Regions. Further, it was clearly recognised that given the low levels of agricultural techniques in communal lands, and given a net population annual growth rate of about 3.1 percent Malawi was moving toward a crisis in land utilisation.

The rates of population density in the various districts of Malawi for the years 1966 and 1977 show that average density rates for three regions increased by 27 percent in the Northern Region, 43 percent in the Central Region and 34 percent in the Southern Region.[23] During this same period agricultural employment increased by more than 300 percent as can be deduced from Table 2.1. From Table 2.1 it can be seen that in the period from 1968 to 1974 agricultural employment rose slowly having risen, in 1974, by 60 percent of the first quarter level of 1968. Employment in the rest of the formal sector increased by about 80 percent during the same period. However, there is a phenomenal increase in agricultural employment in the post-1974 period which saw the expansion of estate farming. By 1977 agricultural employment had increased to 310 percent of the 1968 level and almost 200 percent of the 1974 level. During this same period formal employment in non-agricultural sectors increased only to 190 percent of the 1968 level and only 6 percent above the 1974 level. It is clear that aggregate formal employment was increasing with the larger share going to agriculture. By 1980 agricultural employment was 350 percent of the 1968 level and the rest of the formal sector had increased to 200 percent of the 1968 level. By 1980 the proletarianisation of Malawians had definitely increased by more than twice the level of 1968.

TABLE 2.1

Indices of Agricultural Employment, non-Agricultural Employment, Wages and Food Prices, 1968 to 1980

Year & Quarter	Agri. Empl.	Non-Agri. Empl.	Average Earnings	Low income Food Prices
1968: 1	100 P (50, 790)	100 L (82,053)	100 (K8.06)	
2	89	114	104 L	
3	73 L	117 P	102	
4	87	110	97 L	
1969: 1	110 P	109 L	102	
2	104	127 P	93 L	
3	79 L	114	113 P	
4	96	120	100	100
1970: 1	122 P	121 L	106	101
2	100	136	103 L	94 L
3	87 L	137 P	122 P	97
4	116	122	111	107 P
1971: 1	130 P	128 L	106 L	120 P
2	114	149	107	104 L
3	89 L	150 P	128 P	106
4	118	133	123	114
1972: 1	145 P	142 L	112 L	117
2	124	163 P	114	115
3	108 L	163	121	112 L
4	126	147	126 P	119 P
1973: 1	177 P	158 L	105 L	119 L
2	151 L	180 P	117	122
3	134	178	113	124
4	138	164	147 P	130 P
1974: 1	185 P	170 L	134	140
2	159	187 P	127 L	139 L
3	132 L	185	144 P	147
4	159	174	138	151 P
1975: 1	214 P	178 L	117 L	166 L
2	177	191	141	168
3	149 L	192 P	144 P	168
4	191	181	134	185 P
1976: 1	236 P	185 L	133	188 P
2	198	204 P	139	172
3	173 L	197	143 P	169 L
4	189	186	131 L	174
1977: 1	351 P	177 L	143 L	182
2	302	193	144	170 L
3	300 L	196 L	163	173
4	289	188	166 P	189 P
1978: 1	381 P	191 L	167 L	201 P
2	332	213	193	189
3	304 L	216 P	191 P	180 L
4	314	205	191	190
1979: 1	400 P	206 P	171 L	207 L
2	351	225 P	174	209
3	334 L	208	186 P	216
4	347	211	185	233 P
1980: 1	404 P	214 L	181 L	250 L
2	361	221	185	263
3	328	216	190	275
4	320 L(162,813)	226 P (185,264)	209 P (K16.7)	291 P

Source: Calculation from statistics in the Reserve Bank of Malawi, *Annual Economic Review.*
P: Peak for the quarter   L: Lowest for the quater   K: Kwacha

As many observers have noted, this increase in aggregate and agricultural employment took place at a time when real wages were declining as can be deduced from Table 2.1. Between 1970 and 1974 the urban low income price index had increased by 36 percent for all items and by 44 percent for food items, while average earnings in agriculture had only increased by about 18 percent. Thus real earnings fell by about 18 percent (36 percent minus 18 percent) during this period. Between 1974 and 1978 the same price index had again increased by 36 percent, at the same rate as average monetary earnings showing that real wages remained constant during this period. It is therefore clear that earnings in agriculture could not have been the factor enticing peasants to join wage employment. The increased employment was a direct consequence of a deteriorating situation in peasant subsistence agriculture. It should be noted that Malawi did not experience any serious or persistent droughts during this period.

It is true that the dramatic increase in wage employment coincide with the curtailment of the number of migrant workers going to South Africa starting in 1975-76 as can be seen in Table 2.2. Further, this curtailment meant a drastic decrease in repatriated earnings which were a significant source of additional income for the support of migrant families who remained in Malawi. It is thus possible that the swelling of the wage force was in part from migrant returnees and in part from former dependants of migrants who now saw the need to earn a wage in order to replace the lost support from previous repatriated earnings which were used to purchase nonsubsistence goods.[24] That a prima facie case for the above explanation exists can be seen from Tables 2.1 and 2.2. First, note that after 1974 the fall in the number of migrants going to South Africa is roughly 100,000 which is about the net increase in the agricultural labour force in the post-1974 period over that of the previous period. Second, the number of registered job seekers shows a significant increase after 1974, (Table 2.2). Third, the percentage of vacancies filled also shows a significant increase following the low levels registered from 1972 to 1974.

The crucial question however is why returning migrants would opt for agricultural wage employment as opposed to self-employment in cash crop and subsistence farming. It is well known that those Malawians who were most inclined to migrate were the least likely to go into agricultural wage employment since this form of employment was despised.[25] Further, returning migrants were generally inclined to go into cash crop farming. Thus, the most likely effect of the curtailment of the export of labour to South Africa would have been an expansion of smallholder farming on communal land. Under the circumstances, however, with estate farming expanding at the expense of communal land, all that may have occurred is an increase in population density on cultivable communal land resulting in a generalised shortage of land hence the high population density rates in 1977. In the final analysis, then, it is the deteriorating situation in the peasent sector caused by the expansion of estate farming and the deliberate policy stance to reorient government resources away from smallholder communal agriculture toward estate and "Achikumbe" schemes that forced labour to migrate toward agricultural wage labour.[26]

TABLE 2.2

Number of Registered Workseekers, Vacancies Notified and Filled at Employment Exchanges and Malawian Migrants to South Africa

Year	Registered Work-seekers (workseekers per vacancy)		Vacancies notified	Vacancies Filled (% of vacancies filled)		Contracted Migrants in South Africa at End of Year
1969	15,044	(1.83)	8,201	5,896	(72 %)	53,184
1970	22,139	(1.83)	12,101	9,095	(75 %)	90,642
1971	17,120	(1.23)	13,875	9,673	(70 %)	99,849
1972	23,575	(1.64)	36,863	14,153	(38 %)	123,845
1973	22,986	(.59)	38,874	15,381	(40 %)	123,251
1974	25,364	(.47)	53,784	19,709	(37 %)	68,448
1975	30,734	(.71)	43,060	21,567	(50 %)	2,711
1976	30,266	(.92)	32,894	25,027	(76 %)	-
1977	27,796	(.90)	31,004	25,506	(82 %)	17,443
1978	27,220	(.64)	42,466	34,679	(82 %)	17,981
1979	33,283	(.77)	43,490	38,967	(90 %)	17,87

Source: Malawi Statistical Yearbook, 1980.

An interesting fact that emerges from Table 2.2 is that between 1972 and 1974 Malawi was definitely experiencing labour shortages at the same time that the government was initiating estate farming. During this period only 38 percent of the reported vacancies were being filled. After the curtailment of labour migration to South Africa the rate of filling vacancies increased to an average of 77 percent for the period 1975 to 1980. This rather confirms the much held suspicion that Dr Banda seized upon the Francistown airplane disaster in which hundreds of Malawian migrants were killed as a convenient excuse to replenish the labour force within Malawi.[27] It should also be noted that between 1970 and 1973, during the labour shortage there was a significant increase in agricultural earnings (see Table 2.1) and that after 1974 earnings were roughly constant indicating an easing of the labour situation.

A closer look at the indices in Table 2.1 reveals some interesting cyclical features of the labour market as shown in Diagram 2.1. As would be expected, agricultural employment in each year peaks in the first quarter during the crop-growing season which means that labour is absorbed away from subsistence farming, which is under-taken is done as a part-time duty.[28] In the first quarter also, the non-agricultural employment index also reaches its lowest point in the year indicating a movement of labour from this sector to agriculture. However, agricultural employment reaches its lowest point in the year in the third quarter during the dry season at the same time that employ-ment of non-agricultural labour reaches its peak for the year. Earnings per worker in agriculture reach their peak in the third of fourth quarters when agricultural employment is increasing towards its peak indicating that some increase in wages is still needed to attract labour in spite of its general availability. We also note that the low income food index peaks in the fourth or first quarter reflecting the increased demand for food by the increased number of wage earners. This also suggests that wage earners depend on the market and not on subsistence stocks for their support. These cyclical trends suggest that the opportunity cost of wage employment during the cultivating and growing seasons in terms of foregone smallholder self-employment on communal land is considered so low as to make agricultural employment preferable. In other words the effort-price of self-employment in agriculture is too high and this in spite of the low minimum wages[29] and average earnings in formal employment.

It is tempting to want to suggest that, given the high population density rates, wage employment represents a release of surplus labour such that males engage in wage employment while women and children engage in smallholder crop cultivation. In this case wage income would represent supplementary family income. However, the fact is that in Malawi female wage employment has been increasing as well. Between 1971, female wage employment increased by 270 percent in agriculture and by 62 percent in the non-agricultural sector. Admittedly, female employment started off at very low levels, but the increase is still significantly high between 1973 and 1977 (an increase of 45 percent). The general trend therefore is for adults to move out of the subsistence sector, and this, given the constant real wages in the agricultural sector, can only mean that conditions in the subsistence sector are deteriorating. Thus, both the expansion of estate farming at the expense of smallholder communal agriculture, and the deliberate neglect of this sector by the government have resulted in the forced expulsion of labour from the latter sector thereby providing

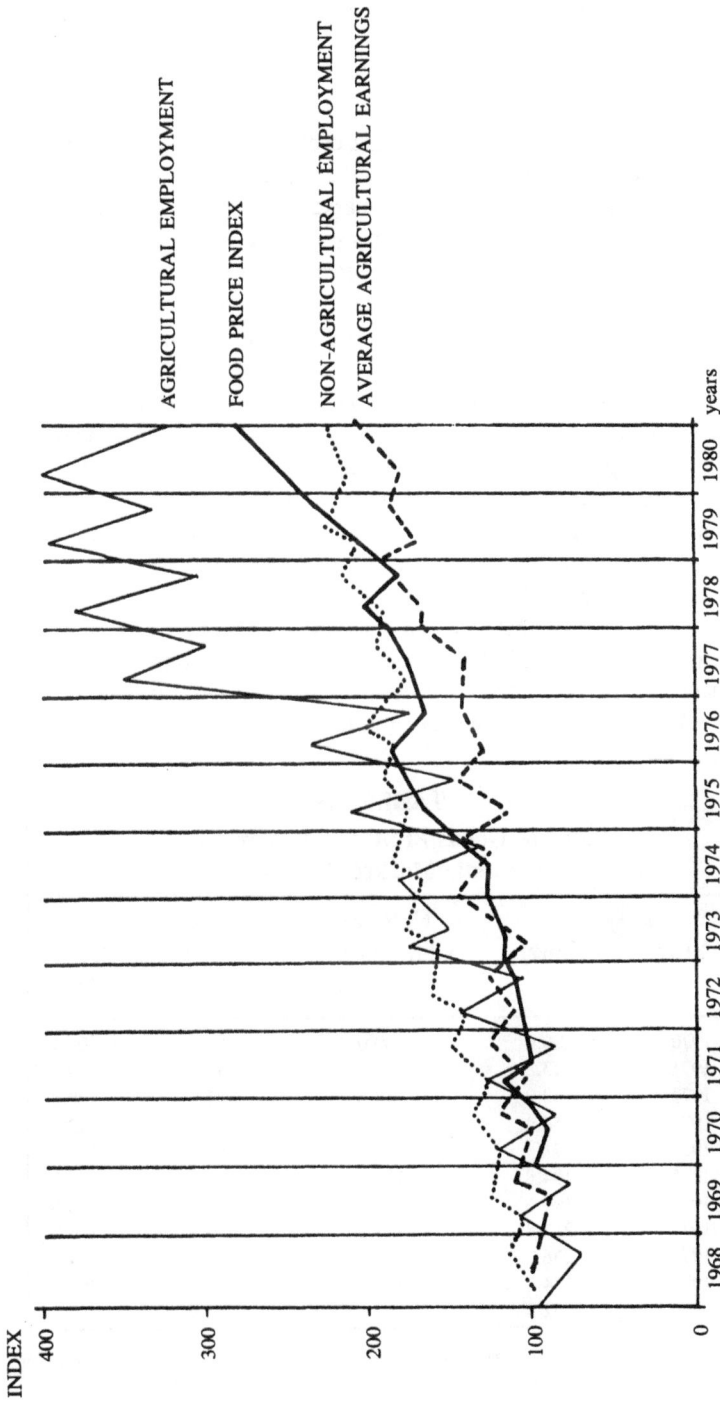

Diagram 2.1   Annual Peaks & Lows in Indices Plotted from Table 2.1

a basis for accumulation on the basis of cheap labour in the former sector.

Food Production

It has been suggested above that the cyclical trends in employment when related to those in the price index suggest that wage employment in agriculture depends on the market for its food. Since the food index is an urban low income one this suggests a dependence on foods that have been processed through the manufacturing sector. Table 2.3 shows the indices of food and manufacturing output excluding exports (weighted 20 percent of total output) and building and construction materials (weighted 18 percent). From Table 2.1 it can be seen that between 1970 and 1981 employment in agriculture increased by about 320 percent and this roughly corresponds to the increase in food output of 375 percent in the same period. We also note from Table 2.3 that food output and total manufacturing output both show dramatic increases after 1974, the very period in which wage employment, particularly in agriculture, dramatically increases. Now, since Malawi does not import food for its wage earners this means that the increases in food production that went hand in hand with increases in employment were produced domestically.[30] But since estates as a rule specialise in cash crops the food must have been produced by smallholder farmers as has been indicated earlier.

TABLE 2.3

Index of Goods Produced Mainly for the Domestic Market (1970=100)

Year	Total	Footwear Clothing and Textiles	Food Beverages and Tobacco	Other Goods
1970	100.0	100.0	100.0	100.0
1971	113.2	118.1	112.3	110.8
1972	120.9	118.8	127.6	108.0
1973	150.4	115.2	169.4	142.4
1974	171.7	120.1	192.1	175.1
1975	194.8	130.2	225.7	187.3
1976	194.6	109.9	238.6	176.8
1977	212.8	126.2	259.8	190.2
1978	236.2	148.3	274.4	234.3
1979	234.7	169.2	273.9	205.2
1970	252.5	149.4	292.9	259.7

Source: Malawi Statistical Yearbook, 1980.

Not only did aggregate food production, and in particular maize, not suffer as wage employment increased, but it actually increased during this same period. Thus between 1970 and 1976 an average of 46,000 tons of maize per year were purchased by ADMARC while an average of 23,000 metric tons were sold between 1973 and 1976. During the dramatic increases in employment after 1976 up to 1982 average annual purchases of maize were 126,000 metric tons while average annual sales were 98,000 metric tons allowing for an average annual surplus of 28,000 metric tons. This indeed is the success that Malawi has achieved, the success of smallholder production of food.

However, this aggregate data is misleading. In the aggregate it seems to contradict the argument that the status of peasants (the smallholder communal producers) has worsened as a result of estate expansion and policy neglect. Such a contradiction even bothered perceptive analysts such a Kydd and Christianssen and arises fundamentally from viewing the structure of agriculture in Malawi as a bimodal one consisting of the dominant estate sector and the smallholder sector.[31] But this, as indicated earlier, is misleading since the smallholder sector is differentiated, and is becoming markedly so. The aggregation of the smallholder sector into one category masks a very fundamental recent development, and that is the development of the "Achikumbe" (the kulak class) as the dominant force in food production. In food policy, Dr Banda's pride and joy has been the success of the "Achikumbe" scheme to which he has diverted government attention and resources that would have otherwise gone to the smaller subsistence producer. It is this group that is increasingly producing food surpluses purchased by ADMARC. It is this same group that has won Dr Banda's accolades; but this group has been succeeding while the condition of the peasant masses engaged in subsistence farming on communal lands has been deteriorating. The Malawi agricultural sector is *trimodal* in a very complementary sense, for the deterioration of the third sector (the subsistence sector) is the precondition for the growth and expansion of the petty capitalist "Achikumbe" sector and the capitalist estate sector and the seeming contradiction we referred to earlier thus disappears[32]

The foregoing conclusion raises the question as to the role that ADMARC pricing policies play. Between 1962 and 1966, the period of concern for the communal small farmer, ADMARC producer prices averaged about 55 percent of the ADMARC selling prices. From 1968 onwards as the emphasis shifted toward estate farming and "Achikumbe" the producer price offered by ADMARC fell from about 42 percent of the selling price in 1968 to about 31 percent in 1978, to the alarm of the World Bank. The smallholder farmer then was exploited both at the point of production and in the realm of circulation.

However, the differential pricing of purchases and sales by ADMARC again differed between "Achikumbe" and communal producers. Since the "Achikumbe" are subsidised by the government in terms of the provision of inputs and extension services which are not available to communal farmers, their (the "Achikumbe") exploitation in food crops is much less. In fact it is through these subsidies that the government has been able to persuade "Achikumbe" to produce maize rather than more valuable cash crops. Further, it would not be surprising to find that the losses in the ADMARC's maize accounts may be a result of these subsidies. It is also true however that ADMARC maize selling prices are longer than what they would otherwise be in order to subsidise urban wage earners. Thus pricing policy is a greater disincentive to communal subsistence farmers who may not be persuaded to produce surplus maize or who may choose to sell their maize surpluses privately and this may have increased the effort-price of farming and further reinforced the need to seek wage employment.

Further, the ability of communal farmers to react to producer price rises which the World Bank has been persuading the Malawi government to undertake, it quite limited. Land-extensive production of food in response to higher prices is strictly forbidden as clearly indicated in the quote at the beginning of this paper. And the intensive use of land is excluded by the lack of resources and government support. The major beneficiaries of producer price incentives on food crops are likely to be the "Achikumbe" who can extend their land by converting communal land into leasehold private land and still maintain the required ratios between cash crops and food crops and who have access to their own and government resources that would allow them to engage in land-intensive cultivation.[33] Thus the World Bank emphasis on price incentives in practice ends up reinforcing existing inequalities.[34]

Social Welfare

If Malawi's food policy were a substantive success and one worth emulating this should be reflected in the health and nutritional status of the population.[35] However, while aggregate food production indices such as those reported by the World Bank give a glowing picture, the social and health indicators of welfare give a markedly depressing view of the situation. In 1980 life expectancy at birth stood at 44 years while the average for low income countries was 50 years. The infant and child mortality rate for 1981 was 207 per thousand while the average for low income countries was 145 per thousand. Now, Tanzania, a country much criticised for its agricultural policy, had a life expectancy of 52 years and an infant and child mortality rate of 120 per thousand, and neighbouring Zambia, also much maligned for its agricultural policy, had a life expectancy of 51 years and an infant and child mortality rate of 124 per thousand.

While Malawi is among the highest in the proportion of government expenditures spent on agriculture (11.5 percent in 1978), roads (11.1 percent in 1978) and other economic services (13.3 percent in 1978) it was among the lowest in expenditures on education (11.8 percent in 1978), health (5.3 percent in 1978) and other community services. As indicated earlier expenditures on economic services are complementary to the structure and functioning of the inequitable agricultural trimodal system. However, the low expenditures on social services reinforce whatever health problems may arise from the skewed access to food which may be reflected in the high infant and child mortality rates since these groups are highly susceptible to nutritional diseases when food deficiencies are experienced in a section of the population. Thus, apart from the fact that food production has quantitatively increased in Malawi, this has not been reflected in a general improvement in the welfare of the population. In fact the country fares significantly worse than its neighbours which are constantly asked to emulate the Malawi "miracle".

Sectoral Linkages and Structure of the Economy

Implicit in Malawi's development strategy is the faith in the Arthur Lewis/Fei and Ranis development model. With relatively high rates of population density especially after the 1970s Malawi has been simultaneously releasing seemingly redundant or less productive labour from communal lands (Lewis' traditional sector) and expanding commerical agriculture and manufacturing (Lewis's modern sector) by absorbing the released labour. This process has been facilitated classically by very low and constant real wages in the modern sector thus allowing for high rates of surplus accumulation in this sector. Further, Malawi even met Lewis' (and, incidentally, Rostow's as well) minimal prescription of attaining an average savings ratio of more than 15 percent of GNP between 1973 and 1979 and a respectable growth rate in GNP took place during this period. One therefore would expect that the economy should have experienced structural change and an increase in the absolute standard of living of the majority of its people.

But in 1980 and 1981 the Malawi economy was in a crisis and growth came to a halt. Not only had significant structural change failed to occur but the standard of living of the majority of the people had not improved either. That the standard of living of the people did not improve is clearly evidenced by the constant real wages in commercial agriculture, where the majority of the formal sector employees were, and where a direct relationship existed between wages in the sector and the standard of living in the subsistence sector. An improvement in the standard of living of communal farmers would have definitely increased real wages in commercial agriculture. Thus the constancy of real wages in Malawi is a clear indication that the standard of living of the majority had not improved.[36] However, this simply means that the benefits of growth were unequal and skewed in favour of a few.

That no major structural change occurred in the economy is evidenced by the continued and expanded dominant role of agriculture in Malawi's economy. It is true that some light manufacturing has significantly grown. But this has been a development primarily representing an extension of agriculture for it has entailed import substitution in processing of primary products and the manufacture of food items from domestic agricultural produce. This form of import substitution is quite limited for its possibilities depend on a low threshold demand (primarily domestic) and the high transportation cost as a component of the prices of competing imports. In Malawi, a further facilitating factor in this type of import substitution was the government's unshakeable commitment to low and constant wages. Thus between 1973 and 1979 the monetary value of manufacturing grew by about 50 percent; and between 1970 and 1979 the volume of manufacturing output had increased from an index of 100 in 1970 to 224 in 1979. However, while the share of agriculture in GDP remained roughly at 40 percent as the dominant sector in both 1973 and 1979, that of manufacturing had declined from 36 percent of GDP in 1973 to 12 percent of GDP in 1979. The decline in the significance of manufacturing was replaced by the growth of the service sector.

Thus by 1979 Malawi was still predominantly an agricultural economy structurally. But this sector had failed to play the key role assigned to it in the Lewis/Fei and Ranis model, primarily because Malawi continued to be inextricably tied to the international economy as a dependent and peripheral appendage. The benefits of the growth of the 1970s were not only unequally distributed but such benefits accrued abroad as well, as a result of Malawi's dependence on the international economy. It is the leakages that result from dependency relations that crucially nullify the operation of the Lewis/Fei and Ranis model in spite of increasing domestic accumulation by the "saving" and "investing" capitalist class. The implications of Malawi's external dependency came to the fore in the economic crisis of 1980 and 1981. Income transfers from the domestic economy to external sources occurred in the form of increased debt servicing and repayment of loans with a bunching up of maturity dates, increased transportation costs, increased transfers for imported factor inputs, loss of foreign exchange from migrant remittances, worsening terms of trade from crucial export crops, the devaluation of the Malawi Kwacha, and limited regional export markets for Malawi's manufactured products. It is not surprising then that in Malawi growth took place without any significant development or improvement in equity.

CONCLUSION

Within the limits of dependency Banda's effective personal control over the economy was subsequent to his having gained absolute control over political affairs in Malawi. He has transferred to the economy the unabashedly dictatorial and paternalistic stance he takes in politics. The capitalism Banda propounds, aspires to, and practises, and for which he is commended by the West has to be seen as a crude and primitive form of accumulation whereby he has transferred the hierarchical and dictatorial relations internal to the workplace in a capitalist firm to the economy as a whole. Thus the economy as a whole is managed in the interest of private accumulation in the same way that management operates a firm. Given his authoritarian control over the political process and his dominance in the economy through Press Holdings, Banda has been able to transform Malawi into a fascist capitalist state.

Malawi does not even have autonomous capitalist interests which can assert themselves independently of Banda. Banda himself is the premier monopoly capitalist. Where foreign private interests are involved they have had to accept partnership with Press Holdings or the government. Where indigenous capitalists are involved they are constantly reminded that their positions are due to the patronage accorded them by Banda. Further, these interests cannot complain much for he has demonstrated an uncanny ability to intervene in the economy on their behalf. He has used his authoritarian powers to control wages, prices, commodity and labour flows, and government fiscal and monetary policy to ensure accumulation by the capitalist class. The only constraints are external relations of dependency.

Public investment expenditures have been oriented towards supporting and subsidising directly productive private economic activities. Thus, expenditures on social investments with a long gestation period or with indirect long-term effects on private accumulation have been shunned. The only economically unproductive expenditures tolerated and even encouraged by Banda himself are those that are politically productive. These are expenditures on political patronage and on symbolic investments which consolidate and legitimise Banda's power and thus stabilise the polity. The ideological justification of the resulting exploitative and inequitable economic structure has been given an interesting puritanical twist by Banda. Banda has made it clear time and again in public and parliamentary speeches that he accepts and encourages the acquisitive instinct in man. He has preached that people have to be prepared to work hard. Thus inequities are explained away as being a result of differential propensities to work or laziness. The poor and exploited are blamed for their condition and are exhorted to emulate the rich. The wealth of the latter is conspicuously reported in the belief that this will provide a demonstration effect to the poor and lazy. Thus, the accumulation which is the consequence of their exploitation is paraded before the masses as an incentive for them to work harder.

It is in the context of the above that agricultural and food policy in Malawi should be viewed. The concern for the masses is an incidental one couched in an illusory and paternalistic puritan ideology. The whole political and economic structure in Malawi is geared against their interests. Foreign exchange and wage employment are the basic prerequisites for accumulation by the few in Malawi and these are the primary goals of agricultural policy in Malawi. Food policy is only important in so far as it reproduces and stabilises the existing social, political and economic order in Malawi and as such it is not a policy at all but a survival strategy. Since many authoritarian regimes in Africa are also poor performers in the economic sphere the "success" and "miracle" Malawi has achieved is perhaps that it has maintained fascism with a modicum of economic growth.

The essence of agricultural policy in Malawi is unravelled once the bimodal model of the agricultural sector is discarded and instead a trimodal one assumed. It is then brought out that the logic of Malawi economic policy lies in the government's ability to manipulate wage policy, labour flows, agricultural price and subsidisation policies, and monetary policies to the maximisation of forced savings which are directed into productive investment. Thus the Malawi government has fairly good control over savings and investment in the economy. In this respect the Malawi model of accumulation is not too dissimilar from that of the apartheid model of South Africa and the former Rhodesia.

It there is a lesson at all to be learned from the Malawi "miracle" it is that the logic of economic growth lies in the implementation of macro policies that maximise internal accumulation of savings and their productive reinvestment internally. The economic failure of the so-called "socialist" or "progressive" states in Africa has been their inability not only to undertake indigenous accumulation but also to invest productively whatever capital funds existed. Thus we have a growing pie that is inequitably shared in Malawi and a shrinking pie that is relatively equitably shared in countries like Tanzania and

Zambia. And the difference is simply that Malawi, like Kenya and the Ivory Coast, has been able to identify the logic of accumulation and growth while the latter countries (i.e. Tanzania and Zambia) have found it elusive and herein lies the lesson of the Malawi "miracle" to failing "capitalist" or "socialist" economies in Africa. The crucial question is whether it is possible to effect forced accumulation and planned productive investment in a manner that ensures that the burden of accumulation and the benefits of investment are equitably shared. Malawi unfortunately has failed on this score.

NOTES

1. Mwakasungula, A.K., *The Rural Economy of Malawi: A Critical Analysis*, Derap Working Papers, The CHR. Michelsen Institute, Bergen, Norway 1984; Mkandawire, Thandika, "Economic Crisis in Malawi", paper presented in Uppsala at the conference on "Africa which way out of the Recession?" September 1982 (published in Jerker Carlsson, et. al; *Recession in Africa* , Scandinavian Institute of African Studies, Uppsala 1983; Kydd, Jonathan and Christianssen, Robert, "Structural Change in Malawi since Independence: Consequences of a Development Strategy based on Large Scale Agriculture", *World Development*, Vol. 10, 5, 1982; and Kydd, Jonathan "Malawi in the 1970's: Development Policies and Economic Change", paper presented at a conference on "Malawi—An Alternative Pattern of Development", Centre of African Studies, Edinburgh University, May 1984. The foregoing works are essentially complementary in their explanation of developmental, and in particular agricultural policy changes in Malawi. However, Mwakasungula and Mkandawire interpret the changes as having definitely worsened the condition of the masses while Kydd and Christianssen equivocate on this and a number of other issues by unnecessarily concentrating on the trees thereby missing the forest at times. The latter two writers go to unnecessary lengths to find collaborative micro statistical evidence when macro data are adequate for the formulation of deductions.
2. With regard to the peasantry there are three issues of importance arising from agrarian policies in Malawi since the 1970s. There is, first, the question of the degree of inequality experienced by the peasantry; second, the impact of agrarian policy on the absolute standard of living has been of concern; and third, the degree to which differentiation has taken place among the peasantry. For a review of different interpretations on these issues see Kydd and Christianssen (1982) and Kydd (1984).
3. See Mkandawire (1984).
4. Thus the World Bank report on *The Development of the Agricultural Sector* (8 May 1981) observes: "Despite the country's extremely low level of development at independence, the Malawi Government opted for an outward-looking strategy of rapid growth based on encouragement of private enterprises, use of the market mechanism, favourable treatment of foreign capital, low levels of protection and wage restraint", p.1.
5. The most notorious forms of labour utilisation in colonial Malawi were *Thangata*, which reduced indigenous labour to forced labour and serfdom, and the "visiting tenancy" scheme which tied the peasantry to a manorial arrangement on the estates.

For a history of the development of agrarian policy and structure in Malawi see Myambo, Simon, "The Shire Highlands Plantations: A Socio-economic History of the Plantation System of Production in Malawi, 1891 to 1938", M.A. Thesis, University of Malawi, 1973; and Vail, Leroy, "The State and the Creation of Colonial Malawi's Agricultural Economy" in Rotberg J., (ed.), *Land, Hunger and Labour in Central Africa*.

6. Dr Banda has been quite adept at combining political repression with a paternalistic populism in economic pronouncements and policies. There is a much needed synthesis of the conjuncture of political and economic factors in Malawi. The perverse synthesis of the political and economic factors is manifested in Dr Banda's dominance and not necessarily in a class dominance, hence our insistence on identifying Dr Banda as the dominant force in this paper.

7. On the attempt by the colonial government to make amends for past malpractices see works by Leroy Vail and Mwakasungula cited above.

8. See Kydd and Christianssen (1982), Kydd (1984) and Mwakasungula (1984) for details on estate development in Malawi.

9. It is Banda's standard practice at rallies not only to brag about his own successes in estate production and empolyment but actually to call upon members of his entourage to do the same as well.

10. Banda has argued as follows: "Under our present system of landholding and land cultivation, no one, either as an institution or as an individual, will lend us money for developing our land because our present methods of landholding and land cultivation are uneconomic and wasteful. They put responsibility on no one. No one is responsible here and now for uneconomic and wasted use of land because no one holds land as an individual. Land is held in common. They say 'everybody's baby is nobody's baby at all. We have to put a stop to this..." Quoted in Pachai, B., *Land and Politics in Malawi 1875-1975*, Limestone Press, Kingston (Ontario), 1978.

11. The legal changes facilitating the individualisation of land are discussed by Mwakasungula (1984).

12. It is however true that the pricing policy with respect to the purchase of peasant produce. and the sale of agricultural inputs involved cross-subsidisation. Nevertheless, as the World Bank report notes, while neither gains nor losses were passed on to farmers the general trend resulted in net gains for ADMARC and thus a net taxation of peasants; see *The Development of the Agricultural Sector*, World Bank (8 May 1981).

13. "Progressive" or "Master Farmers" had the image of being collaborators of the colonial regime and were despised by nationalist movements. In the post-independence situation they are similarly identified as Dr Banda's sycophants and beneficiaries. While the govenment's idea is that they should be admired and emulated by peasants as models to which they should aspire, the communal peasants themselves view them with suspicion.

14. Quoted in Mwakasungula, *The Rural Economy of Malawi*.

15. Kydd and Christianssen seem to think that since the World Bank-supported "integrated rural development schemes" are being extended into a national policy they are overshadowing the "Achikumbe" scheme. Thus on this point they are sceptical of Ghai and Radwan's argument ("Growth and Inequality: Rural Development in Malawi", in *Agrarian Policies*) contending that there has been increased differentiation among the peasantry as a result of the "Achikumbe" programme, hence their insistence on viewing the Malawi agricultural structure as a bimodal one. While detailed data on peasant differentiation are hard to come by it is quite clear that Banda favours the "Achikumbe" programme and that a proportionately

larger amount of output comes from this class. Banda has been dragged into "integrated rural development" which still remains a token welfare programme. The seriousness with which the "Achikumbe" scheme is treated is quite clear from the Ministry of Agriculture's *Guide to Agriculture Production in Malawi* 1980/81 and 1981/82. In the preface to the 1981/82 issue Banda off-handedly and tokenly commends communal farmers for cooperating in the manner suggested by the rural integrated schemes. See Mwakasungula (1984) on the importance of the "Achikumbe" scheme.

16. On differences in strategy between the Malawi government and the World Bank see Kydd and Christianssen (1982) and Kydd (1984).

17. The World Bank opposition to large-scale private farming in Malawi cannot be taken seriously. Kydd and Christianssen exaggerate the importance of the World Bank's difference with the Malawi government. The World Bank has had a good record in supporting and promoting large-scale private farming. Their opposition to large-scale farming is usually directed at state-owned farms. This is clearly brought out for instance in their preference for large-scale commercial farming in Zimbabwe and their reluctance to support small-scale private resettlements. The concern for small-scale farmers by the World Bank is essentially a reaction to government large-scale farming which they frown upon (the Berg Report) and a belated and token bow to the lobbying of basic needs supporters.

18. Dr Banda's view of self-sufficiency refers to the satisfaction of absolute basic necessities. In none of his speeches or policy statements has he ever countenanced the need actually qualitatively to improve the condition of the peasants. Banda's perennial refrain at rallies that his people at least have food to eat, a hut that does not leak, and a shirt on their back means just that and no more

19. Dr Banda's model of capitalism is that of Somoza, Marcos and the former Shah of Iran. It is a personalised fascist capitalism.

20. The interdependence of the three objectives is often blurred by comments such as the following: "The smallholder subsector accounts for over 85 percent of all agricultural production, meets the country's demand for food staples (maize, beans, groundnuts, sweet potatoes and rice) and provides some export surplus. However, the bulk of the country's agricultural exports come from the estate subsector which has functioned as a principal earner of foreign exchange and an engine of growth for Malawi's development." (*The Development of the Agricultural Sector*, p. vi). Such statements make no mention of the fact that the growth of estates has been largely financed by surpluses from smallholder and subsistence peasant produce and that the cheap labour that is the basis for their efficiency is extracted from a degenerating subsistence sector. The engine of growth in Malawi has been the cheap labour extracted from the subsistence sector and the surpluses extracted from peasant farmer—the estate sector has been the beneficiary.

21. The 1981 World Bank reports on agriculture and employment both observe the "the development of the social sector received relatively less attention until recent years", p. 1 of *The Development of the Agricultural Sector*, and p (i) of *Malawi: Employment Aspects of Economic Development*.

22. Estate employment increased from 42,000 to 148,000 between 1969 and 1978 at the same time that the overall employment in the economy was increasing, that the "miracle" in smallholder agriculture was supposed to be taking place, and that wage rates were constant and artificially low. By the end of the 1970s the World Bank estimated that 95 percent of the usable land (which is 38 percent of the total land areas) had been used up. The increased labour force participation and the consequent increase in wage employment could have only been symbolic of deteriorating subsistence conditions.

23. See *Malawi Population Census, 1977*, Vol. 1-2, Government Printer, Zomba, 1980.
24. The number of migrant returnees is considerably larger than is suggested by data on South Africa since after independence the number of returning Malawians from all over Southern Africa increased substantially.
25. It is well known that Malawi migrants greatly despised wage employment within Malawi even if, incidentally, they were likely to be similarly employed in Zimbabwe or South Africa. It was seen as a clear sign of failure for a returning migrant to end up in agricultural wage employment.
26. The overcrowding created by returning migrants does not require micro statistics to prove. It is clear from the numbers involved given fixed land for which adequate density data exist. Whether or not returning migrants went directly into wage employment is less important than the fact that overcrowding created a crowding-out or bumping-out effect on labour such that some labour had to be released out of the communal lands to seek wage employment.
27. See Leroy Vail's article on the problem of capital and labour shortages implied by estate farming since the colonial days. The preoccupation with manipulating policy so as to ensure adequate supplies of labour is something Banda has adeptly inherited from the colonial past. Incidentally, if the disaster was a fitting reason for curtailing labour migration to South Africa one is at a loss to explain why labour exports were resumed albeit at a lower rate—the risk of death was still there.
28. This cyclical utilisation of labour lowers productivity on communal land. This was further reinforced by the increase in female labour-force participation.
29. Minimum wages are very low in Malawi and were constant between 1969 and 1973 when they averaged about 35t per day and between 1973 and 1980 when they averaged about 40t per day (slightly more than the same equivalent in United States cents). The agricultural sector has generally been exempted from minimum wages.
30. Among low-income African countries Malawi has the least dependence on food aid imports. See Table 24 of *Accelerated Development in Sub-Saharan African*, World Bank, 1981.
31. Insistence on a bimodal structure of agriculture in Malawi is a serious misreading of the situation by Kydd and Christianssen. Anyone familiar with the real situation in Malawi does not require statistics to appreciate the significance and importance of the "Achikumbe" scheme in rural differentiation and in the role attached to it by the government and ADMARC.
32. Dr Banda clearly recognises the critical importance of the "Achikumbe" programme for political legitimisation. It gives the peasants the possibility of partaking in capitalism in the same way small businesses are promoted by conservatives in the United States thereby masking the larger inequalities due to monopoly capital.
33. It should be noted that acceptance of government assistance requires compliance with government suggestions on crop mix—another crucial way in which the government intervenes at the point of production.
34. Kydd and Christianssen lament the lack of statistics on private sales by peasants and seem to want to give these sales undue weight. Such sales can only be done at neighbourhood markets and as such they are likely to be small. If attempts are made to transport large quantities of crops long distances, the possibilities of being apprehended are quite high and so are the political consequences. ADMARC has effective monopsonistic power over peasant produce.
35. The writer would have liked to get data on the incidence of nutrition-related health problems but was unable to do so. A review of Malawi's performance in the satisfaction of basic needs is given in the World Bank report, *Malawi: Basic Needs*, 8 May 1981. While its own data on various indicators of social welfare in Malawi

is patently unflattering and in fact depressing, by making tortuous statistical comparisons showing that Malawi indices are average either for Sub-Saharan black Africa or with respect to what would be predicted given its per capita income (even if most indices are below the average for all developing countries), the World Bank report observes: "This orientation [Malawi's economic policy] has succeeded in giving Malawi perhaps the highest growth rate in Africa while meeting the basic needs of the population at least as well as most other countries in the region."

36. Here again Kydd and Christianssen (1982) and Kydd (1984) equivocate unnecessarily on the improvement of the peasants because of lack of detailed data. This is unnecessarily hair-splitting since aggregate data clearly allow for the deduction that the impoverishment of the peasants increased. Unless Malawian peasants are totally irrational increased labour force participation in agriculture in the face of declining constant real wages and in a sector where they are simply transferring known skills which could be used for their own employment can only be a result of a deterioration in the alternative of subsistence farming. Detailed data would be interesting but superfluous.

BIBLIOGRAPHY

Arhin, K. et. al., (ed.). *Marketing Boards in Tropical Africa*, Routledge and Kegan Paul, London, 1985.

Carlsson, J. (ed.). *Recession in Africa*, Scandinavian Institute of African Studies, Uppsala, 1983.

Ghai, D. and Radwan S. (ed.), *Agrarian Policies and Rural Poverty in Africa*, ILO, Geneva, 1983.

Kydd, J. and Christianssen, R., "Structural Change in Malawi since Independence: Consequences of a Development Strategy based on Large Scale Agriculture", *World Development*, Vol. 10, N° 5, 1982.

Kydd, J., "Malawi in the 1970s: Development Policies and Economic Change", paper presented at a conference on "Malawi—an alternative pattern of development", Centre of African Studies, Edinburgh University, May 1984.

Malawi Government, *Guide to Agriculture Production in Malawi*, 1980/81, 1981/82, Malawi Government.

Malawi Government, *Malawi Population Census, 1977*, Volumes 1 and 2, Government Printer, Zomba, 1980.

Malawi Government, *Monthly Statistical Bulletin*, April 1982, Government Printer, Zomba, 1982.

Malawi Government, *Reported Employment and Earnings, Annual Report, 1975*, Government Printer, 1976.

Malawi Government, *Malawi Statistical Yearbook, 1980*, Government Printer, Zomba, 1981.

Mkandawire, T., "Economic Crisis in Malawi", paper presented in Uppsala and subsequently published in Carlsson, *Recession in Africa*.

Mwakasungula, A.K., *The Rural Economy of Malawi: A Critical Analysis*, Derap Working Papers, The CHR Michelsen Institute, Bergen, Norway, 1984.

Myambo, S. "The Shire Highlands Plantations: A Socio-economic History of the Plantation System of Production in Malawi, 1891 to 1938", M.A. Thesis, University of Malawi, 1973.

Overseas Development Administration, *Malawi Manpower Review, 1980*, ODA, London, 1980.

Pachai, B., *Land and Politics in Malawi 1875 to 1975*, Limestone Press, Kingston (Ontario) 1978.

Reserve Bank of Malawi, *Economic Review*, 1969 to 1984.

Rotberg, J. (ed.), *Land, Hunger and Labour in Central Africa*.

World Bank, *Malawi: Basic Needs*, 8 May 1981.

World Bank, *Development of the Agricultural Sector*, 8 May 1981.

World Bank, *Malawi: Employment Aspects of Economic Development*, 8 May 1981.

World Bank, *Accelerated Development in Sub-Saharan Africa*, 1981.

3. THE STATE AND AGRI-BUSINESS IN THE SWAZI ECONOMY

Patricia McFadden

INTRODUCTION

The agricultural policies of most African countries are a persistent reminder of how capitalist interests continue and will continue to dominate the future of the continent and its people. In Southern Africa in particular, the structure of agriculture in countries surrounding South Africa is directly related to the historic hegemonic influence of South Africa in the region, and to the growing importance of the region for global imperialist interests.

Since the late 1960s agriculture within the region has, through deliberate and carefully planned policies, been reoriented to follow a certain pattern—that of export-oriented production. Under the supervision of multinational capital, and with the direct collaboration of the neocolonial state, countries like Swaziland and Malawi have emerged in the 1980s as major exporters of commodities like sugar, citrus, tea, tobacco, cotton etc. In Swaziland, for example, in the early 1970s, "taken together, these export-oriented primary commodity industries accounted for slightly less than half of total output", and between 1970-74, "total real output may well have increased at an annual average rate in excess of 6 percent. The main upward propellant of the economy was undoubtedly the agricultural and forestry industries producing for export".[1]

The role of the state in providing economic and political conditions conducive to surplus extraction in agriculture is central to an understanding of the economies of the region. In Swaziland, the state opened up the economy to total domination by multinational and South African capital, and it is the multinationals which dictate economic policy. It is this *apparent* inability of the state to determine economic policy which makes the case of Swaziland peculiar and specific in many ways. In fact, the state plays a very definite role within the economy, related to labour, land, marketing of certain commodities, provision of infrastructural facilities, and the provision of a "suitable" investment climate.

The operation of capital is further facilitated by the existence of the Customs Union which allows for the free and unrestricted flow of capital and manufactured goods (machinery and production inputs) within Botswana, Lesotho, Swaziland and South Africa. Between 1970 and 1973, the total value of equipment and materials imported into Swaziland from South Africa, rose by 40 percent. (See Table 3.1).

TABLE 3.1

Imports of Equipment and Material for Investment (in E '000)

	1970	1971	1972	1973
Machinery and Transport Equipment	9,158	11,700	11,239	13,664
Metal Manufactures	2,521	1,328	1,554	2,061
Metals	1,069	1,596	1,613	1,913
Construction Materials	1,474	1,510	1,849	2,168
Total	14,222	16,134	16,255	19,806

Source: Central Statistical Office, Mbabane, Swaziland, 1974.

The early 1970s saw the introduction and implementation of policies which aimed at restructuring the economies of the countries of the Southern African region along lines suitable to the interests and needs of the major imperialist countries. Southern Africa, and South Africa in particular, had assumed vital importance for the international capitalist system since the discovery of gold and other strategic minerals at the end of the last century. During the 1960s as an expression of "confidence" in the ability of the South African regime to bring to fruition an economic and political system which offered the highest rates of profit through the fascistic repression of the cheapest labour force in the world, multinational capital poured into the South African economy at an unprecedented level.[2]

This not only served to raise productivity in industry and manufacture within South Africa through the introduction of new technology and expertise, but the entire region was affected by this upsurge in economic activity generated by the "boom". Through an emphasis on the "capabilities" of the state in the region, multinational capital began actively to change the economic structure of what had previously been "suppliers" of cheap black labour to the centres of capitalist production located in South Africa. The main focus was on agriculture, a sector which had been neglected for half a century, but whose efficient integration into the international agri-business structure was vital for the continued profitable operation of world capitalism.

Latin America and Asia had been brought under control through the "Green Revolution", and the importance of Southern Africa in economic and political terms required a more efficient organisation of the vast agricultural potential of the region.

Agribusiness in the Third World means an integrated food system that extends from farm to factory to consumer—from food production to the manufacture of farm implements and pesticides to food processing and food marketing. Besides linking agriculture to industry, agribusiness also means that agricultural production increasingly resembles industrial production, in the application of technology to control nature, and increase productivity and in the use of wage labour... the spread of agribusiness to the Third World also entails a central role for the multinational agribusiness corporations that dominate the food system (internationally).[3]

Through the internationalisation of agri-business to the Third World, the most crucial issue which preoccupies the millions of poor people in Africa, Asia and Latin America—FOOD—comes under the complete control of multinational corporations. Agri-business "not only determines how food is produced and distributed, but it also shapes the lives of millions of third world people who depend on agriculture for their livelihood".[4] Agri-business is in fact the transfer of a particular system of production and the social relations of inequality which characterise the capitalist system.[5]

A country like Swaziland, which has an excellent agricultural potential and an abundant and cheap labour-force, experienced huge inflows of capital directly through loans and "aid" grants from agencies like the IMF and the World Bank, and indirectly through United Nations development projects and technical assistance schemes. According to the Swaziland Government *Economic Review, 1970-74*, the 1970s showed a clear restructuring of the Swazi economy, and while about a quarter of the total national output came from agriculture and agricultural processing, by the end of the 1970s, this share had risen to approximately 30 percent of the total national output.

By 1979, agriculture was generating 31 percent of gross domestic product, contributing more than 70 percent of national export earnings and providing employment for nearly 70 percent of the total indigenous work-force. The combined value of citrus, cotton and sugar output increased by 223 percent between 1971 and 1976, while total production increased by 39 percent and these three crops earned 93 percent of the total value of crops grown on freehold land.[6] The Barclays Bank report entitled *An Economic Survey of Swaziland—1979* described commodity agriculture thus:

Farming practices are highly mechanised and market oriented. Output per hectare is relatively high, accounts for 60 per cent or total agricultural production and is growing at a rate of about 5 per cent per annum.[7]

There were huge increases in acreage under sugar cane, citrus (oranges, grapefruit and pineapples), timber and "cotton production and exports were some three times greater in volume in 1974 than they had been in 1970".[8] Cotton, which is produced mainly by small peasant producers, increased from 2,500 metric tonnes in 1971-72 to 5,200 metric tonnes, an increase of over 100 percent. Manufacture, although relatively small, saw an increase of approximately 11 percent. The development of export-oriented primary commodity industries centred mainly around sugar, citrus and wood-pulp, and

from 1972-74... several new manufacturing enterprises were established. These were almost all industries relying on inputs of semi-manufactures and on the export of finished products to the South African market.[9]

As far as investment was concerned,

it is known that there was a net annual inflow of official capital through the period (1970-1974) and in most, if not all years, there was also some inflow of private capital for direct investment... In 1971-1972, the total volume of fixed investment, both for replacement of existing assets and to create new assets, amounted to about E18.3 million. This was about 22 per cent of gross domestic product,[10]

and most of this went into commercial agriculture (international finance capital) and/or into infrastructural development (South African capital), i.e. into the surfacing of main roads leading from the main centres of agriculture to South Africa. Waged employment also increased dramatically over this period, especially in agriculture, with the female component increasing almost two and a half times over a period of nine years. There was also an increase in those industries and sectors which have a direct relationship to the agricultural sector—manufacturing, processing, construction and transportation, and communication (see Tables 3.2 and 3.3).

TABLE 3.2
Structure of Employment

Year	Agriculture Forestry	Manufacturing	Construction	Transport Communication	Total *
1970	18,319	5,383	2,328	1,877	42,426
1971	20,840	5,837	2,537	2,050	47,051
1972	24,332	6,512	3,629	2,280	53,856
1973	23,655	7,360	3,950	2,688	57,032
1974	28,069	7,547	4,421	2,483	62,061
1975	28,407	8,998	3,341	2,540	64,405
1976	28,520	8,216	3,075	2,566	66,215
1977	26,377	8,411	4,081	2,768	66,225
1978	27,152	8,743	7,909	2,934	71,256
1979	28,573	8,873	8,223	3,012	73,767
1980	-	-	-	-	-

* The totals include figures on Mining & Quarrying: Electrical/Gas/water; Trade/Restaurants/Hotels; Financial Institutions; Community Social Services.

TABLE 3.3
General Levels of Employment

	1970	1971	1972	1973	1974	1975	1976	1977	1978	1979
Total	42,426	47,051	53,856	57,032	62,061	64,405	66,215	66,225	71,256	73,767
Males	35,068	35,953	42,609	44,769	48,746	50,011	47,559	47,266	54,017	56,767
Females	7,358	11,698	11,247	12,263	13,585	14,394	18,656	17,572	17,239	17,000

Source: Labour Department Annual Report, Central Statistical Office, Swaziland Government, Mbabane, 1980.

LAND POLICY—A POLITICAL MECHANISM OF CONTROL

Land is the source of all reproduction and life. Human society has moved through history and made advances, within the confines of each epoch, through its knowledge of the most important sources of food and wealth—the land. It is the process of production and our relationships with each other. It is in the process of harnessing nature, of creating wealth and thus moving society forward, that people enter into social relationships with each other. It is the process of production and our relationship to it which determine our relations to one another and to society as a whole. Or, in the words of Marx:

> In the social production which men carry on they enter into definite relations that are indispensable and independent of their will; these relations of production correspond to a definite stage of development of their material powers of production.[11]

The reliance upon the land for almost everything we need, from food, clothing and shelter to minerals with which to make tools or weapons, has underpinned all the struggles in human history.

Colonialism—the domination by imperialism of precapitalist societies in what is now called the Third World—was implemented essentially through a labour policy which recognised the crucial importance of land. To clear the way for the development of new and different relations of production—commodity relations—a land policy which not only alienated most of the land from the subsistence producers, but also redefined their relationship to the land, was necessary. Prior to the colonisation of Southern Africa, the Swazi, as one of several communities in state formation, had begun to settle down and establish a definite pattern of agricultural production.

They cultivated mainly sorghum and millet, and later grew maize which has become their staple food. Murdoch makes a very interesting assessment of precolonial land settlement among the Swazi and of the types of land which they chose to cultivate. The Swazi, who are organised along clan lines, initially occupied what were the most fertile areas of the country.

"The remaining 15 original clans, the main body of the Swazi tribe (sic) settled around 1750 in the cool middleveld sector of good soils, although capitals (royal residences) tended to be south-east..."[12] of the country. He goes on to show how in fact the five most fertile areas of the country today were used by the Swazi before they were carved up among the concessionaires from the mid-1880s.

Thus all five choicest arable areas with mean annual rainfall exceeding 30 inches, had been tribe (sic) or clan nuclei by 1850. Swazi are pastoralists but have not been purely pastoralists for two or more centuries and their crop husbandry—first millet and sorghum then maize—has been concentrated on what are, or were in the virgin state, better than average soils. Land use has long reflected land capability.[13]

Land allocation was determined by status—the king, princes, chiefs and headmen having access to the best and largest pieces of land, and demanding tribute labour from their ''subjects''. Individual (male) land rights were defined on the basis of communal or traditional tenure.

An individual's rights to land are derived from his membership of a community, from his position in the political organization. Rights to allocate land to others are similarly based on the allocator's political position.[14]

But under colonialism the right to land, arable land upon which the peasantry could reproduce itself, was determined and defined by how much labour capitalist enterprises need and when this labour was required. Land alienation became one of the main means of labour extraction from the rural homesteads into the mines, plantations and factories of the capitalist economy.

The Swazi were pushed onto one-third of the country's total land surface, 35 little pieces of poor land scattered around the country in convenient proximity to the huge white-owned farms.[15] These were known as Native Reserves, reservoirs of abundant and cheap labour, and after ''independence'' in 1968 it was called Swazi Nation Land. Until 1900, the Swazi were self-sufficient in food production, but after land alienation, productivity began to fall, and has never risen in any substantial terms, for the last 80 years.[16] ''Land shortage, mainly caused by the massive European appropriations, not only created a growing landless population, it also increased the oppressive and exploitative character of chieftaincy.''[17] By 1940, there were approximately 25,000 landless ''peasants'' out of a population of 180,000. The migration of male labour into the mines and plantations of South Africa was initially encouraged by the colonial government and the traditional rulers, the latter working in close collaboration with the colonial state for mutual benefits.

As Kowet stresses, labour migration allowed chiefs to keep control over land distribution, providing the ''safety valve'' without which the traditional systems would have faced immediate collapse.[18] The colonial state preserved most of the structures of the precapitalist political system—the king was given the status of "Paramount Chief" and his status was upheld as the "ruler" of the Swazi people—the peasants and emergent working class. Here we can see the basis of later collaboration between the oppressive classes in Swazi society— the traditional monarchy, the new petty-bourgeois elements and multinational capital.[19]

The traditional political system participated actively in the enforcement of taxes, which served to coerce labour out of subsistence production and into wage employment. Land became once again a tool of repression and control over the producing classes.

Since the chiefs controlled the population through their land allocation functions, the participation of the chiefs was of crucial importance particularly when the payment of taxes also included those who were not eligible for land i.e. the non-heads of households.[20]

The chief used to call the people to "tax-camps" and those who refused to attend were liable to a fine of ten pounds. Non-attendance could also mean several other possible forms of punishment, the most dreaded being banishment from one's area of residence. The question of access to a piece of land, even if it was non-arable and underused because male labour had been extracted from the household labour pool, became a powerful weapon during and after colonialism, in the control and repression of labour, especially in agri-business enterprises like the sugar, citrus and timber industries.

The specificity of Southern Africa is directly tied to the forms that labour assumed in the development of capitalism in the region. Migrant labour, of which Swaziland was an important supplier, shaped the political economy of the region and fundamentally affected the character and levels of consciousness of the black proletariat in the sub-region. Swazi workers were manipulated by the colonial state into entering commodity production, at excessively low rates of pay, while the state refused to recognise them as members of and heads of households. Land alienation had deprived the peasants of the best land, and because their techniques of production were so backward, the shortage of land made reproduction even more difficult. The situation was aggravated by the absence of male labour and by the continually falling subsistence levels in the rural homestead.

The migration system enabled the colonial state to treat black workers as "single" men, claiming that their families were adequately catered for in the rural areas, and therefore workers were and still are paid dismally low wages which did not include the reproduction of their families. Extensive literature on this issue has clearly exposed the myth about "the little piece of land back home" theory and shown it for what it really is—an excuse by the capitalist system to shift the responsibility of reproducing the family of the worker and the next generation of the labour force onto the shoulders of the peasants and workers, and especially onto the women, who remained behind in the rural areas.

But in Swaziland, state policy is still based on this myth that the families of workers should be able to reproduce themselves on the little scattered plots of land, of approximately 2.5 hectares, in the countryside, in spite of all the evidence to the contrary. One of the clearest refutations of the above claim that adequate means of reproduction are available in the countryside for subsistence producers, is that every single attempt to increase productivity on Swazi Nation Land has failed since the land expropriation of 1904.

RURAL DEVELOPMENT SCHEMES—PEASANTS PRODUCING FOR CAPITAL

Having created the conditions which pushed large numbers of male workers onto the labour market, the colonial state was faced with the consequent decline in the ability of the peasants to survive. Constant reference is made in the colonial reports to the "laziness" and the "apathy" of the Swazi. Although some recognition was made of the reasons behind the decline of productivity on the Native Reserves, it was always quickly qualified with claims that the peasants were not trying hard enough or were simply incapable of producing surplus because they were "irrational" and "inherently lazy".

The first government-sponsored scheme aimed at improving subsistence production was the Swaziland Native Land Settlement Scheme, begun in 1945. Based on a system of leasehold tenure, this scheme was meant to resettle families who were described as "squatters" on white-owned farms or on land which had been declared "Crown Land" (the British Crown), thirty years earlier. There were about 22,000 "squatters" on white-owned farms and about 5,000 on Crown Land. The initiators of the scheme admitted that "of the total area of the reserves, no less than one-third, containing over two-fifths of the cattle and one-half of the population has more than four times the number of cattle the land can support".[21] They also admitted that "deteriorating land fertility and the steadily diminishing area available to the average family have caused cultivation to be pushed on to steep slopes so that yields have fallen and erosion has increased".[22] But, they claimed that "there is little doubt that insecurity of tenure is one of the main causes for the agricultural improvidence and apathy of the African and the consequent misuse of the land".[23]

There were essentially two reasons why the scheme failed, and these were related to inherent contradictions in the scheme, but they seem to have eluded both the colonial government and later the neocolonial regime. In spite of all the efforts by the peasants to increase production on the settlement, they failed, and a census revealed one of the reasons for this failure. "The census disclosed that 40 per cent of the families have no cattle, or, with a few exceptions, any other stock, while 10 per cent have less than 13 heads."[24] Furthermore,

> the majority of settlers who produced poor crops are those
> who own too few or no cattle at all, and are unable therefore
> to make sufficient kraal compost for their needs, and in addi-
> tion are compelled to plant too late in the season, as they
> generally cannot hire draught animals from those settlers who
> have stock, before the latter have completed their own plan-
> ting operation.[25]

Second, just as the scheme was beginning to make some headway, multinational capital stepped in to take over the land for the large-scale timber forests. Peak Timbers decided that they needed the land upon which the scheme was being developed, and the colonial government agreed that "the acquisition of the Piggs Peak Settlement by Peak Timbers for afforestation ensures that the land will be put to its most correct and suitable use".[25] So, the people were told to leave the area with whatever they could carry, to be resettled on "better" land in another part of the country. The company undertook to "compensate on a very generous basis the the settlers who [were] required to move. For this purpose, £10,000 was made available"[27] for 158 families. The scheme never recovered and was formally abandoned in 1954.

This was the first example of many to come, where the interests of capital were placed above the needs of the people, and resettlement of Swazi peasants and the families of workers to make way for the development of agri-business concerns has become part of the "development" programme of the neocolonial state.

The next scheme was the "Master Fare Scheme", begun in 1963, but which did not prove much of a success either, for want of suitable new recruits, according to Low.[28] It had aimed at changing the Swazi peasants into "yeoman" farmers, and also at "accelerating the introduction of a cash crop economy to supplement and replace traditional subsistence farming in the African areas".

The colonial government once again blamed the unsuccessfulness of the scheme on the traditional tenure system, which it argued was not suitable for "development"!

> Government nevertheless considers that there is a strong case for a system of leasehold tenure on a limited number of specialised Swazi schemes for the production of high value commodities, mainly under irrigation, which will not only create a new class of prosperous Swazi farmers but will also test and demonstrate the value of the system of private ownership.[29]

Thus, property ownership, that essential element of capitalist production, became the basis for the recruitment of a kulak class within the context of the Swazi Settlement Schemes, which were established "on land set aside for the purpose together with a limited area for suitable title deed land to be purchased for settlement".[30] The settlements were meant to achieve the following—to produce "high value" crops like sugar, pineapples and cotton, and to encourage the rearing of good stock on the basis of leasehold tenure, under the auspices of the state and multinational capital. The state undertook to advance credit loans to those rich peasants who had enough collateral in the form of cattle and other stock, to establish the Swaziland Credit and Savings Bank for this purpose, and to develop strong cooperatives to provide inputs and technical assistance to the farmers.

As far as the sugar industry was concerned, and as a future "partner", in the scheme, the state made an encouraging suggestion.

> It is in the political and economical interests of the country as a whole that more Swazi be directly involved in sugar production. While government is prepared to play its part in promoting Swazi sugar growing, it considers that the industry should, in its own interests and as soon as the economics of the industry so permit, take the initiative in this regard and play the major role in the development of suitable schemes.[31]

Little wonder that the colonial state was anxious for the industry to take advantage of the schemes, since the largest agricultural and especially sugar growing concern in Swaziland was the Colonial Development Corporation.

By 1968, there were approximately 120 Swazi outgrowers in the CDC-supervised smallholdings. "As a general rule, three-quarters of the smallholdings are devoted to cane, and the remaining land is used for homestead and other crops..."[32] The 1969 Swaziland Livestock and Agricultural Department *Report* described the scheme thus:

> A sector of the agricultural economy where fundamental innovation has been introduced is on Settlement Schemes, notably at Vuvulane and Mpetseni. In 1968 at Vuvulane the 120 leaseholders (60 on 16-acre and 60 on 8-acre farms) produced 48,000 tonnes of sugar cane worth R200,000, 141 tonnes of seed cotton... and some maize, beans, potatoes and vegetables which were not all centrally marketed.[33]

There were also about 40 outgrowers in a pineapple scheme attached to the Libby's factory.

These kulaks have become an important source of raw cane and pineapples, produced by hundreds of Swazi workers under atrocious conditions of employment and living, who are not the responsibility of either the companies who receive the sugar and pineapple quotas, or the state, which turns a blind eye to the conditions on the schemes. It is interesting to note that the last Prime Minister of Swaziland, recently deposed in a "bloodless coup", was one of the larger outgrowers, supplying one of the sugar mills in the south-west of the country, and he still retains his "plot", which is about 60-acres of former Swazi Nation Land.

In 1971, 28 new settlers were granted 10-acre farms, and in 1972, a further 34 were holding plots of approximately 145 acres each.

> This brings the total number of smallholdings leased by Swazi farmers (in 1972) to 197, excluding the five 60-acre areas owned by large-scale farmers. Each of the new farms was planted with 10 acres of sugar cane, leaving approximately 4 acres for other crops.[34]

Another project, begun in 1959, as part of the Colonial Development Corporation's integration of the traditional monarchy into the young sugar industry, began as an area of 129 hectares, managed by CDC on behalf of the Swazi National Council, until the post-independence period when it was taken over by the Ministry of Agriculture. Out of this estate, came the basis for the establishment of the "Sihoya Farmers Settlement" established in 1969. "The settlers on this settlement were evacuated several years ago from the area now occupied by the Estate. They each have now three acres of sugar cane."[35] The entire project came under the management of Tibiyo Taka Ngwane

TABLE 3.4

Sihoya Sugar Project, 1969-75

Season	Tonnes cane cut	Sucrose	Tonnes sucrose	Ha. cut	Tonnes per ha.	Av. ag. cut/Mn.
1969/70	8,445	12.84	1,084	108.9	94.4	13.2
1970/71	9,413	14.04	1,322	109.3	82.5	13.0
1971/72	11,686	13.67	1,598	112.8	11.0	14.1
1972/73	12,775	12.95	1,716	129.9	11.5	12.9
1973/74	10,718	11.96	1,282	116.7	91.8	11.7
1974/75	12,800	13.35	1,800	110.1	116.3	-

Source: *Annual Report* of the Ministry of Agriculture, Swaziland Government, 1974.

in 1976.* The project seems to have been very successful in cane production, especially (see Table 3.4).

These are just a few of the many schemes which show the kind of collaboration between the neocolonial state and multinational capital in the production of raw agricultural commodities using very cheap labour on land which is formally for the use of the Swazi people, especially the peasants. The majority of peasants' and workers' families have remained on those parts of Swazi Nation Land which has not been declared "development" land.

> The promise that technical aid programmes will, of themselves, result in substantial increases in production or the "Commercialisation" of the Swazi farmer enables planners to take the easy narrow view and consider agricultural development in isolation from the broader, often conflicting economic wage (employment) and political (land tenure) issues.[36]

---

* The Tibiyo Fund, in which every Swazi is a shareholder although without receiving any dividends, has a varying stake in all sectors of Swazi industry. Its functions include not only promoting economic development and extending the Swazi stake in economic activity, but also preserving national customs, promoting education and purchasing and holding land for the Swazi nation. The value of Tibiyo's capital assets at the end of April 1980 was almost E38 million and their income in the 12 months to that date totalled E9.2 million. The source of the funds is the proceeds of the king's mineral rights made over to the fund.

So far, most of the attempts at improving subsistence production have met with failure, for the simple reason, that the peasant household, as an economic unit, is not structured to produce more than subsistence, and the disturbance of the division of labour within it, through the migration of able-bodied labour (both male and female) into wage employment in commodity agriculture, has undermined the viability of this unit, and only a total transformation of the relations of production can solve the problem in the countryside.

It has been necessary to discuss the historical relationship between the colonial state, the monarchy and the producing classes in Swazi society in order that the current relationship between multinational capital, the neocolonial state and the traditional monarchy on the one hand, and the workers and peasants on the other, may be clearly understood. This relationship will be brought out more clearly in our discussion of the manner in which these oppressing forces collaborate to oppress and exploit the Swazi workers and peasants.

AGRI-BUSINESS—SUGAR, CITRUS, TIMBER, TOBACCO AND COTTON

Private tenure land presently occupied approximately 30 percent of the rural land area of Swaziland. Although the collection of statistics on private land use was discontinued in 1978, the figures in Table 3.5 are still relevant to an understanding of the situation presently prevailing, since land distribution and use patterns have not changed very radically.

Private tenure land can be divided into two categories—privately owned white farms, which average about 800 hectares, and company farms or estates, which run into thousands of hectares. But for the sake of expediency, we shall lump these two land categories together and only make a distinction in reference to private tenure farms as Individual Tenure Farms (IFTs). In 1981, these farms which are owned by freehold or concessionary title, covered about 750,000 hectares out of a total land area of 1,727,116 hectares. The rest of the land came under Swazi Nation Land. "IFTs produce all the citrus and forestry products, most of the sugar cane and pineapples, about 40 percent of the cotton and a relatively small amount of maize."[37] Most of the tobacco (80 percent) and cotton (48 percent) is produced on SNL by petty commodity producers.

Sugar
Swaziland is a major producer of sugar, and since the opening of the third sugar mill in 1980, the combined capacity of the sugar industry

TABLE 3.5

Land use—1977-78 (hectares)

Land use category	Whole country[1]	Swazi Nation Land	Individual Tenure Farms[2]
Cropland of which:	193,616	137,924	55,692
crops fallow	159,322	108,652	50,670
	31,294	29,272	5,022
Grazing Land of which:	1,199,256	788,715[5]	410,541
natural veld	1,101,407	788,715	312,692
improved	97,849	-	97,849
Commercial Forests	97,702	-	97,702
of which: Pines	71,118		71,118
others	26,584	-	26,584
Other Farm land[4]	37,215	4,855	32,360
All other land	199,327	67,039[5]	132,288[6]
Total land	1,727,116	998,533	728,583

1. Excludes urban areas of approximately 9,300 hectares.
2. Including Sihoya Swazi Nation Land Sugar Project.
3. Includes all unallocated communal grazing land and mountains and hills on Swazi Nation Land.
4. Includes areas of farm buildings and services.
5. Includes Purchase-Land for RDSs and other SNL resettlement schemes.
6. Land unused or used only for traditional Swazi agriculture.

Source: Central Statistical Office, Mbabane, Swaziland.

is now 405,000 metric tonnes of milled sugar. In 1980, sugar exports comprised 40 percent of total export earnings, and in 1981, sugar and molasses exports earned the country E69 million, about 70 percent of the total value of sales.

The main sugar growers are subsidiaries of some of the world's largest multinational corporations, and the main foreign company in Swaziland is the Commonwealth Development Corporation (formerly the Colonial Development Corporation).** The first trial planting of sugar cane in Swaziland began in 1957, and by the end of 1958 the

---

** CDC, the Commonwealth Development Corporation, is a British statutory corporation established by an Act of Parliament in 1948.

first mill was being constructed on the banks of the Usutu river, one of the four large rivers cutting across the country on their way to the Indian Ocean.

Mhlume Sugar Company Limited was founded under the auspices of the CDC in 1958. The Swazi Nation (i.e. the traditional monarchy) was made a 50 percent share partner, and this enabled the CDC to lay the basis for future collaboration between the company and the traditional political system on all matters relating to the recruitment and control of labour and access to land.

The CDC have more investments in Swaziland than anywhere else in the world on a per capita basis. In addition to Mhlume, they own the Swaziland Irrigation Scheme and Vuvulane Irrigated farms. They also have interests in forestry and in industry.[38]

In addition, the CDC operates as a manager of the Mananga Agricultural Training College, which trains middle-level managers for the sugar industry in Central and Southern Africa. Through the outgrowers schemes, the CDC has been able to gain access to more land and expand its operations in milling without incurring the costs or the risks involved in the cultivation of the necessary amounts of cane. All in all, the CDC has made fantastic profits since its establishment. By 1980, there were about 19,000 hectares of land under cane, all irrigated, and the CDC had control over much of this.

The next big sugar producer in the country is the Ubombo Ranches Company, also established in 1958, and now owned mainly by Lonrho. This company owns 60 percent of the shares, and the remaining 40 percent are held by the Swazi government. In 1979, the mill at Ubombo Ranches had an annual capacity in excess of 140,000 tonnes of sugar. This particular company witnessed extensive strike action by workers in 1963 and again in 1979, which was brutally put down in both cases by the police and the army. A large quota of raw cane is delivered by private white growers, and by Swazi outgrowers located around the mill. The Sivunga Sugar Estate, in which Tibiyo Taka Ngwane has a 40 percent share equity, is managed by Ubombo Ranches. "Ubombo Ranches provides cutting and haulage facilities, tractor hire, irrigation and water, extension services, seed and fertiliser supplies."[39]

The outgrowers schemes have certainly been very productive and profitable more so for the companies than for the outgrowers, mainly because the companies do not have to bother about labour and the attendant "problem" under very poor working conditions and low wages. But, the development of a kulak class has come into conflict with some of the "traditional values" over which the traditional monarchy and the chiefs have assumed custodianship. Two consequences, which are clear indications of differentiation among the peasantry, are seen as a threat by the traditional political system. One of the objectives of the outgrower schemes was to use the (successful) outgrowers as examples to other Swazi peasants who had remained on SNL. Thus, "there was no provision for purchase of land, only leasing. The legal position of the lease is for 10 years with one possible renewal. It is also agreed that heirs may inherit a valid lease."[40]

One result has been that some of the lessees did not give up their homestead rights on SNL and therefore have retained their pasture and crop land rights in their chieftaincies. Due to the high price of sugar especially during the early half of the 1970s,

the profits have been sufficiently large that in some instances the lessee has become a landlord. Thus, instead of smallholder cultivation, the lessee hires all the field labour thereby performing neither a management nor a labour function and earns income merely from having a lease.[41]

To this the traditional authorities object, and feel threatened that loyalties to traditional authority are being eroded. Quite rightly so, but their class position makes it virtually impossible for them to see the contradition in the system.

There is even more outcry about the effects of the schemes on those outgrowers who do not live on Swazi Nation Land and have not pledged allegiance to any chiefs. These are the outgrowers on the 8-16-acre plots, who grow cane on 75 percent of the plot and use the rest of the plot for homestead space and for food crops. When the Swazi government takes over the ownership of such plots, these "tenants" become liable to eviction, but have nowhere to go since they are unattached to any chiefs. Once again a dilemma, whose solution does not favour the outgrower at the end of the day. It is interesting to note the effects of the outgrower schemes on peasant producers in other countries like Kenya, for example.

The third sugar producer in Swaziland is a corporation made up of several multinational subsidiaries, but managed by Tate and Lyle, the big British multinational, which also has a share in the company. This is the Royal Swaziland Sugar Corporation (Simunye), incorporated in 1977 with an authorised share capital of E40.1 million, and costing E120 milllion by 1980 when production began.

The majority of the share capital is being subscribed by the Government of Swaziland and the Tibiyo Taka Ngwane, who will each contribute E13 million. The Federal Republic of

Nigeria (E4 million), Tate and Lyle (E3.5 million), the German Development Bank-DEG (E2 million), Coca-Cola Corporation—USA (E1.7 million), Mitsui and Company Limited (E1.5 million), Commonwealth Development Corporation (E1 million) and the International Finance Corporation, an affiliate of the World Bank (E0.4 million), are the other shareholders. The balance of the finance required (E81 million) has been obtained by means of loans and export credit finance. The estimated cost of the project is E121 million.[42]

This project alone involves about 9,000 hectares of land and the mill was expected to have a productive capacity of 126,000 tonnes by the end of 1983. The deal was that the Swazi government would not only raise a large part of the share capital, but would also provide the land for the cane to be grown upon, and of course once again hold the whip over the work-force, to maintain conditions "conducive" to investment. The marketing and pricing of sugar are undertaken solely by the companies involved and the Swazi state has no influence at all at this level, except to try to get as good a price as is possible within the European Community (EC) through the Lomé Agreements. The market price of the product is determined by the international market forces in the final analysis, but state pressure is useful in getting the companies certain concessions.

For example, until 1964, sugar sales were integrated into the South African industry, and the sugar and molasses were transported by road and rail to South African terminal points. In 1961, South Africa declared itself a Republic, thus making it "an embarrassment" for the British companies to continue selling to South Africa. Therefore, in 1964, Swaziland was admitted into the Commonwealth Sugar Association, with an annual quota of 85,000 tonnes. When Britain joined the EEC in 1974, it became more convenient for Swaziland (and the CDC) to be a signatory to the 1975 Lomé Convention, which meant a commitment on the part of the companies to supply a quota of 116,000 tonnes of white sugar annually to the EEC. "The EEC in its turn, [has] agreed to purchase the sugar and guaranteed a c.i.f. price within the range obtaining in the community."[43]

In the 1979-80 season, when the price of sugar was beginning to fall to disturbingly low levels, Swaziland was granted an increased quota from 85,000 to 93,000 tonnes under the "hardship" allocation in the EEC. "The EEC quota which gives Swaziland a preferential price, is at its maximum of 120,000 tonnes. For the balance the country is subject to the variations of the world price."[44] This means that approximately 200,000 tonnes, since 1981, must be disposed of every year outside of the guaranteed quota, and although the United States and Canada have been taking a large part of this, most of the remaining sugar has to be sold on the open market. Thus, as the price of sugar continues to fall, the revenues to the Swazi government have dropped as well.

> Due to the sharp fall in world sugar prices, the sugar levy dropped to E1.4 million from E4.0 million during the year 1981-82. Taxes realised from companies (and individuals) also decreased from previous years' level, with the former falling from E7.3 million to E3.8 million while the latter decreased minimally to about E0.9 million.[45]

In addition, the Swaziland government agreed to a clause in the Simunye sugar project agreement, drawn up by the International Finance Corporation, that it would guarantee the level of working capital of the company. Therefore, the government has the obligation to inject money/capital into the firm once the working capital falls below the acceptable level required for the operation of the company. In this way, the company is assured of operating at a non-loss level even if the price of sugar falls, as it did in the early 1980s. Because of this clause, the Swazi government had had to pay the company instead of receiving revenues from it. Consequently, the fall in state revenues mentioned above.

Part of the land upon which the Simunye project is built was unallocated Swazi Nation Land and another part, the Ngomane area, was occupied by about 550 families (approximately 5,000 people) who were instructed to move from the area and threatened with forcible eviction and prosecution if they refused to obey the king's order. In fact, the king, who has since died (August 1982) came out on the local television station to warn the peasants that the state would send bulldozers in to demolish the homes of those peasants who behaved like landowners and who were refusing to move. An excellent example of the collaboration between the state, the traditional monarchy and the companies, in expropriating and oppressing the producing classes in the society. Most of the able-bodied men and women from the area concerned now work on the cane fields which were once their gardens and crop fields, and the old and the very young live in slums on the edges of the sugar cane estate.

In 1964, the Swaziland Sugar Association was formed, which controls all the sales of sugar, both locally and on the world market. "The Association co-ordinates sugar research through its extension committee and undertakes testing at mills under the supervision of its Cane Testing Service Technical Committee."[46] By 1978-79 the value of sugar produced had risen to E69.1 million, from E10.5 million in 1968-69, and the value of molasses had also risen from E0.5 million in 1968-69 to E1.1 million in 1978-79.

Citrus

Citrus is unique in that it occupies only 5-6 percent of the crop area, contributes 12-15 percent of the value of crop sales, and is concentrated on eight farms. Seven of the eight farms have their own packing houses and the industry is considered efficient by world standards. This component of the sector is totally owned by non-Swazis.[47]

The citrus industry in Swaziland had been operating since the 1950s, mainly in the production of oranges and grapefruit. These citrus fruits were planted as far back as the 1920s, mainly as private orchards, and although it has become an important industry, supplying Libby's with most of its oranges and grapefruit quotas, the fruit is still grown mainly on large orchards and some estates in the middleveld of the country where the temperature is mild. By 1978-79 total production had reached 717,519 tons of which 421,643 tons were exported. As can be seen from Table 3.6, the industry enjoyed reasonably stable production levels for the years 1969-1979.

TABLE 3.6

Swaziland—Citrus Production and Trade with the Republic of South Africa

Year	Production	Exports	Local Sales[1]
		(tons)	(tons)
1969-1970	623,827	311,679	233,323
1970-1971	637,086	293,291	276,578
1971-1972	709,489	302,456	250,596
1972-1973	666,646	308,305	338,518
1973-1974	734,537	300,105	318,374
1974-1975	659,894	346,406	324,613
1975-1976	712,219	380,158	225,831
1976-1977	613,033	385,129	267,731
1977-1978	745,472	366,291	209,958
1978-1979	717,519	421,643	261,152

1. Refers to sales via the Citrus Board, other direct sales may have occurred.

Source: RSA, Citrus Board, *Annual Report*, 1978-79, cited in Swaziland Agricultural Sector Memorandum, Main Report, December 1980.

All marketing and pricing of citrus fruit (oranges and grapefruit) is done by the Swaziland Citrus Board, which is in fact part of the South African Citrus Board, and which was formed in 1967 to undertake this function.

> By agreement and with the approval of the Swaziland and South African governments, the Board markets its fruit through the South African Co-operative Citrus Exchange, which is responsible for marketing policy, and distribution, promotion and sale of fruit, both in South Africa and in Swaziland's main overseas markets... The Board is the sole marketing agency for citrus produced for export by growers who have more than 50 trees.[48]

All fruit is marketed without distinction of origin, except in those countries which have trade embargoes with South Africa. Through this joint scheme, South African fruit is sold on markets which would not otherwise be accessible to it, and this is justified by the claim that:

> Swazi production is insufficient to justify its own marketing organization and by using the Republic of South Africa organization, higher prices are acquired than could otherwise be obtained.[49]

Also, the Lomé Convention allows fresh grapefruit access free of customs duties, and fresh oranges get a 93 percent reduction in tariffs. Therefore, South African growers are able to use Swaziland to sell their fruit, and the South African state scores (indirectly) against embargoes put up as a protest against apartheid. The political implications of such collaboration are embarrassing, to say the least, especially since Swaziland is already considered a "political skunk" in the region, because of its open friendly relations with South Africa both politically and economically.

The pineapple processing industry, owned by Nestlé (80 percent) and by the CDC and the Swaziland government (10 percent each), is the only citrus processing industry in the country. First planted in the early 1950s,

> the area planted has increased steadily and has spread beyond the Malkerns Valley. Pineapples have proved a most satisfactory crop... because they are hail resistant. Dry land cultivation methods are employed and yields have considerably improved by hormone treatment and better planting materials. Yields of 100-125 tonnes per hectare per cycle have been obtained.[50]

As a privately run concern, the industry experienced financial problems and had to be temporarily taken over by the Swazi government in the early 1960s. In 1968, Libby's took over the ownership and management of the industry, introduced high-speed processing machinery, expanded the factory area and increased the land under pineapples. The firm grows about 65 pecent of its own pineapples and the rest is supplied by outgrowers and individual white farmers in the Malkerns Valley. One of the outgrower schemes is the Mphetseni Scheme.

> In this scheme Libby's Inc. developed land for pineapples and constructed residences which the settlers can buy or lease, with payment from the profits of pineapple production. Libby's production manager specifies the kind and timing of agricultural operations, which if not undertaken will be performed by the Libby's company's own field labour. Libby's retains the right to evict lessees if provisions of the lease are persistently broken. Profits are alleged to be relatively high.[51]

Between 1968 and 1978, the value of canned fruits (pineapples, oranges and grapefruit) rose from E0.6 million to E8.2 million, but the conditions of employment and living of the hundreds of workers (95 percent of whom are women) who slog in the fields and in the factory, are barbaric and inhuman.[52] The manipulation of labour legislation by the company, and the "cooperative" attitude of the government Labour Department, have made it possible for profits to increase steadily while wages have continued to fall in relative terms. Through various forms of repression, the labour force has been kept in control, in a constant state of fear and intimidation. The operations of the Libby's company, with the collaboration of the Swaziland government, are another example of the callousness of agri-business in Swaziland and in the Third World.

Timber

The timber industry, which also began in the 1950s, is today a very large and profitable enterprise. Owned and controlled mainly by British capital (Courtaulds and the CDC), this industry has grown to occupy about 97,000 hectares of man-made forests on freehold land, and 65 percent of this area is on two plantations. The total land area planted with commercial trees is 120,000 hectares, 23,000 of which are on former Swazi Nation Land. There are about 100 plantations, and the deep fertile soils and the humid near-temperate climate of the Highveld and parts of the Middleveld provide ideal conditions for the growing of trees such as conifers and gums. Further, in Swaziland, pulpwood is harvested, on average, after 16 years, as compared with 40 years in northern Europe.[53]

The two major companies involved are the Peak Timbers (Pty) Limited and Usutu Pulp Company. The latter is 50 percent owned by Courtaulds and 50 percent by CDC. The Peak Timbers Company is also a British company, incorporating some South African capital, and it produces about 270,000 tonnes of logs per annum, and operates two saw mills. It exports to Mauritius, Reunion, the Persian Gulf, South Africa and Zambia. Wattle bark is exported to South Africa, on the basis of a quota which is controlled by the South African Wattle Growers' Union. The Usutu Pulp Company has a pulp mill at Bhunya, with a capacity of 150,000 tonnes and uses about 2,500 tons of pine wood a day. It produces unbleached Kroft pulp for export to Italy, Japan, Korea and Venezuela.

In spite of the huge areas of land which this industry has occupied and the huge profits it has made—in 1978 woodpulp brought in E27.1 million and sawn timber raised E5.5 million—the companies are still anxious to expand because they claim that "there is ample land available for forestry development". Yet, Swaziland has to buy all paper products from South Africa, because these needs are not catered for by the local industry. The conditions of living and employment of the work-force are terrible, and this industry has a very high accident and fatality rate, especially in the forest where many workers are injured or killed through carelessness, mainly on the part of the company.

The marketing and pricing of this product are also outside the government's control or influence, and like the sugar and citrus industries, the timber industry only declares what they think are "reasonable" profits, and the government, which is unable and also unwilling to question the accuracy of such figures, continues to accept the manner in which these companies operate in the country.

Besides sugar, citrus and timber, there are three other corps which are grown for commercial purposes in Swaziland, although under conditions which differ from those discussed above. These are cotton, tobacco and maize. These three crops are grown mainly on SNL and partly on privately owned white farms. The main crop, cotton, which is a dry-land crop, is grown mainly by petty commodity producers in the Low and Middle veld. In 1978-79, there were 4,000 small growers producing 48 percent of the national crop on 10,130 hectares of SNL. Because it is labour-intensive and therefore can use family labour, in addition to not requiring irrigation, cotton production has become one of the means by which the peasants and families of the workers can supplement their meagre subsistence levels.

But it is in the marketing and pricing of this crop that we observe the relationship between the Swazi state and the South African Marketing Boards. As in the case of tobacco and maize, which are also grown mainly on SNL, the pricing and marketing of cotton are done by organisations which are controlled and dictated to by the needs of South African industry. In the case of cotton, the Swaziland cooperative movement cooperates with the Swaziland Cotton Board, which works closely with the South African Cotton Board in determining the value and price of cotton lint, seed and other by-products of the crop. Marketing is done under the South African Marketing Agreement. ''This agreement deals with crop take-up, allocations to spinners, grading, coding and classification, specification and price and many other aspects of the marketing of cotton.''[54]

The Swaziland Cotton Board was formed in 1968, and operates on the basis of the 1967 Cotton Act (N° 26 of 1967), and one of the main functions of the Board is to undertake research on cotton improvement and to advise the government on matters affecting the cotton industry. The Swaziland Cotton Cooperative provides peasants with fertiliser, seeds, crop-spraying, picking and storage. ''The Cotton Board is responsible for the orderly marketing of the cotton crop from the producer to the ginner.''[55] Until 1981, the crop had three marketing outlets, two in the Transvaal and the local ginnery at Matsapa. A new cotton ginnery is being built at Big Bend by Guinness Peat International which should allow Swaziland to process its entire crop locally. An oil seed crushing mill is also beng constructed as part of the project, which is in all costing E3.5 million and will be run as a new joint venture company—Swaziland Oil Seed Mills Pty Limited. Partners in the joint venture are Guinness Peat International, the Swaziland government and the Swaziland Cotona Cotton Corporation.[56]

It is quite apparent that the Swaziland interaction with RSA Statutory Boards is a dependency relationship. The prices of maize, tobacco, and cotton in Swaziland are very much a function of the RSA price.[57]

It is also very apparent that the Swazi economy is predominantly dominated by British capital, a typical characteristic of most former British colonies. British economic interests have come to dictate the political importance of Swaziland for the British economy and for the other imperialist powers which dominate the Southern African region. (See Appendix to this chapter for a list of British companies in Swaziland.)

INVESTMENT POLICY

Swaziland has a very "favourable" investment policy, in addition to the necessary "political stability" which has made it a haven for multinational capital. With an "excellent" labour-control record, Swaziland is highly recommended by most of the World Bank and international finance organisations.

> Up to 1979, capital inflows, in which direct investment featured prominently in that year, have resulted in an overall balance of payments surplus and in an addition to international reserves. In 1979, however, a small overall deficit of £1.4 million was recorded.[58]

The Tropical Africa Advisory Group (TAAG) Report goes on:

> Other advantages include access by Swaziland goods to both Black and South African markets, quotes for certain agricultural products to the European Economic Community (EEC) and duty free access by locally produced goods. Membership of the Customs Union with South Africa, Botswana and Lesotho allows the unimpeded sale of goods to these signatories, although there is a common external tariff in force on imports from outside the Union. Exchange control regulations apply only to transfers between Swaziland and countries outside the Rand Monetary Area, but investment activities generally encounter no difficulties with transferring or raising capital. Unlike many other African nations, there are no foreign exchange problems. The repatriation of dividends and interest is freely permitted, subject to a witholding tax of 15 percent and a tax of 12.5 percent respectively, and expatriate workers are allowed to remit up to one-third of their gross income.[59]

By 1979, United Kindom investment in the Swazi economy stood at around £200 million.

Besides the above advantages, there are a host of investment incentives offered, ranging from training allowances and depreciation allowance on machinery to special incentives for development, when certain industries are granted "pioneer status" and thus qualify for additional income tax concessions. All prospective companies have to do is to write for a copy of *A Guide for Investors in Swaziland* to have access to some of the highest profits in the world.

But surplus value is not possible without a labour force to produce it, and here once again Swaziland offers the highest benefits to multinational capital without the "fuss" of labour problems. "In general, wages in Swaziland are low, even when compared with those of other African countries"[60] and in agriculture, they are about the lowest anywhere in the region, including South Africa. Wage rise demands have constantly met with threats and repression from the companies and the government. The share which the Swazi government has in the companies is in fact remuneration for the role the state plays in keeping conditions "conducive" for foreign investment, through repressive legislation and the use of the police and the army to quell any "disturbances". In 1979, with a labour force of about 73,000 there was only one factory inspector for the whole country. The industrial accident rate is very high, in fact, it is one of the highest on the continent, and most of the accidents occur in agriculture.

TABLE 3.7

Accidents in Wage Employment, 1973-81

	1973	1974	1975	1976	1977	1978	1979	1980	1981
Total Reported	1,208	1,059	1,209	1,396	1,316	1,327	1,629	1,370	1,375
Fatal Cases	23	21	26	18	23	33	38	20	37

1. Total active working population in 1980 was 82,000.
2. Sugar industry had 28 percent of the accidents (378 accidents and 4 fatalities) and 10 percent of the deaths in 1981.
3. Agriculture generally had 68 accidents and 5 fatalities (outside of the sugar industry) in 1981.

Source: Annual Report of the Labour Department, Swaziland Government, Mbabane, 1980.

1978 was a particularly bad year, as was 1979, with 33 and 38 deaths respectively, and the Government *Annual Labour Report* made the following statement:

33 deaths at work in one year is a very high figure in a country like Swaziland with only half a million people. It is a
known fact that Swazi workers are willing and eager to sweat
on the job, and there is a matter of simple justice that
employers should not ask them to bleed for their
willingness.[61]

But, in the absence of any trade union activity, the state has effectively used certain traditional ideological and political structures to
undermine workers' struggles and to militate against any organised
forms of worker resistance. This has been possible by transferring
one of the traditional functionaries into the work-place. He is known
as the *Ndabazabantu*—"the people's business"—and he represents
both the interests of the state (and the traditional political system) and
the interests of the companies. He is appointed by the traditional
monarch.

Such a decision is taken by the Swazi National Council and
the official looks after national as well as workers' interests.
He advises management on issues involving Swazi tradition.
It is usual for management to abide by requests made by the
king's representative, but should there be any disagreement,
the matter must be referred to the Deputy Prime Minister's
Office or the Department of Labour. The king's representative usually asks for a group of workers to be elected to
a consultative committee to whom he can refer in need. This
system is known as "Libandla".[62]

All workers' grievances are directed to the *Ndabazabantu* who
derives his authority and status directly from the traditional political
system which until recently was personified in the king. In this way,
the companies do not have to deal with workers' grievances and the
Ndabazabantu can easily identify any "trouble-makers" who can be
fired without much ado. The system also enables the state to use constant threats of "banishment", a practice of social and political
intimidation characteristic of the traditional political system. Because
the majority of workers' families live in the rural areas, on Swazi
Nation Land, workers are constantly threatened with loss of the
"right" to land, which has become a privilege, if they cause any
"trouble" for the company. They are also reminded that the companies are providing them with jobs, so they should not "spoil for
themselves" by agitating and demanding better wages and living
conditions.

Historically, as far as colonial history is concerned in Southern

Africa, the Swazi authorities have always taken the line of least resistance to domination and they collaborated with the British and the Boers in the repression of some of the other African peoples in the region. This "peaceful" attitude, as it is officially called, has become something of pride for the authorities, and they constantly remind the workers that "the Swazi are a peaceful people".

The collaboration between the neocolonial state and the traditional monarchy on the one hand, and the capitalist firms is a very important example of how these three oppressing classes have formulated a system of social and political control using backward forms of culture and ideology as part of the state repressive mechanisms.

In spite of all this repression and exploitation, the producing classes in Swaziland have not ceased to struggle against the capitalist system and the structures which represent it. The forms of struggle are covert, but they represent a resilience and strength which gives one the confidence that in the future the Swazi people will be able to live in a free and just society.

NOTES

1. Swaziland Government, *Economic Review, 1970-1974*, pp. 2-3.
2. First, R. et al., *The South African Connection*.
3. Burbach, R & Flynn, P., *Agribusiness in the Americas*, Monthly Review Press, London, 1980, p. 13.
4. Ibid., p. 13.
5. Burbach and Flynn, op. cit. Dinham, B. and Hines, C., *Agribusiness in Africa*, E.R.R., 1983.
6. Patrick, D.H., *An Economic Survey of Swaziland, 1979*, Barclays Bank, Swaziland.
7. Ibid., p. 5.
8. Swaziland Government, *Economic Review, 1970-74*, p.4.
9. Ibid., p. 6.
10. Ibid., pp. 11-12.
11. Marx, K., "1859 Introduction" *A Contribution to the Critique of Political Economy*, Chicago, 1981, p.11.
12. Murdoch, G., *Soils and Land Capacity in Swaziland*, Ministry of Agriculture, 1970, p. 243.
13. Ibid., p. 264.
14. Huges, A.J.B., *Some Swazi Views on Land Tenure*, p. 256.
15. McFadden, P., "Women in Wage-labour in Swaziland: A Focus on Agriculture"; Fransman, M., "Labour, Capital and the State in Swaziland, 1962-1977"; Booth, A., "The Development of the Swazi Labour Market", in the *South African Labour Bulletin*, Focus on Swaziland, Vol. 7, No. 6, April 1982.
16. McFadden, P., "Food Dependency in Southern Africa: the case of Swaziland", 1982.

17. Kowet, D.K., *Land, Labour Migration and Politics in Southern Africa: Botswana, Lesotho and Swaziland*, Uppsala, 1978, p. 74.
18. Ibid.
19. Fransman, M., *The State and Development in Swaziland, 1960-1977*, Ph.D. Thesis, Sussex University, June 1978.
20. Kowet, D.K., pp. 90-91.
21. Scott, P., "Land Policy and the Native Population in Swaziland", *Geographical Journal*, 1951, p. 444.
22. Swaziland Livestock ad Agricultural Department, *Report*, 1950, p.2.
23. Swaziland Native Land Settlement Scheme (1945), p.10.
24. Ibid. (1945), p. 12.
25. Ibid. (1948), p. 4.
26. Ibid. (1948), p. 3.
27. Ibid. (1948), p. 1.
28. Low, A., "Farm-household Theory and Rural Development in Swaziland", University of Reading, Department of Agricultural Economics and Management, 1982, p. 2.
29. Swaziland Government, *A Policy for Agriculture*, 1966, p. 2.
30. Ibid., p. 3.
31. Ibid., p. 11.
32. Patrick, D.H., op. cit., p. 9.
33. Swaziland Livestock and Agricultural Department, *Report*, 1963, p. 3.
34. Swaziland Government, *Annual Report of the Ministry of Agriculture*, 1972, p. 36.
35. Ibid., p. 46.
36. Low, A., op. cit., p. 89.
37. Swaziland Agricultural Sector Memorandum, Main Report, 1980, p. 2.
38. Tropical Africa Advisory Group (TAAG), *Fact Finding Mission; Swaziland*, 1981, p. 11.
39. Patrick, D.H., op. cit., p. 11.
40. Swaziland Agricultural Sector Memorandum, Main Report, 1980, p. 64.
41. Ibid.
42. Patrick, D.H., op. cit., p. 9.
43. Ibid.
44. TAAG, op. cit., p. 11.
45. *Swaziland Observer*, 5 March 1983.
46. Patrick, D.H., op. cit., p. 10.
47. Swaziland Agricultural Sector Memorandum, Main Report, 1980, p. 33.
48. Patrick, D.H., op. cit., p. 10.
49. Swaziland Agricultural Sector Memorandum, Main Report, 1980, p. 98.
50. Patrick, D.H., op. cit., p. 14.
51. Swaziland Agricultural Sector Memorandum, Main Report, 1980, pp. 63-64.
52. McFadden, P., "Women in Wage-labour in Swaziland : A Focus on Agriculture", *South African Labour Bulletin*, Vol. VII, No. 6, April 1982.
53. Patrick, D.H., op. cit., p. 14.
54. Ibid., p. 12.
55. Swaziland Agricultural Sector Memorandum, Main Report, 1980, p. 100.
56. TAAG, op. cit., p. 12.
57. Swaziland Agricultural Sector Memorandum, Main Report, 1980, p. 100.
58. TAAG, op. cit., p. 5.
59. Ibid.
60. Patrick, D.H., op. cit., p. 24.
61. Swaziland Government, Labour Department, *Annual Report*, 1978, p. 16.
62. Patrick, D.H., op. cit., p. 25.

APPENDIX

Local Company	British Shareholder
Industry	
Havelock Asbestos Mines (Swaziland) Ltd	—Turner & Newall Ltd
Libby Swaziland (Pty) Ltd	—CDC
Mhlume (Swaziland) Sugar Company	—CDC
Yeopac (Swaziland) Ltd	—CDC
Royal Swaziland Sugar Corporation Ltd	—Tate & Lyle Group/CDC
Shiselweni Forestry Co. Ltd	—CDC
Swaziland Irrigation Scheme	—CDC
Swaziland Oil Seed Mills (Pty) Ltd	—Guinness Peat International
Swaziland United Transport Ltd	—British Electric Traction Company
Ubombo Ranches Ltd	—Lonrho Group
Usutu Pulp Company Ltd	—CDC/Courtaulds Group
Finance Management	
Barclays Bank of Swaziland Ltd	—Barclays Bank International
Commonwealth Development Corporation	—CDC/ODA
EDESA (Swaziland, Pty Ltd	—Barclays Bank International/Union Castle Steamship Company Ltd
Lonrho Sugar Corporation	—Lonhro Group
Standard Bank Swaziland Ltd	—Standard Chartered Bank Group
Insurance	
Bowring & Minet (Swaziland) Pty Ltd	—C T Bowring & Co.
Swaziland Insurance Brokers (Pty) Ltd	—Price Forbes Bland Payne/AR Stenhouse & Partners/Hogg Robinson Ltd.
Auditors	
Cooper and Lybrand	—Cooper Bros. & Co.
Peak Marwick Mitchell & Co.	—Peak Marwick & Co.
Arthur Young and Co.	—Arthur Young & Co.

Source: TAAG Fact Finding Mission: Swaziland, 1981, Appendix IV.

4. STATE POLICIES AND FOOD PRODUCTION IN TANZANIA

L.A. Msambichaka

INTRODUCTION

Tanzania's agriculture is not yet developed and the level of production is still very low. It has enormous land resources and a favourable climate conducive to growing both temperate and tropical crops. About 46 million hectares are estimated to be suitable for agriculture, of which about 13.5 percent is under cultivation.

The government's pledge in 1964, that Tanzania should be self-sufficient in food by 1981 has not been attained. The pledge was directed towards the production of "international" staples which include maize, rice and wheat. Other food crops such as bananas, cassava, potatoes and yams were marginally considered by planners.

During the second independence decade (1970-80), Tanzania was importing cereals almost annually. In 1974, 1975 and 1980, Tanzania had to spend approximately one-fifth of its export earnings for importing only three "preferred" cereal grains, namely: maize, rice and wheat.

Despite the above anomaly, the importance of agriculture to Tanzania's national economy is indisputable. About 85 percent of the population resides in the rural areas and 85 percent of the economically active population is engaged in agriculture. It provides food for the greater part of the population and raw materials for the processing and manufacturing industries. Agriculture is the major source of foreign exchange earnings which are needed for importing both capital and consumer goods. Between 1973 and 1980 it contributed on the average (exports of raw and processed products) 95 percent of the annual export earnings. In addition, agriculture contributes about 40 percent (1966 prices) of the Gross Domestic Product (GDP).

In view of the importance of this sector, there is no doubt that the state is interested in the activities taking place in it. The state therefore guides or influences the nature of the activities with the assumption that they will benefit the state and the society as a whole.

This paper attempts to observe the impact of state policies[1] on Tanzania's economic development and food production in particular. It is expected that this observation will highlight some reasons as to why the food sector has stagnated throughout the decades. The discussion is organised by first examining the characteristics and functions of agriculture. We are of the opinion that this will act as a drawing board indicating whether major changes have taken place since independence both functionally and featurewise. In a similar manner section three attempts to discuss the consequences of various state policies on human and economic development as well as food production. It is our considered opinion that socio-economic development and food production are closely inter-related. This means that any policy which affects socio-economic development does implicitly or explicitly affect food production. The paper ends by trying to search for reasons supporting the importance of having a defined food policy in Tanzania.

FEATURES AND FUNCTIONS OF AGRICULTURE

During Colonial Days

The colonial government brought about many changes in the rural economic activities and the people's concern to produce food. Agriculture became synonymous with crop production.

Principal export crops included sisal, cotton, coffee, tea and tobacco. In the early years of this century, rubber was also grown. These were the crops which the colonial government emphasised, and policies were designed and directed towards enhancing these crops. Food production did not appear on the agenda of economic activities which were to be emphasised. The traditional practice of concentrating on food production was virtually undermined.

As a consequence of emphasis on export crop production large-scale settler farms were established where men were required to work as wage labourers. The demand by the government for each male citizen to pay poll(head)-tax forced many men to look for possible ways of earning cash income. This was possible by working in urban areas, in settler farms or in their own farms if they grew cash crops. Women were for a long time left alone in the villages because males had to work on estates, plantations or in towns. As a consequence the role of women in food production slowly gained prominence.

Unlike during the precolonial era when the main farm activity was food production, the introduction of export crops in some parts of Tanzania meant double work for the farmer. He had to work much harder than before in order to produce food for household consumption, food for the growing non-farm population and cash crops for the export markets. Farmers who were situated in areas which had a favourable climate for food growing concentrated most of their time on exports crops. The traditional function of agriculture of producing food was thus changed. Production of crops was essential, for feeding the indigenous population was treated as a peripheral activity. Export crop production as a source of money income and raw materials for the metropolitan states became the major obligation of the agricultural sector.

As a consequence of this emphasis Tanganyika was turned into a colony which was a source of raw materials for metropolitan Europe. Huge national resources in terms of labour, finance, and technology were wilfully invested in the production of raw materials in order to support the parent industries in industrialised and rich nations. Employment opportunities in the metropoles were also established as a result of having assured sources of raw materials and market outlets. At home the emphasis on export crop production was accompanied by the acquisition of the best and most fertile soils and inter-rural differentiation started to emerge because of ecological differences. However, the problems of food appear to have been less than during the period since the attainment of independence.

The Independence Era
The independence era did not witness any radical change in agriculture in terms of characteristics and functions. However, the functions expanded at least theoretically and its features continued to vary from one region to the other. In terms of functions, agriculture was expected to provide food for the nation, employment for the economically active population and raw materials for the domestic industries and to generate foreign exchange. The government continued to put priority on export crops and the same cash crops dominate foreign trade to date.

On the other hand, the independence period continued to be characterised by fewer varieties of foodstuffs. Whereas during the precolonial period there existed a variety of food crops, today the nation relies on very few of them. No substantial effort has been made to revive some of the food crops which were abandoned during the colonial period. A few traditional food crops such as cassava and bananas managed to survive the extinction, presumably because of their high yields per hectare. Production of crops like millet and sorghum has been neglected despite their ecological suitability simply because they are less palatable and have a low export value. And yet, cassava, bananas, sorghum and millet contribute about 40 percent of the human energy supply of foodstuffs consumed in Tanzania.[2]

The methods of farming used are still archaic. Primitive farming methods still dominate agricultural practices at the household level. Some of these practices have been very unproductive and destructive of the landscape. The poor farming practices include the burning of the area to be cultivated, shifting cultivation, cultivation on steep slopes and river banks and the continuous cultivation of the same areas with the same crop for many years (the non-existence of crop rotation). About 85 percent of the land is cultivated by hoe, and only 10 percent is cultivated by using ploughs and tractors. Just as prior to independence, the hoe is the dominant farm implement used by most farmers in Tanzania. There are a few regions like those in the north such as Arusha and Kilimanjaro and those of the cotton belt such as Mwanza, Shinyanga and Musoma, where a plough is an important farm implement.

In terms of rural transportation network the situation is poor. The difference between regions is considerable. Whereas Kilimanjaro region has a road density of 116.1km per 100km^2, Lindi region registers a density of 11.8.[3] In addition to these discrepancies there are also suggestions that the quality of rural road networks which expanded rapidly in the 1960s and 1970s has deteriorated.

THE IMPACT OF STATE POLICIES ON SOCIO-ECONOMIC DEVELOPMENT AND FOOD PRODUCTION

General Analysis: Impact on Socio-economic Development

Tanzania with an area of 945,087km^2 and a population of about 20 million (1983) has experienced a number of state policies during the colonial and independence era. The objectives and emphasis during the two periods have been different. In both cases state policies towards agriculture could be explained by looking at a number of variables among which are the utilisation of land resources, the cropping patterns and the structure of exports. We will turn to the specific policy variables in the next section.

Since the beginning of the century the best land in Tanzania has been actively put into the production of sisal, coffee and cotton. Food production areas have always been on the marginal lands. Past literature also indicates that during colonial days, in areas such as Kilimanjaro and Kagera regions, the growing of bananas together with coffee was prohibited despite the crop being the main staple food of these regions. Bananas were only allowed to be grown as wind-breaks and as a source of mulching material for coffee. Similarly, in Mwanza region, cereals were not allowed to be grown before cotton since they were considered to harbour insects and pests which would later attack the cotton crop. These measures were taken when it was clearly understood that late planting of cereals such as maize would lead to poor yields and consequently low farmers' incomes.[4]

The independence era does not appear to have made any significant change in the emphasis of agricultural policy on export crops. The newly independent Tanzania continued to place its best and enormous resources in the production of export crops. This is observed in budget allocations, research activities, extension service, distribution of inputs, prices incentives etc.

Despite this emphasis, export production figures are disappointing. In the mid-1960s and early 1970s production of most of the major export crops except tea and coffee started to fall and this continued in the 1980s (see Table 4.1). For example, the 1966 peak record of cotton production (86,200 tons) has never been reached again. Likewise, peak production levels of sisal, cashew nuts, tobacco and pyrethrum have never been attained since 1962, 1974, 1977 and 1975 respectively. All these crops belong to the group of traditional export crops and practically determine the foreign exchange situation of Tanzania (see Table 4.2). As can be observed from the table, sisal was the chief foreign exchange earner during the early years of independence. Between the mid-1960s and the mid-1970s cotton and coffee changed positions, and coffee has taken the leading role.

In the case of food marketing the institutions have been changing hands. During the colonial period most of the food marketing was undertaken by private traders except for wheat which was handled by the then National Wheat Board. In an attempt to eliminate the middleman so that farmers can realise more from their farm produce, the government in 1962 established the National Agricultural Products Board (NAPB). The Board was given the responsibility of controlling and regulating the marketing of maize and later maize flour, paddy and rice. In 1973 the activities of the NAPB were taken over by the National Milling Corporation (NMC).[5] As from July 1984, NMC has been buyer of last resort and deals more with the enlargement of the Strategic Grain Reserve (SGR). The entire purchase of cereals from the farmers is carried out by the respective regional cooperative unions.

It is anticipated that cooperatives will collect the crops from farmers timely and late payments will be minimised. These have been some of the major complaints of the farmers since 1976 when cooperatives were disbanded.

The second independence decade also marked the beginning of decline of official food purchases (see Table 4.3). Volumes of preferred staples started to decline in 1977-78 (except maize) and that of drought staples in 1978-79.

TABLE 4.1

Production of Major Export Crops in Tanzania 1961-83 (in '000 Tons)

Year	Coffee	Cotton	Sisal	Cashewnuts	Tea	Tobacco	Pyrethrum
1961	24.9	30.2	200.9	40.0	3.2	0.5	1.4
1962	25.7	33.0	219.5	59.0	4.0	0.4	1.9
1963	28.7	48.6	214.4	41.5	4.0	4.0	2.3
1964	33.8	45.2	203.9	56.7	4.5	0.2	2.3
1965	34.2	56.2	210.1	64.6	4.3	1.7	3.7
1966	52.0	86.2	218.0	83.3	6.8	5.1	4.4
1967	40.5	60.8	225.0	84.3	7.2	7.8	6.7
1968	51.5	62.9	220.0	117.6	8.0	7.3	4.8
1969	46.1	56.8	197.0	113.5	8.8	11.6	3.9
1970	49.7	60.6	209.0	111.2	8.5	11.1	2.3
1971	46.7	54.5	202.0	107.3	9.2	12.0	2.7
1972	46.7	64.5	181.0	113.8	11.6	13.1	4.3
1973	52.4	60.0	157.0	125.4	13.3	12.7	4.0
1974	47.5	49.1	155.0	145.1	12.3	18.3	3.3
1975	42.3	39.1	138.0	119.0	13.9	14.2	4.7
1976	52.1	57.6	120.0	83.7	13.0	17.9	3.9
1977	54.2	40.4	114.0	97.6	15.2	18.3	1.3
1978	48.7	47.0	105.0	68.4	18.5	17.1	2.9
1979	51.9	60.5	92.0	57.1	17.5	17.3	1.6
1980	47.8	58.4	81.0	41.4	17.3	16.9	1.6
1981	67.3	53.8	86.0	47.0	16.4	16.1	2.0
1982	54.8	44.2	73.7	44.2	15.5	13.6	1.0
1983	53.8	42.9	60.0	32.5	17.5	13.6	1.6

Source: *Hali ya Uchumi wa Tanzania* (Various Years); URT, *National Agricultural Policy*, Dar es Salaam, October 1982, p. 191.

The four southern regions of Tanzania, currently known as the big four (Iringa, Mbeya, Rukwa, Ruvuma), continued to be neglected despite their enormous agricultural potential with favourable climate and almost sure rainfall for growing all major staples consumed in Tanzania. It is only recently that policy-makers have publicly declared that these regions are the food surplus production areas of Tanzania. Other regions of Tanzania, like the central dry zone have remained poor despite having fertile soils and better communications than the regions in the south. Frequent droughts are singled out as a major drawback in these zones.

The second independence decade also witnessed enormous imports of cereals. While huge volumes are recorded, import volumes per unit price are continuously declining, a reflection of the increasing

TABLE 4.2

Major Agricultural Export Contribution to Total Export Value in Tanzania

Year	Total Export Earnings T.Shs.M.	Coffee	Cotton	Sisal	Tobacco	Tea	Cashew-nuts	All six Exports Crops
					Percentage Contribution			
1961	1,056.7	12.8	12.9	26.6	0.2	2.5	3.4	58.3
1962	1,119.3	11.7	13.2	28.1	0.1	2.9	4.2	60.3
1963	1,371.3	9.9	15.6	33.1	0.1	2.3	2.9	63.9
1964	1,530.9	14.4	12.9	28.6	0.0	2.0	4.3	62.3
1965	1,402.5	12.2	17.4	20.4	0.7	2.2	5.9	58.8
1966	1,785.2	16.9	19.6	13.1	0.9	2.5	5.8	58.9
1967	1,667.7	14.3	15.1	12.0	2.0	2.6	6.1	52.1
1968	1,717.7	15.4	16.5	9.2	2.3	2.6	6.5	52.5
1969	1,792.4	14.3	13.1	8.9	1.9	2.7	7.6	48.6
1970	1,851.9	16.9	13.3	9.7	2.4	2.4	7.4	51.9
1971	1,988.8	11.4	12.3	5.7	2.2	2.5	7.4	42.5
1972	2,344.8	12.3	14.3	6.2	2.1	2.3	7.4	44.6
1973	2,607.7	18.9	20.4	8.5	2.1	2.1	6.7	51.1
1974	2,944.4	12.7	16.1	15.7	2.9	2.4	8.2	58.1
1975	2,848.9	17.0	10.5	10.6	3.1	2.8	7.7	51.8
1976	3,886.0	33.0	15.8	6.2	4.8	3.6	5.3	68.6
1977	4,545.4	40.9	11.9	5.0	4.6	3.9	6.0	72.4
1978	3,685.4	35.4	11.4	6.0	6.0	4.6	6.2	69.6
1979	4,484.3	27.1	10.9	5.9	3.3	3.7	5.1	55.9
1980	1,467.5	26.9	8.6	5.9	2.4	4.4	4.4	52.6

Computed from : Bank of Tanzania, *Tanzania : Twenty Years of Independence (1961-1981), A Review of Political and Economic Performance*, p. 278.

unit price of the grains. (Figure 4.1). The low import volumes per unit price as observed in Figure 4.1 are the ones which Tanzania has to battle with. Food imports per US$ 1,000 are declining annually and when compared with other variables the situation is equally alarming (see Figure 4.2). In 1980, all four comparisons were moving upwards. The food import bill took more of foreign exchange earnings than the previous four years. Cereals constituted a larger part of the food imports.

The impact of all these on the economy and well-being of the people is adverse. The government's ability to continue importing food in order to meet local demands is seriously being undermined and a favourable condition for malnutrition is being prepared. The cause of such huge imports lies solely in the perverse agricultural

TABLE 4.3

NMC Purchases of Major Crops in Tanzania (in '000 tons)

Year	Maize	Rice	Wheat	Sorghum	Bulrush Millet	Finger Millet	Cassava	Pulses	Sugar
1966-67	112.9	23.5	28.0	-	-	-	-	-	-
1967-68	104.3	5.1	33.1	-	-	-	-	-	-
1968-69	127.5	4.2	24.6	-	-	-	-	-	-
1969-70	54.1	59.8	21.0	-	-	-	-	-	-
1970-71	186.0	62.7	34.0	-	-	-	-	-	-
1971-72	43.0	45.9	56.7	-	-	-	9.1	14.3	18.9
1972-73	106.4	49.0	46.8	0.6	-	-	14.3	-	105.1
1973-74	73.8	39.9	27.9	1.7	-	-	18.4	-	96.2
1974-75	23.9	15.2	14.4	1.9	2.5	-	17.8	-	103.2
1975-76	91.0	12.2	24.2	2.9	2.2	-	17.4	-	112.1
1976-77	127.5	14.9	23.0	10.1	6.4	4.5	20.2	n/a	99.2
1977-78	213.2	35.5	35.1	33.6	14.4	22.3	36.9	40.3	135.6
1978-79	220.4	34.5	27.5	58.6	16.5	23.4	63.8	58.3	128.6
1979-80	161.5	30.8	26.5	20.7	1.3	15.5	44.2	59.4	118.0[1]
1980-81	103.8	13.4	26.9	19.3	0.3	1.2	7.5	48.6	122.0[1]
1981-82	85.1	14.1	2.1	-	-	-	-	9.1	-

1. Calendar year 1980, 1981.
Source: MDB, *Price Policy Recommendations for the 1981-82 Agricultural Price Review.*

policies which have been followed throughout the decades.

This brief observation underlines the fact that both during the colonial era and after independence, no state policy was seriously geared towards food self-sufficiency nor improvement of the human environment. Efforts have never been made to ensure that the nation is adequately supplied with enough food domestically. The country has a large area which can be cultivated (46 million hectares) compared to its small population of 20 million people.

Given the country's favourable climatic conditions and soils, the country is capable of producing most of the required foodstuffs locally.

It is therefore beyond people's imagination that Tanzania should be experiencing food problems so frequently.

Specific Policy Variables and their Impact on Food Production

This part of the paper attempts to discuss the impact of various state policies on food crop production. We will mention a few, which in our judgement appear to have contributed to the country's failure to produce enough food and achieve breakthroughs in agricultural productivity.

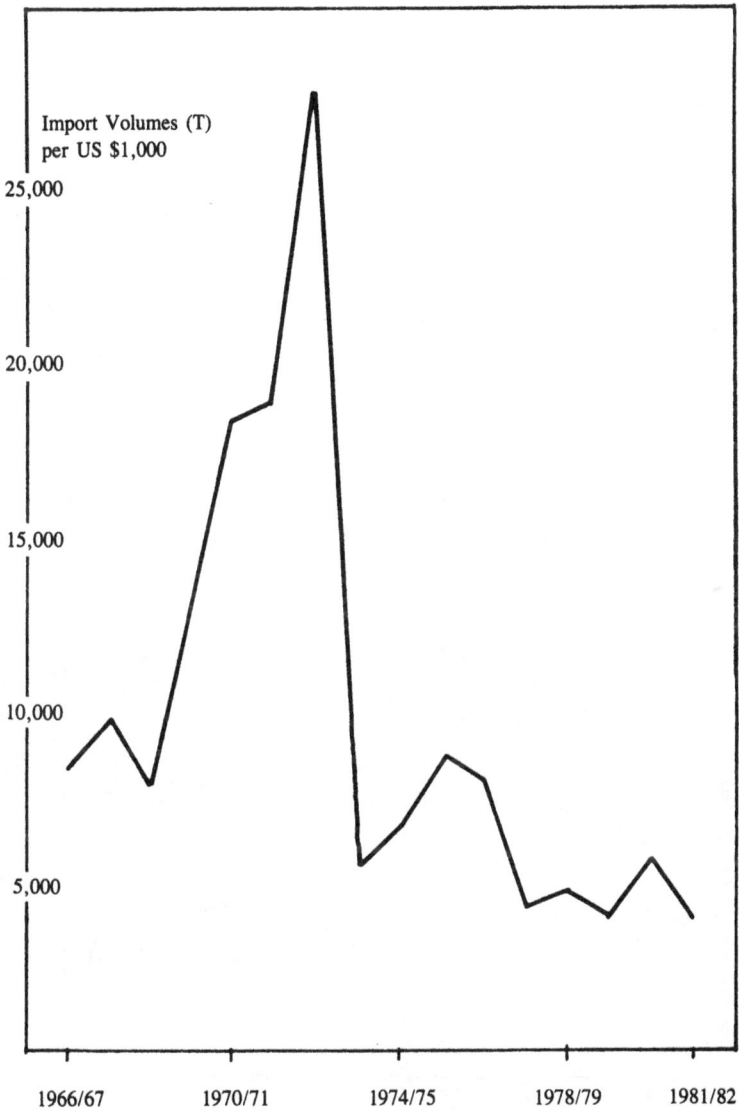

Figure 4.1   Trends of Food Imports per US$1,000

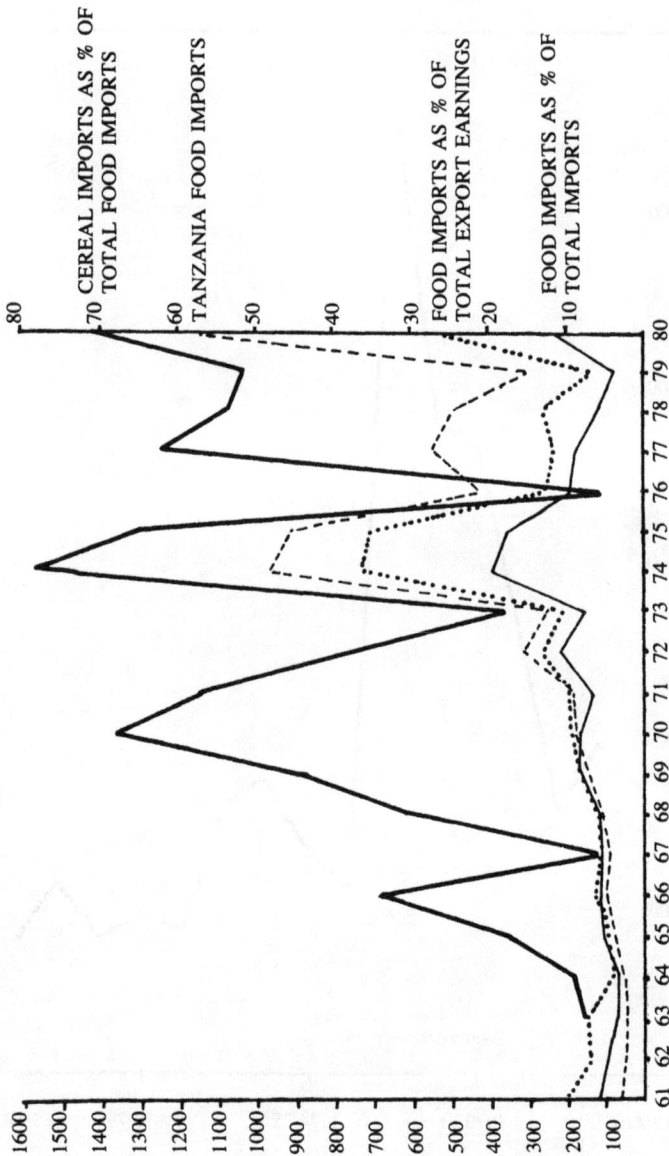

Figure 4.2   Food Imports in Tanzania (000 tons) compared to total Export Earnings & total Imports (in %)

Land use and land allocation. Since prior to independence, allocation of land has always favoured the development of export crops. For instance, sisal, which is a drought-resistant crop and could be grown in marginal areas, is produced in the best alluvial soils which are well watered and could be used for the production of cereals, pulses, fruits and vegetables. The Kilimanjaro region, which is blessed with the richest volcanic soils in Africa with excellent rainfall and water from various rivers flowing all the year round, grows principally coffee. This region could very well produce maize, fruits, vegetables and livestock products especially milk.

When land was allocated for large-scale wheat farming in Harang district, Arusha region (Appendix 1), the primary purpose was to serve the urban population and not to diversify food sources and incomes for the indigenous population.

In view of these policies, allocation of land to both rural and urban dwellers, for the purpose of promoting food production in the country, has not been given due consideration.

Education. Education plays an important role in production. It is a factor of growth and productivity and a medium of faster adoption and diffusion of modern techniques and practices in agriculture.

The importance of farmers' education was recognised soon after independence. As a result, literacy campaigns were launched all over Tanzania and in 1974 the Universal Primary Education (UPE). Farmers' Training Centres (FTCs) were established in various regions of Tanzania to provide some of the basic knowledge on the use of ox-ploughs, fertilisers, pesticides, crop rotation etc. For reasons best known to decision-makers, during decentralisation (in 1972), FTCs were transferred to the Ministry of Education where they served multiple functions including the teaching of ideology. As a result farmers were left without any educational institution of their own, where practical problems could be discussed and examples of modern farming demonstrated.

In terms of agricultural development and food production in particular the abolition of FTCs was an error. FTCs provided the broad-based type of farmers' education which is crucial in production. And a prosperous agricultural development requires that kind of education for the farmers. This is what Tanzania is currently lacking. Tanzania needs to increase the quality of its farmers if per capita food output is to increase rapidly.

Manpower. In terms of manpower utilisation, the food sector has throughout the decades been undermined. About 85 percent of the economically active population (9,192,970 in 1978) are engaged in agriculture and more than 23 percent of the wage-earning group are engaged in production of estate crops.[6] Sisal alone contributes about 9 percent of the wage-earning labour force.

The efforts made by the government and party leaders through campaigns in order to promote production of export crops are enormous. The army of extension personnel has throughout the decades been directed towards export crops, mainly coffee, cotton, tea and tobacco. Most of the (elite) farmers who have some knowledge about modern agriculture are concentrated in the export crop sector.

Smallholder efforts which are dedicated to growing coffee, cotton, tobacco, tea, cashew nuts and pyrethrum are considerable. Unfortunately these efforts cannot be computed. But it is likely that more than 50 percent of the total farmers' efforts are devoted to production of export crops. It means therefore very little of farmers' efforts are invested in food production.

All what we are trying to argue here is that previous agricultural policies have for many years neglected the food sector even in terms of allocation of human efforts, both qualitatively and quantitatively.

Rural transport. Transport is one of the most important factors in the process of agricultural development. It is only through roads, railways and waterways that agricultural inputs and products can be moved from one place to the other. Research findings indicate that a good transportation network alone can contribute about 15 percent of the total agricultural production. For this reason, investment in agricultural transportation is a great asset to the entire development of both exports and food crops. Yet in Tanzania, the railways (Dar-Mwanza and Dar-Moshi) and ports (Tanga and Mtwara) were developed in order to facilitate the development and export of cotton, coffee, sisal and oilseeds. Evidence shows that when the groundnut scheme which was run by the Overseas Food Corporation in Nachingwea failed, the railway line which was constructed primarily to serve the scheme was uprooted.

The construction of railway networks in Tanzania seems to have been played down in favour of roads. Yet the cost per ton road mile of freight of moving bulky goods such as inputs is very high. For instance, it is cheaper to move an equal volume of fertiliser from Dar-es-Salaam to Kigoma by rail (T. Shs. 200/- in 1979) than Dar-es-Salaam to Sumbawanga (T. Shs. 1,000/- in 1979) by road although the two areas are situated at an almost equal distance from Dar-es-Salaam.

The road transport charges are also prohibitive in transporting food crops.[7] At one time it was argued that it is cheaper to ship maize to Dar-es-Salaam from America, which is almost 10,000 km away, than to transport maize from Songea which is just 800 km away. There is also the road density factor. As discussed earlier, the discrepancy between regions is still very big. Given the current high road construction costs (T. Shs. 0.5-1.5 million per km of gravel roads and T. Shs. 4.0 million per km of permanent roads), it is unlikely that the difference between regions will be rapidly reduced in the near future.

As a consequence, Rukwa and Ruvuma regions which are favourable for the production of food crops such as maize, beans and millet and whose potential has remained untapped for many years will continue to be deprived of the economic benefits of railway transport.

Much will have to be done in the area of transport, if farmers are to be motivated to produce more.

Agricultural investment. Agricultural investment in Tanzania has always given preference to export crops. This pattern of investment, which can be traced back to the colonial days, is observed in the recent development budgets as shown in Table 4.4. For eight years, the Ministry of Agriculture expenditure for cereals ranged between 5.9 percent (1975-76) and 35.6 percent (1980-81) of the development budget. Allocation to export crops ranged between 15.5 percent in 1974-75 and 55.6 percent in 1981-82. A substantial amount of the development budget was invested in sugar development, a semi-food crop which is mainly intended to cater for the urban population.

At the household level, the present economic difficulties have forced many middle income earners to reduce their cash requirement for investment. Average farm returns are low and when compounded by poor crop harvests, farm saving has been difficult. Farmers who used to employ workers have abandoned the practice and now depend on family labour. Rich farmers fail to maintain their farm equipment, such as tractors and vehicles, because of lack of spare parts, fuel, oil etc.

The impact of these problems on production is the reduction of hectarage cultivated and consequently less total food output. This is what Tanzania is currently facing.

Credits. Modern inputs and improved farming practices are important in increasing productivity. But inputs need capital and most farmers in Tanzania have very low incomes. They cannot afford to buy inputs, equipment and the like without the provision of credit.

Research conducted in four regions of Tanzania in 1975 revealed that about 60 percent of the farmers produce less than what an average farmer can produce. It means many food producers need credit

TABLE 4.4

Expenditure on Major Crops in the Ministry of Agriculture Development Budget in Tanzania 1974-75—1981-82

	1974-75	1975-76	1976-77	1977-78	1978-79	1979-80	1980-81	1981-82	Total Constant 1969 Prices
Total Expenditure (Million T. Shs.)	233.0	351.6	436.3	349.8	371.5	605.0	472.9	394.5	1,144.6
Out of which:									
Sugar Crops %	77.6	52.0	61.3	38.0	35.4	25.4	23.1	15.7	43.7
Export Crops %	15.5	42.1	23.5	51.9	39.3	49.9	41.2	55.6	38.6
Cereals (NMC & NAFCO) %	6.9	5.9	15.2	10.1	24.0	24.0	25.6	28.1	17.3
Oilseeds %	-	-	-	-	1.3	0.7	0.7	0.6	0.3
Total Percentage	100	100	100	100	100	100	100	100	100

Source: Ministry of Agriculture Development Budget 1975-76 to 1981-82.

facilities in order to improve the current poor food outputs. Aggregate yields growth rates of major food staples for the past 22 years (maize, rice, wheat, sorghum and millet) are declining and so are the per capita food outputs.[8] Traditionally, credit institutions in Tanzania have favoured cash crop producers more than food crop producers. Tobacco, tea, coffee and cotton have been the major beneficiaries of the facilities even after establishing the Tanzania Rural Development Bank (TRDB), (now Tanzania Cooperative and Rural Development Bank—TCRDB).

The TRDB was established in 1971 in order to provide equal service to small-scale producers of both export and food crops. It paid more attention to export crops, especially tobacco, to the point of being nicknamed the "Tobacco Regions Development Bank". The introduction of the National Food Credit Programme (NAFCREP) in 1977-78 opened a new phase of activities for the TRDB. The Bank started disbursing credits for food crops such as maize, sorghum and wheat. Nevertheless, there does not seem to be a significant change in the levels of food production since the bank started to issue loans for the purpose. In fact from 1978 to 1982 aggregate food production growth rates were declining at the rate of 12.20 percent per annum.[9] There is probably a need of looking into the current credit policy so that the farmer can use the extended resources much more productively.

Research. Modern agriculture cannot take place in Tanzania without adequate and proper research. It is research which has turned many countries into food surplus growing regions. Yet ever since colonial days, research activities have been carried out with the purpose of improving export crop production. The current available evidence shows that most of the oldest established research institutions and centres in Tanzania are dealing with research on export crops. For instance, Mlingano Research Station for sisal, Lyamungu for coffee, Ukiriguru for cotton, and Nachingwea for cashew nuts. These are some of the most famous and best research stations in East Africa.[10]

Research on food crops is a recent phenomenon and has not taken full grip. Many food crops such as rice, bananas, cassava, round-potatoes and sweet potatoes have been left to the whims of the farmers. Only a few crops such as maize, wheat and beans have been given some research attention for quite sometime. Despite these exceptions, the beneficiaries of the research findings constitute neither a large percentage of the smallholders nor a sizeable crop hectarage.

Equal emphasis has to be given to the research on food crops if the current low food production is to be improved.

Extension service. Dissemination of findings to the farmers needs the services of extension agents. During the early years of independence the role of extension agents was actively supported. However, their role was mainly directed towards promoting export crops such as coffee, cotton, tobacco, tea etc. The role of extension agents in food crop production was minimal. This was further weakened by the decentralisation programme of 1972, when a number of agricultural officers were posted into regions to carry out administrative duties. In 1978, about 16 percent of 7,836 agricultural professionals were employed by export crop authorities.[11] Unfortunately, there were no agricultural professionals employed specifically to serve the food sector. This situation continued until the government decided in 1983 to place all extension officers under the Ministry of Agriculture.

Just like the export sector, the food sector also needs the infusion of modern agricultural technologies. This role can best be performed by extension agents.

Use of biological inputs. Inputs occupy a central place in modern agriculture. In Tanzania, it has taken long to recognise the advantages of modern inputs. Currently, the amount of fertiliser used per hectare is about 20 kg. When compared with developed countries which use more than 100 kg per hectare, the amount is very small. It is also estimated that about 50 percent of the fertiliser is channelled into coffee, cotton, tobacco and tea. This type of unequal distribution is being carried out when food crops could as well benefit from its miraculous impact on yields. For example, research conducted by MDB indicates that average yields of paddy in smallholder plots could be raised from the current levels of 0.4 tons per hectare to more than 3.0 tons, only if improved seeds and fertiliser were used.[12] Research findings also indicate that proper application of fertiliser can increase production up to about 60 percent.

Likewise the use of improved seeds is very low, as is use of pesticides and fungicides. In 1978-79 only about 9 percent of the total crop seeds used in Tanzania were improved seeds. Average consumption of pesticides, fungicides and herbicides is around 320 gm per hectare.[13] It is therefore not surprising that 25-30 percent of the harvested food crops do not reach the table because of vermin, pests and diseases.

The importance of biological inputs for food production has been recognised by the government at a time when the food sector is in severe difficulties and the world market prices for inputs are high.

Irrigation. Irrigation is an important factor in food production. It increases crop intensity and ensures crop harvests. Unfortunately, this kind of agricultural practice has been left to develop without the government's guidance.

There are more than 933,000 hectares which are suitable for irrigation but only 144,000 hectares were under irrigation in 1980. About 83.6 percent of the irrigated area is under smallholders and the rest under large-scale private and public farms.

Lack of emphasis on irrigation farming existed before independence and was carried over thereafter. This was taking place while having a division or a full government department responsible for irrigation.

The explanation as to why the government has not given due emphasis and support to irrigation farming over the years is difficult to find. There was an irrigation campaign in 1974 which ended without any record. In 1978 the ruling Party (Chama Cha Mapinduzi-CCM) central committee appealed to the government to start large-scale and small-scale irrigation projects. The Party guidelines of 1981 emphasised the importance of irrigation farming. Nothing significant seems to have taken place ever since.

Surveys conducted in some regions of Tanzania indicate that paddy yields can increase from 400 to 2500 kg per hectare if put under irrigation.[14] Despite the appeals and survey evidence on the positive impact of irrigation on yields, neither large-scale nor small-scale irrigation seems to be taking place. The water resources are in abundance but the country continues to depend entirely on erratic rains for food production. This is amazing, specially when it is known that the rains are unreliable. What is more amazing is the fact that even traditional irrigation methods have been abandoned and the government just let the system rot. Something needs to be done at least as far as simple irrigation is concerned.

Mechanical technology. Tanzania has had a demoralising experience in mechanised farming both prior to and after independence. First there were the Kongwa and Nachingwea groundnut schemes of the 1950s and then the cooperative tractor service of 1964. Both programmes failed miserably. Since then, Tanzanian policy-makers have considered mechanisation as an omen and the hoe as the appropriate farm tool for transforming Tanzania's agriculture and food production in particular.

Research conducted in various parts of the world and in Tanzania indicates that mechanised modes of production can enormously expand output per person and hectare. Different research findings indicate that through mechanised farming, paddy yields per hectare can increase threefold and the number of man-hours spent per hectare be reduced by about 97 percent. Similarly, maize yields per hectare can increase approximately ten times and the expended man-hours be reduced by almost 92 percent.[15] These advantages have never been exploited. Instead policy-makers have always warned against the use of tractors because of the foreign exchange constraints. To argue that tractors should not be widely used because they consume considerable foreign exchange appears odd in light of the fleets of cars, trucks and heavy ammunition which are imported annually. It really depends on "how" and "where" the tractor is used.

On the other hand the emphasis on the hoe has not provided the country with a better food situation. Aggregate indicators of major staples for the period 1961-82, show either a low growth rate or a decline.[16]

Tanzania could have done a better service to the nation if emphasis had been placed on careful planning and gradual introduction of mechanisation. The rejection of mechanisation (tractorisation), on the pretext of foreign exchange constraints, has adversely affected the expansion of productive capacities in agriculture and more so in food production.

Producer prices. Case studies undertaken in Tanzania indicate that pricing policies have an impact on production.[17] Sisal production (a non-peasant crop) has declined consistently since 1964 because of the big fall in the world market prices. There is also enough evidence that small farmers respond to changes in relative prices. For example, ratios of export crops prices to food crops prices (Table 4.5) first showed improvement between 1974-75 and 1975-76. Despite the low ratios, it appears that the numerator changed slightly faster than the denominator. This change was in favour of the export crops. The overall impact of the price changes on marketed output of the major staples was rather ambiguous. There was a sudden upsurge of grain purchases in 1977-78 and 1978-79 (see Table 4.3). Thereafter purchases have consistently declined, despite the marginal increases in export prices. The enormous food purchases in 1977-78 and 1978-79 might have been caused by a switch of farmers to food crops.

At this point we would like to mention that the phenomenal increase in the purchase of the so-called "new crops" (sorghum and millet) was a costly venture. They had neither a domestic nor an international market. The result was the costly overstocking of the National Milling Corporation (NMC) godowns with the less demanded grains but a shortage of the major staples.

This is one of the pieces of evidence of how pricing policy can induce agricultural production and how bad pricing policy can misallocate resources. Additionally, the sudden increase of drought crops testifies to the farmers' quick response to prices. Why the country has not been able to attain food self-sufficiency appears to be partly explained by the pricing policy.

	TABLE 4.5		
	Ratio of Export Crop Prices to Food Crop Prices		
Year	Food Crops Weighted Average Price	Export Crops Weighted Average Price	Ratio of Exports to Food Crop Price
1971-1972	u/a	1,045	u/a
1972-1973	171	919	5.4
1973-1974	157	873	5.6
1974-1975	163	698	4.3
1975-1976	200	955	4.8
1976-1977	237	1,331	5.6
1977-1978	289	880	3.0
1978-1979	260	839	3.2
1979-1980	223	779	3.5
1980-1981	177	695	3.9
1981-1982	170	640	3.2

Source: Marketing Development Bureau, *Price Policy Recommendations for the 1982-83 Agricultural Price Review: Summary.*

Notes: a) Deflation using National Consumer Price Index (NCPI).
b) Weights used are value of official purchases.

Rural transformation programmes. We do not have supporting data as to the impact of the following programmes on food production in Tanzania : a) the Village Settlements' Scheme which started soon after independence; b) the Ujamaa Village Programme of 1967; and c) the Government's Villagisation Policy which was announced in 1973. Through these programmes the government expected that the advantages of economies of scale and modern farming in food production would be easily acquired. Unfortunately these expectations have never been attained. The expensive Village Settlements' Scheme which started in 1964 was shelved two years later. The Ujamaa Village Programme which was conceived in 1967 does not seem to have delivered the expected goods. Likewise the Villagisation Policy of 1973.

The overall production outputs in villages is far from encouraging. Less food is produced today in villages than at the time of independence. Something ought to be done in order to improve the productivity levels in the villages. This will ensure adequate food supplies in the villages and more for sale.

Government policies. We would like to wind up the discussion by looking at the policies as a drawback to food production in Tanzania. Ideally, policies towards the food sector are expected to stimulate production. This has unfortunately not been the case. Throughout the decades the government has devoted more efforts to designing policies which can enable them to collect more revenue from the farmers than supporting growth and expansion of the sector.

As things are now, farmers cannot influence policies which are favourable to food producers and they cannot easily acquire farm implements, inputs and above all consumer goods. Consequently, farmers have decided to take refuge by producing for household consumption and selling their food in local and parallel markets. This has meant less food purchased through official marketing channels and an unstable food situation.

LOOKING AHEAD:
THE RATIONALE FOR FOOD POLICY EMPHASIS

In this section we attempt discussing the rationale of emphasising food production and the need for having a well-defined food policy. To facilitate discussion we try and present the arguments looking at the impact of food crops on both economic and human development as follows:

Food and Economic Development.

Socio-economic development entails improvement of the nutritional levels of the population, greater stamina for work, greater labour productivity and ultimately greater income. Adequate food supplies to the people are therefore a key to socio-economic development.

Economic development in any country is closely linked with increased productivity. Rising labour productivity, despite technological advancement, has its roots in the adequate provision of food for the working population. Food is necessary in generating the required energy while working. For instance in almost any large installation where substantial human energy is required, food concession is given top priority. Since physiologically Tanzanians are no different from other human beings, it means they also need adequate food if they are to provide optimum productivity. And exactly this is what is happening right now in the food sector.[18]

The development of the livestock industry greatly depends on a strong food sector. This is because food which is excess to local demand can always be converted into livestock feed with an eventual increase in the supply of meat, milk and eggs. This is the major advantage of food crops contrary to export crops which cannot be channelled into the livestock industry in the event Tanzania cannot export all that it has produced. Additionally, food crops currently face no competition from synthetic food products.

A stable food production ensures a much more reliable source of income (especially now that food prices are almost competing with oil prices) than export crops. This is important in increasing farmers' incomes, which ultimately results in growth of the national market for nonagricultural products.

Food and the State of Health of the Population
Since independence, there has been no policy document which spells out that the government is obliged to ensure that there is adequate food for all the people, all the year round. This kind of assurance to the population which forms the basis of a sound food and nutritional policy for the people has been lacking.

Tanzania's population is poorly fed and experiences a low standard of living. About 35 percent of the population is said to be malnourished and more people go to bed hungry.[19] The most deeply affected group includes the children, lactating women, the disabled and the aged. Studies indicate a high incidence of protein energy malnutrition (PEM) which is closely related to the high mortality rates for the under-fives. Whereas in 1960 protein energy malnutrition was recorded to be 30 percent, in 1983 it averaged 60 percent.[20] It is not quite certain whether malnutrition has really doubled or more mothers are now reporting to the Maternal Child Health (MCH) clinics. Whatever the argument made it is likely that a large segment of the population is experiencing severe difficulties in maintaining the nutritional levels of the 1960s. The economy has been severely damaged and the food problem is more acute now then before.

The 1978 census indicates that the infant mortality rate stands at 137 per thousand which means about 90,420 babies lose their lives in Tanzania annually.[21] The causes of most of these deaths are rooted in malnutrition. Those who manage to reach the productive age group have very little opportunity of growing to their full physical stature and intellectual ability. The prescription to this erosive effect which suffocates the nation lies entirely in the change of existing policies which have given low priority to food crop production. Malnutrition is disastrous because it saps the veins of power of the nation's development and blocks advancement.

The current situation of malnutrition is a reflection of how the nation is badly fed. Thus malnutrition in Tanzania will only disappear if the people's effective demand for food is adequately covered. This problem can be tackled best by producing enough food domestically and keeping adequate reserves. This is possible because land is not a limiting factor. The country has an abundance of fertile soils and an army of manpower resources which could be channelled into food and livestock production easily.

The state of health of the nation will continue to be in a precarious situation if the country continues to promote export crops at the expense of food crops. A rational agricultural policy should attempt to promote food production first. This guarantees an acceptable human environment and state of health of the entire population. Hence it extinguishes malnutrition.

CONCLUDING REMARKS

This paper has attempted to discuss very modestly the impact of state policies on socio-economic development and food production from before independence to the present day. In order to understand the situation better it was desirable to discuss very briefly the features and functions of agriculture during the era of colonial administration and after independence. By so doing it has been clear that both the features and functions of agriculture have basically not changed even after achieving constitutional independence.

State policies have, even after independence been directed towards export crops. As a consequence of the emphasis on export crop production food crop production has been adversely affected. This has led into rural poverty, malnutrition and acute food shortages which in turn have been mitigated by enormous imports of food supplies.

There is an indication from the paper that increased food production in the Tanzanian peasant agriculture would necessarily require a dramatic change in the emphasis of state policies towards: land use, farmers' education, manpower, transport, investment, credit, research etc.

Finally, the paper contends that Tanzania's socio-economic development is predicated upon the improvement of the nutritional levels of the population. Adequate domestic food supplies are an urgent and indispensable appeal to policy-makers and producers. Therefore, as a matter of life and death, the nation as a whole and people as individuals should devote all their efforts to increasing food production and support state policies which are directed towards improving the food sector. If this is not taken seriously, there is every reason to believe that the state of health of the Tanzanian citizen will continue to deteriorate.

NOTES

1. The discussion on "States Policies" as it appears in this paper basically refers to agricultural policies.
2. Msambichaka, L.A. and Amani, H.K.R., "Food supply problems and consumption levels in Tanzania", paper presented to the workshop on Food Self-sufficiency in Sub-Saharan Africa, Dar-es-Salaam, May 1984, p. 26.
3. ILO, *Basic Needs in Danger; A Basic Needs Oriented Development Strategy for Tanzania*, 1982, p 130.
4. Semiti, G.A., "Cash Crops Versus Food Crops and Their Comparative Effects on Human Environment", *Viertel Jahresberichte*, Nr. 79, March 1980, p.95
5. Wagao, J., "State Control of Agricultural Marketing in Tanzania—1961-1976", ERB Paper 82.7, p.4.
6. URT, *Population Census, Preliminary Report*, 1978.
7. Tanzania Food and Nutrition Centre (TFNC), *Data Report*.
8. Msambichaka, L.A. and Amai, H.K.R., op. cit, p.12.
9. Ibid., p.5.
10. Malima, V.F., "Agricultural Research in Tanzania", paper presented at the Agricultural Policy Seminar held in Arusha, March 1983.
11. United Republic of Tanzania, *Tanzania National Agricultural Policy*, Ministry of Agriculture, 1982.
12. Tanzania Food and Nutrition Centre (TFNC), op. cit.
13. Khamisi, S. "The Role of Extension in National Agriculture Policy", address to the Ministerial Meeting on National Agriculture Policy, Arusha, 1983.
14. Tanzania Food and Nutrition Centre (TFNC), *Data Report on the Food and Nutrition Situation in Tanzania 1974-75—1979-80*, Dar-es-Salaam, 1982.
15. Marketing Development Bureau (MDB), *Price Policy Recommendations for the 1981-82 Agricultural Price Review*.
16. Msambichaka, L.A. and Amai, H.K.R., op.cit.
17. Ellis, P., op. cit.
18. Msambichaka, L.A. and Amai, H.K.R., op. cit.
19. Ibid., p.51.
20. Maleth-Lema, "Hunger Kills 150 a Day in Tanzania".
21. URT, op. cit.; Mascarenhas, A., "Food and Nutrition in Tanzania's Development", Keynote Address to the Symposium on "Food and Nutrition in Tanzania's Development—Issues of the 1980s", 1983.

BIBLIOGRAPHY

Bourenane, N., "State Policies, Agricultural Development and Food Production in Africa", CODESRIA Research Proposal, 1983.

Ellis, P., "Agricultural Pricing Policy in Tanzania 1970-1979; Implications for Agricultural Output, Rural Incomes and Crop Marketing Costs", ERB Paper 80.3.

ILO, "Basic Needs in Danger; A Basic Needs Oriented Development Strategy for Tanzania", 1982.

Maleth-Lema, "Hunger Kills 150 a Day in Tanzania, A wicked Distortion of Facts", TFNC (mimeo).

Malima, V.F., "Agricultural Research in Tanzania", paper presented at the Agricultural Policy Seminar held in Arusha, March 1983.

Marketing Development Bureau (MDB), *Price Policy Recommendations for the 1981-82 Agricultural Price Review, Summary and Price Proposals*, Dar-es-Salaam, September 1980.

Mascarenhas, A., "Food and Nutrition in Tanzania's Development", Keynote Address to the Symposium on "Food and Nutrition in Tanzania's Development—Issues of the 1980s", 1983.

Msambichaka, L.A., "Food Grain Shortfalls in Tanzania 1961-1981: A Retrospective Assessment", ERB Paper 81.3.

Msambichaka, L.A., Nduly, B.J. and Amai, H.K.R., "Agricultural Development in Tanzania: Policy Evolution, Performance and Evaluation", in *The First Two Decades of Independence*, Gottinger Tageblatt GMBH & Co., 1983.

Msambichaka, L.A. and Amai, H.K.R., "Food Supply Problems and Consumption Levels in Tanzania", paper presented to the Workshop on Food Self-Sufficiency in Sub-Saharan Africa, Dar-es-Salaam, May 1984.

Semiti, G.A., "Cash Crops Versus Food Crops and Their Comparative Effects on Human Environment", *Viertel Jahresberichte*, Nr. 79, March 1980.

Tanzania Food & Nutrition Centre (TFNC), *Data Report on the Food and Nutrition Situation in Tanzania 1974-75—1979-80*, Dar-es-Salaam, 1982.

Tanzania National Agricultural Policy, Ministry of Agriculture, October, 1982.

URT *Population Census, Preliminary Report*, 1978.

Wagao, J., "State Control of Agricultural Marketing in Tanzania 1961-1976", ERB Paper 82.7.

Yambi, O. "Monitoring the Nutrition Situation (with specific reference to Iringa Region)", paper presented at the Symposium on "Food and Nutrition in Tanzania's Development—Issues of the 1980s", Dar-es-Salaam, 1983.

APPENDIX 4.1

Purchase of Major Food Crops in Tanzania, 1966-67—1981-82

Crops	1966-67	1967-68	1968-69	1969-70	1970-71	1971-72	1972-73	1973-74	1974-75	1975-76	1976-77	1977-78	1978-79	1979-80	1980-81	1981-82
Preferred staples	164.4	165.7	156.3	134.9	323.0	168.0	226.3	161.3	61.0	131.7	169.4	292.0	291.3	228.6	147.0	128.3
Maize	112.9	104.3	127.5	54.1	186.4	43.0	106.4	73.8	23.9	91.1	127.5	213.2	220.4	161.5	103.8	85.1
Rice	23.5	5.1	4.2	59.8	62.7	45.9	49.0	39.9	15.2	12.2	14.9	35.5	34.5	30.8	13.4	14.1
Wheat	28.0	33.1	24.6	21.0	43.0	56.7	46.8	27.9	14.4	24.5	23.0	35.1	27.5	26.5	26.9	2.1
Drought staples	-	-	-	-	-	9.1	75.2	23.0	22.2	44.7	41.2	107.2	162.3	81.7	28.3	-
Sorghum	-	-	-	-	-	-	0.6	1.7	1.9	2.9	10.1	33.6	58.6	20.7	19.3	-
Bulrush Millet	-	-	-	-	-	-	-	-	2.5	2.2	6.4	14.4	16.5	1.3	0.3	-
Finger Millet	-	-	-	-	-	-	-	-	-	-	4.5	22.3	23.4	15.5	1.2	-
Cassava	-	-	-	-	-	9.1	14.3	18.9	17.8	17.4	20.2	36.9	63.8	44.2	7.5	9.1
Pulses	-	-	-	-	-	-	-	-	-	-	n/a	40.3	58.3	79.4	48.6	-
Sugar	-	-	-	-	-	-	105.1	96.2	103.2	112.1	99.2	135.5	128.6	118.0a	122.0a	-

a Calendar year 1980, 1981.
Source: MDB: Price Policy Recommendations for the 1981-82 Agricultural Price Review.

APPENDIX 4.2

**Production and Area of Mechanised Large-Scale
Wheat Farms in Hanang District**

Year	Area (Hectares)	Production (Metric Tons)	Yield Tonne/Hectare
1971-72	3,628	3,869	1.06
1972-73	2,717	1,257	0.46[a]
1973-74	3,298	666	0.20[a]
1974-75	810	408	0.50
1975-76	2,894	3,948	1.36[a]
1976-77	6,529	9,381	1.43
1977-78	8,834	11,153	1.26
1978-79	9,530	17,103	1.79
1979-80	10,648	17,280	1.62[a]
1980-81	16,234	25,225	1.55[a]

[a] Drought years + full implementation of new practices in later years.

Source: Young, R., *Canadian Development Assistant to Tanzania*, The North-South Institute, 1983, p. 66.

APPENDIX 4.3: Ratio of Export Crop Prices to Food Crop Prices

	1969-70	1970-71	1971-72	1972-73	1973-74	1974-75	1975-76	1976-77	1977-78	1978-79
A1/C1	3.25	3.5	3.79	3.5	2.76	2.06	1.32	1.34	1.32	1.93
21/C2	1.75	1.57	1.75	1.63	1.60	1.58	1.03	1.07	0.93	1.37
A1/C3	1.60	1.60	1.60	1.60	1.60	1.33	1.03	0.89	0.90	1.31
A1/D1	-	-	-	3.03	1.82	1.87	1.37	1.19	1.12	1.64
A2/C1	17.86	19.42	21.92	23.23	19.06	11.2	13.85	20.63	13.29	10.65
A2/C2	9.62	8.71	10.12	10.79	11.03	7.78	10.80	16.50	9.42	7.54
A2/C3	8.77	8.86	9.22	10.60	11.03	7.27	10.80	16.50	9.42	7.54
A2/C3	8.77	8.86	9.22	10.60	11.03	7.27	10.80	13.75	9.04	7.24
A2/D1	-	-	-	20.13	12.58	10.18	14.4	18.33	11.30	9.05
A3/C1	3.79	3.93	4.43	4.21	3.34	3.90	2.44	2.39	2.56	2.70
A3/C2	2.02	1.76	2.05	1.95	1.93	2.22	1.90	1.91	1.81	1.92
A3/C3	1.86	1.79	1.87	1.92	1.93	1.88	1.90	1.59	1.74	1.84
A3/D1	-	-	-	3.65	2.20	2.63	2.53	2.12	1.72	2.98
A4/C1	10.71	10.65	11.88	10.58	8.33	8.42	5.17	5.04	4.72	5.32
A4/C2	5.77	4.78	5.48	4.91	4.82	6.48	4.03	4.03	3.34	5.77
A4/C3	5.26	4.85	5.00	4.82	4.82	5.47	4.03	3.36	3.21	3.62
A4/D1	-	-	-	9.17	5.5	7.65	5.37	4.48	4.01	4.52
A5/C1	1.64	1.77	2.00	1.85	1.67	1.1	0.83	0.86	1.29	1.29
A5/C2	0.88	0.79	0.92	0.86	0.96	0.85	0.65	0.70	0.92	0.92
A5/C3	0.81	0.81	0.84	0.84	0.96	0.71	0.65	0.58	0.88	0.88
A5/D1	-	-	-	1.6	1.1	1.00	8.87	0.78	1.10	1.10
A6/C1	14.36	18.81	21.83	20.31	16.97	14.12	9.23	9.89	9.45	9.81
A6/C2	7.73	8.43	10.08	9.43	9.82	10.86	7.20	7.91	6.69	6.95
S6/C3	7.05	8.58	9.19	9.26	9.82	9.17	7.20	6.59	6.42	6.67
A6/D1	-	-	-	17.6	11.2	12.84	9.6	8.79	8.03	8.34

Notations:

A1 — Cashew-raw
A6 — Tobacco-Wetleaf
A2 — Coffee-clean
C1 — Maize
A3 — Cotton-seed
C2 — Paddy
A4 — Pyrethrum-dried flowers
C3 — Wheat
A5 — Tea-green leaf
D1 — Sorghum/Bulrush Millet.

Source: Computed from various MDB Reports.

5. STATE POLICIES AND AGRICULTURE IN MAURITIUS

Raji Virashawmy

INTRODUCTION

Within the international division of labour, Mauritius has, since the beginning of the century, been assigned the role of a plantation economy, along the same lines as the islands of the West Indies. This role, as a consequence, has transformed the whole island into a vast sugar cane field, making the production of sugar for export the centre of gravity of economic and social life. All other developments, all other sectors, all other state policies have been, and still are, to some extent, subservient to King Sugar.

The logic of the world market and that of the plantation system as it has evolved in Mauritius make the island extremely vulnerable to world market sugar prices (now at half the costs of production). Climatic conditions add substantially to the state of vulnerability. It is rare that a good harvest coincides with good world market prices. The future of the world sugar market is all the more uncertain given the recent development of isoglucose (synthetic sugar) and the expansion of beetroot sugar in Europe, both of which are rivalling cane sugar from the Third World countries.

The world market, which has, over half a century reinforced the mechanism for the reproduction of cane sugar in Mauritius and elsewhere, is today itself threatening the very reproduction of the system.

In such a plantation society, capitalist relations of production are highly evolved. History, the type of colonisation which made Mauritius more of a settler colony, the nature of the penetration of the world capitalist system, and the existence of a local elite transformed by the colonial power into an agrarian and commercial bourgeoisie are so many factors which explain the stage of evolution of capitalism inside Mauritius.

The nature of a settler colony, the indirect rule of the British colonial system, the small size of Mauritius and historical factors, go in one direction to explain, at the same time, why local capital is highly present on the plantations and in the related enterprises, and why foreign capital has been marginalised in equity ownership.

However, these facts should in no way give the impression that Mauritius is not well integrated into the world capitalist system. On the contrary, it is very much so. As an example, external trade constitutes some 65 percent of GDP. But, such an integration takes indirect forms of dependence in the world capitalist system, less through equity, more through marketing, technological and consultancy services and through the purchase of inputs.

Mauritius now has a population of one million, with a rate of natural increase now standing at 1.4 percent. It is a relatively small island of 1,860 km², with a shortage of natural resources. There are practically no mineral resources on the land, although the potential exists for mineral resources on the sea bed. At the moment, Mauritius finds itself in a situation where self-reliance needs to be redefined, not in terms of the absolute and relative size of imports and exports, but more in terms of internal integration, the extent of local value added, diversification of the economy and agriculture, diversification of external trade and resistance to the new policies of the IMF and the World Bank.

Social democracy is fairly developed given the existence of a relative separation of economic power from political and ideological power, the institution of free regular general elections, a free and independent press, free trade unionism, and a complex set-up of lobbying groups coming from the organised classes and ethnic groups.

In a new settler colony, where nation-building poses itself as a priority problem, the existence of various ethnic groups originating from various continents calls for prudent policies which, at all times, must balance the delicate class-ethnicity articulation. Some 50 percent of the population are of Indian descent and of Hindu faith, 17 percent of Indian descent but of Islamic faith, 30 percent are of mixed blood and are mostly Christians. Finally, there are about 3 percent Chinese, most of whom are Christians.

Mauritius obtained its independence in 1968, following 158 years of British colonisation and, prior to the latter, 89 years of French colonisation.

The agricultural policies followed in the 1950s and 1960s are mere products of the colonial strategy.

COLLUSION AND CONFLICTS

The 1950s and 1960s were two decades of the struggle for

independence and, at the same time, of transition which the British colonial power needed in order to impose a neocolonial solution on the independence issue. The 1960s, in particular, were much more of a transitional decade, leading to neocolonisation.

The structure of the colonial plantation economy is simple, it rests on three poles of growth : a) the export sector (cultivation and processing of sugar cane for export); b) the import sector with its network of wholesale and retail trade channelling the imports all over the country; c) the economic and social infrastructure needed to support both the export and the import sector (shipping, harbouring, insurance, banking, roads, electricity, hospitals, state administration).

The state policies vis-à-vis agriculture have been explicit in both government official reports and political statements: ''The sugar industry is the backbone of the Mauritian economy'', has been a constant slogan used by many leaders. As such, no government in power has so far ever managed to go against sugar's ''established right''.

State revenues are largely dependent on sugar export revenues, in a direct or an indirect and roundabout way. Export tax and levies are imposed on sugar exports. They constitute 23.69 percent of total export revenues, bringing in some Rs400 million in 1983 to the state.[1]

Second, there are various levies and taxes on imported goods, which are in the last resort bought with earnings from sugar exports. In the financial year 1981-82, import taxes constituted 31 percent of total government revenues. Import and export taxes and levies constituted some 50 percent of total state revenues.[2]

The state also obtains its revenue from taxation of companies and individuals. Taxes are levied on private companies at the rate of 60 percent on undistributed profits. Since 50 percent of the land is owned by sugar companies and, secondly, since all sugar processing is carried out by 21 sugar mills owned by companies, company taxation on the sugar companies constitutes a substantial source of revenue to government. Lastly, the taxes are also levied on distributed profits, that is on dividends obtained by shareholders of the sugar companies. Taxes are also levied on wage earners and salary earners of the sugar industries in the form of income tax. Taxes on individuals, that is income tax, are levied on a progressive scale, ranging from 5 percent to 70 percent of taxable income.

Only ten years ago the export of sugar and its by-product, molasses, constituted up to 90 percent of the country's total export revenue. Today, with export industrialisation well on its way, the share of sugar revenue in total export revenue has dwindled to 58 percent.[3] However, this highlights the fact that the sugar sector has deep roots in the economy of the country in the sense that it uses local inputs, local technology, local raw materials. It has a low import content compared to most other sectors of the economy. Hence its net ability to bring in foreign exchange in this country is substantial. The consumption of imported intermediate inputs amounts to only 3 percent of the total intermediate inputs required.[4]

Secondly, equity ownership of sugar businesses is highly local. Foreign capital is marginal. Therefore, repatriation of profits is minimised. Only three mills out of the 21 are owned by a foreign company, namely, LONRHO Sugar Corporation Ltd and its affiliates. They owned 9.3 percent of the total harvested cane acreage in Mauritius and the three mills process nearly 15 percent of the country's total sugar cane.[5] However, this does not mean, despite the heavy presence of local capital, that Mauritius is not dependent on the world economic centres. Mauritius, as a Third World country, participates in unequal exchange but in the specific case of Mauritius, the transfer of surplus to the centre takes place in an indirect form. The existence of a local bourgeoisie inevitably calls for a sharing of the surplus between the centre and the local bourgeoisie, albeit on a highly unequal basis.

The sugar sector is also a vast source of employment in the country. The medium-sized and large sugar holdings and their processing plants and ancillary activities employ at present 24 percent of the labour force.[6] This proportion is higher if we include the small sugar cane planters, the metayers (that is those who run or cultivate land rented from landowners) and if we include casual and agricultural workers working for small planters.

Further, given the nature of the plantation system, the indirect employment created as a result of direct employment within the sugar sector is substantial. A good number of ancillary industries have evolved around the processing and cultivation of sugar cane. Banks, research institutions, secondary and tertiary education, public administration and parastatal bodies, insurance, import and export firms etc. are all geared directly or indirectly to the production of sugar for export.

On the one hand the material benefits derived by the state bureaucracy from its collusion with the plantocracy are enormous: state revenue, foreign exchange, employment creation. The state bureaucracy has much to gain by supporting and perpetuating the plantocracy. Hence it is not surprising that the colonial state and the postcolonial state have in both their avowed policy and their practice consciously attempted to increase sugar production to its maximum even if this has been to the detriment of other groups and other sectors.

The plantocracy has, on the other hand, much to gain through supporting the state bureaucracy and developing a class alliance with it. The plantocracy needs the state to look after the economic and social infrastructure needed for the plantation economy (bridges, roads, electricity, water, harbour, docks and so on).the Reconstruction and Development Programme 1960 to 1965 clearly indicated the type of capital expenditure incurred by the state. Infrastructure related directly or indirectly to the sugar sector took almost 40 percent of the total budgeted capital expenditure.[7] In addition, the state provides social infrastructure (hospitals, schools etc.). The state has to provide the necessary legislation to maintain law and order in the interest of the plantocracy: regulations regarding strikes and wage negotiations, labour laws, and also the implementation of the laws through its repressive apparatus which includes the police, the army, the prisons. The state is called upon by the plantocracy to provide the necessary credit on a short-term basis, as well as on a long- and medium-term basis. The Agricultural Bank was set up in 1936 with the objective of providing medium-term and long-term loans specifically to the sugar industry.

Moreover, Professor Meade's report on the economic and social structure of Mauritius indicated that between 1963 and 1968 the sugar industry absorbed the bulk of gross capital formation in the country. Investment in the sugar industry made up 13 percent to 17 percent of the Gross National Product.[8]

As early as 1939, a system of minimum wages was established by the state in order to alleviate labour unrest, especially in the sugar plantations. A minimum wage advisory board was set up. As from 1948, trade unions appeared on the scene of industrial relations : with the rise of trade unions collective bargaining became a practice and whenever deadlock occurred, the matter was submitted to arbitration tribunals which would make awards that were binding on both the employers and employees. The marriage of convenience between the state bureaucracy and the plantocracy became clear.

STATE POLICIES AND IDEOLOGY

During the colonial period the state could not be more than a mere instrument of the colonial power. It helped the plantocracy to accumulate capital by providing the social and economic conditions necessary to that purpose. However, within this state, the plantocracy assumed the status of a secondary partner. Over the years, constant struggles were waged against the plantocracy by other social classes, spearheaded by the agricultural workers. The independence movement originated from the countryside, and expressed particular class-ethnic contents.

Gradually, as a method of cooperation, the colonial power had to offer other classes positions within the state. By the end of the Second World War the state represented the interests, although in different proportions, of various classes and ethnic groups within the country. The advent of more democratic and regular general elections, the enlargement of universal suffrage, the organisation of small agricultural cultivators and the agricultural workers, and the rise in political consciousness of the people were important turning points in the history of Mauritius. Gradually, the state had to give room to almost all the social classes. By the 1960s, and especially after independence, the state became gradually the instrument of the "petit bourgeoisie" rather than that of the plantocracy. Nevertheless, the petit bourgeoisie running the state and trying to protect the interests of small producers could not contest the overall interest of the plantocracy for reasons of convenience. Although not an instrument of the plantocracy after independence, the petit bourgeoisie of the state bureaucracy continues to serve the existing economic base. However, over time, it has relatively separated political power from economic power.

Today, powers and counter-powers are well diversified in Mauritius and are very well articulated. Although the state bureaucratic apparatus has control over the administration, the army and the police, it has relatively less control over the judiciary which for long had remained the pillar of the plantocracy. The state bureaucracy has also less control over the press as a source of power. The press has always been independent although it represents various interests, ranging from those of the plantocracy to those of the bureaucracy and the intelligentsia. The church, which was long under the control of the plantocracy, has gradually admitted representatives of other social classes within its administration.

The separation of powers and diversity of powers inevitably led to some changes in agricultural policies in the country in the 1960s and 1970s. They helped the small cultivators to stay and perpetuate themselves and obtain enough guarantees and protection from the state not to be completely wiped out by the giant sugar companies. Changes in the power structure within the state also brought changes in the strategies of agricultural development. While production for export of sugar continued on a large scale and was even encouraged on a wider scale, some form of diversification of agriculture was also encouraged. Agricultural diversification since the 1960s found much echo in government statements regarding agriculture.[9]

However, diversification of agriculture meant different things to different people. To some it meant a self-reliant agriculture catering for the local market, reduction of food prices and less dependence on imports. To others it meant agricultural production for export but away from sugar. The last definition seems to have obtained much support in the 1960s. The logic behind the strategy was made clear; Mauritius is an open economy, its small size and its relatively poor local resource base point to the search for foreign exchange as a key in development; in this respect, tea production for export was encouraged as a means of diversifying away from sugar. By the end of the 1970s tea exports represented almost three percent of total exports. Hence, to government at that time diversification meant a means to obtain foreign exchange and expressed a concern for the balance of payments rather than a policy move towards self-reliance.

Food production as it has evolved in Mauritius cannot be seen separately from the role assigned to this country by both the colonial division of labour and the existing international division of labour. In the colonial period and even shortly after independence, Mauritius was seen more as a relay economy in between the centre and the periphery, or rather a particular form of peripheral country. Mauritius had no peasantry and no vast amount of land to produce its food itself in order to reproduce its labour power. Hence, within the colonial division of labour, it had do draw its cheap staple food (rice) from other colonies—namely from India and other Asian countries. Cheap wage goods like textile products were also obtained from other colonies namely from Asian countries. Only when labour power was reproduced cheaply could it become possible to produce sugar cheaply for export to Europe (and in the past to Australia, India and even to South Africa). Intermediate products like fertilisers and other inputs like machinery and technology were imported from Europe. So Mauritius was engaged in a particular form of colonial division of labour whereby a triangular trade—Europe-Mauritius-other colonies—developed. So, given the small size of the country and given historical conditions in which the plantation system developed, it has been difficult for Mauritius to produce its own staple food. However, certain non-staple foods like vegetables and fruits have been produced locally to satisfy local consumption. It should be noted that these foodstuffs could be allowed to develop as long as they did not conflict with export agriculture.

EXPORT AGRICULTURE VERSES LOCAL FOOD PRODUCTION

During the colonial period and immediately after independence, the world market price mechanism, combined with state intervention and established rigid institutions and the power structure, prevented any substantial local food production that would jeopardise export agriculture.

However, some local food production took place in the shadow of the established right of sugar. Some vegetables and pulses were allowed to grow in between the rows of sugar cane immediately after harvesting before the cane tops had reached certain heights. Or, they were grown at the end of the cane cycle (average of eight years), between uprooting of old ratoons and replanting of virgin ratoons.

At times, when sugar prices are really bad, sugar cane is replaced by new non-sugar crops. However, this occurs only in fields where the ratoons have reached replacement periods. Instead of new cane ratoons being planted, non-sugar crops are tried until sugar prices pick up again.

In addition, any crop, vegetables or fruits or pulses that might compete with imported foodstuffs and endanger the interests of the big import firms, are discouraged through various methods. It will be noted that the import business is also linked with the sugar industry and constitutes, in the plantation society, through interlocking capital, part and parcel of the power structure of the plantocracy.

Under the colonial structure, export agriculture established its primacy and everything else was subservient. However, under a new state and a new power structure in the civil society, one may raise the question whether or not the two could be made complementary. This important issue is taken up in a subsequent section.

THE NEW STATE AND THE NEW VISION
FOR AGRICULTURE

At this point in the history of Mauritius and given the context of world capitalist crisis, the future of the plantation economy itself is being questioned and is threatened. The possibility of its reproduction as a system is also at stake.

The export of sugar from Mauritius finds its way into two main markets: the European Community, which through the Lomé Convention absorbs 500,000 tons of Mauritian sugar, i.e. nearly four-fifths of total sugar exports from this country, and the rest of the world market. As regards the EEC market which is a guaranteed market with respect to price and to quota, the price increases have been meagre over the past years following the sugar boom prices of 1974. The price increases obtained from the European Community usually follow the agricultural policy of the Common Market countries. At the time of writing, the EEC has announced a zero price increase for the coming crop. This is all the more dramatic given that the terms of trade have drastically turned against Mauritius by over 50 percent.[10] As regards the world market prices, they fluctuate tremendously and most of the time remain extremely low, only reaching a high level once every ten years following its own cycle. It has been estimated that the present world market price represents half the cost of production.[11] The future seems to be bleak for the sugar plantation economy in general, given that both the world market and the EEC market are narrowing themselves more and more. The EEC has already become a net exporter of sugar. If it is still taking the 1.4 million tons of sugar from the African, Pacific and Caribbean countries, it is doing so merely for political motives. The over-production of beet sugar inside the EEC countries themselves is a reality today. The strong lobbying of beet producers in Europe represents another obstacle for sugar cane producers. In addition, the emergence of artificial sweeteners like isoglucose and highcorn syrup, more and more widely used in consumption and industries, seriously threatens cane sugar.

Thirdly, the recession in the capitalist countries has heavily narrowed the demand for sugar, so much so that the narrowing export revenue derived from sugar constitutes a real problem to the balance of payments of Mauritius. The balance of payments problem is aggravated by the world currency crisis. The stunning rise of the dollar in the early 1980s reveals a new form of immediate and utter exploitation of the Third World countries by the international economic and financial system. The dollar rose from Rs 7 a dollar in 1980 to Rs 13 in 1984. In other words the value of the dollar vis-à-vis the Mauritian rupee increased by 85 percent. It should be noted that half of Mauritian imports are paid for in dollars and these include the basic essentials like wage goods, rice and flour. Second, external debt servicing which constitutes nearly 17 percent of export revenue is almost totally paid in dollars. The rise in the price of imports has greatly reduced, absolutely and relatively, the value of imports and hence has reduced government revenue from import duties.

The balance of payments problem is always coupled with government revenue. Budgetary problems and balance of payment deficits are two key elements that have finally put Mauritius on its knees vis-à-vis the IMF and the World Bank. In negotiating its standby agreement with the IMF, Mauritius had to submit itself to various conditions imposed by this capitalist institution. One of the bitter conditions has been the ceiling imposed on the amount of subsidies which the government gives to imported rice and flour. The policy of the IMF is to reduce the subsidy altogether so much so that the food prices of rice and flour reaching the consumers will be increased. Already, with the rise of the dollar, rice and flour have been costing the government of Mauritius much more. Given that the subsidies could not be increased, flour and rice prices had to be increased by 33 percent. The World Bank has also imposed its conditions before granting loans for projects under the Structural Adjustment Programme. These conditions are explicit in the *Statement of Development Policy*, a paper of government intention that reflects the point of view of the World Bank. The World Bank in the *Statement of Development Policy*, made it very clear to the government of Mauritius that its Structural Adjustment Programme will have to rest more and more on two poles : export industrialisation and tourism. Diversification of agriculture for export or for local consumption will have to play second fiddle. The sugar industry in that new strategy designed and shaped by imperialism is given an entirely new role. Once the centre of gravity of economic life, the sugar industry is called upon to play the role of springboard to enable the emergence of the new sectors, tourism and manufacturing for export. The sugar industry in its new role must be able to provide the existing and acquired infrastructure of the plantation economy. Land, factories, management of large-scale enterprises, banks, insurance, harbours, roads, electricity, water etc. which were in the past developed for the production of sugar, must now serve the needs of the two new sectors.

THE REPRODUCTION CRISIS OF THE PLANTATION SYSTEM

The plantation as an economic system is finding increasing difficulties in reproducing itself. The historical trends seem to point to its phasing out. The machinery and plant and other elements of constant capital used in the cultivation, processing and distribution stages of sugar production are not being replaced due to lack of capital funds and lack of faith in the future of this sector.

As an example, the purchase of fixed assets for the four-year-period 1979 to 1982 gradually declined from Rs 131.8m in 1979 to Rs 65.6m in 1982.[12] The purchase of fixed assets which amounted to Rs 391.2m over the four-year period fell just short of the depreciation charges provided for in the account of Rs 402m.[13]

At the time of writing, three sugar estates have reached such a level of cashflow problems that even the month-to-month wages and salaries of staff and workers can no longer be ensured. The state has been called upon to intervene on several occasions to guarantee it vis-à-vis a state-owned commercial bank (the SCB Ltd) to provide further overdraft facilities. Of course, these means are temporary and would only postpone the crisis, if not aggravate it, by deepening the indebtedness of the sugar estates.[14]

In addition to the sugar estates, the small sugar cane producers, cultivating some 35 percent of the sugar cane land on an average holding of 0.7 hectares, are meeting much more difficulties to ensure even the process of simple reproduction. The prices of chemical fertilisers, insecticides, transportation and rent have gone up, while sugar cane revenues have not kept pace.

The reproduction of labour power has also been at stake. It depends on imported food. The staple foods (rice and flour) are imported and are totally paid for in dollars. The revaluation of the dollar has inflated the prices of those commodities. Given the stringent condition of the IMF not to increase subsidies for these two products, the local price has gone up. Milk and meat are also mostly imported. So are basic inputs (like maize) going into the local production of most food commodities.

Moreover, local production of both imported foodstuffs and their substitutes have met severe obstacles both because of the existing power structure and, at times, of the resistance of the population.

The approach of looking to local food production has been too technicist and not global.[15]

On the side of agriculture, measures offered by the World Bank and IMF as a remedy to alleviate the balance of payments have also been too technicist. To those institutions, it is a matter of replacing rice and flour as eating habits by substitutes, which could technically be produced locally, namely potatoes. To local bureaucrats and technocrats, these measures have appeared plausible, and even a white paper was pompously produced for "national debate".[16]

To the masses, replacing rice and flour by potatoes was, in terms of caloric and protein supply, in terms of practical cooking and in terms of price per calorie, unacceptable. This change in eating habits had a class content. What might appeal to the higher income groups, would be catastrophic to the masses.

The size of the country, the availability of agricultural land in relation to competing and alternative uses, the type of soil, supply of irrigation and natural water, the absence of a peasantry and the level of agricultural wages, are objective realities that call for a different notion of self-reliance. Local cultivation of rice is almost out of the question. Given the draining nature of the Mauritian soil, only 4,000 hectares have been identified as suitable for such cultivation. Secondly, comparative advantage, even seen in a dynamic perspective, can hardly establish a reasonable justification for local rice cultivation after having compared local costs and world market prices.

The world market price is, of course, not a basic criterion because it can only reproduce extroverted development. However, delinking local prices from world market prices is justifiable to the extent that it is economic in terms of labour time used. The comparative advantage theory holds true if, through external trade, the country is economising materialised labour.

As regards international trade, Marx made it clear that an underdeveloped country with low labour productivity "may offer more material labour in kind than it receives, and yet thereby receive commodities cheaper than it could produce them".[17]

THE NEW VISION OF THE STATE

In a period of rapid industrialisation the power structure within the state is bound to change to keep in line with changes in the economic base. The plantocracy of the sugar sector is losing both the ideological and the political power it had had since the beginning of last century throughout the British colonial period. Its economic base itself is being threatened as the sugar industry faced with changed world market conditions is under threat. The plantocracy is now in a weaker position vis-à-vis the state.

Industrialists are emerging as a dominant class and their alliance with the state bureaucracy is now more and more obvious. In this changing reality, members of the plantocracy have two options which it can put at the disposal of local food production: a) it can exercise planning in such a way and offer such conditions as to make local food production complementary to export agriculture; b) it can articulate industrialisation with development of local food production. One articulation will have to be the industrialisation of agriculture itself, raising productivity to its highest level. This alone should offer abundant and cheap food, a sine qua non for reproducing labour power cheaply and for ensuring that the whole industrialisation process is not nullified.

In fact, the policy statement of the Ministry of Agriculture which offers a more explicit background to the policy behind the Sugar Authority Bill provides for a constant one percent annual reduction of total acreage under sugar cane without reducing the total volume of sugar produced. This policy is aimed at releasing land for food production.[18]

The success or failure of a meaningful food production programme in Mauritius will not depend on technical matters, but, first and foremost, on the outcome of the class struggle presently waged between the progressive forces and the reactionary plantocracy. Fortunately, on this specific issue in Mauritius, imperialism is out of the battlefield.

The sugar plantocracy has two alternatives: a) perish as a dominant class; b) transform themselves into industrialists and participate fully in the industrialisation process. Some are already doing so and have invested sugar profits in other activities and have used existing sugar estates' infrastructure to put up manufacturing enterprises.

The government's declared policy is to establish a state agency (the Sugar Authority), which, short of nationalising all the sugar complexes, will have effective powers for the exercise of overall control over and the formulation of policies for the development of the sugar industry.[19] These powers are wide indeed as the Sugar Authority (in its draft form) should have its say in the production process, marketing, taxation, wage policy, workers' participation in ownership and management and use of by-products.[20] The Sugar Authority should in other words, wipe out the economic power base of the plantocracy.

The plantocracy is reacting fiercely against the state's action in establishing such an Authority. It has sought the backing of imperialism via the World Bank and IMF, which obviously would not support an industry which is becoming more and more meaningless in the present international division of labour. It has tried to win the support of the working class through its trade unions. This has also failed. Its last outposts of support are a section of the press and a section of the church. It could not even win the support of all the main factions of the local capitalist class.

The establishment of a Sugar Authority which will withdraw effective power from the plantocracy is a historical event. It is a key link in getting the country out of its colonial structure.

Apart from reviving the sugar sector so that it can channel surplus and resources to accelerate industrialisation, the state should, for the first time, monitor effectively a process of agricultural diversification. The state will have had control over land, water and other resources.

NOTES

1. Koenig, H., Press Conference, *L'Express*, 4 April 1984.
2. CSO, *Bi-annual Digest of Statistics*, December 1983, Central Statistical Office, Government of Mauritius.
3. Ibid.
4. Edouard, *Input-Output Table for Mauritius*, Ministry of Economic Planning, Government of Mauritius, 1984.
5. Virashawmy, R., "Transnational Corporations in African Agriculture: The Case of Mauritius", CODESRIA, Dakar, November 1982.
6. CSO, op. cit.
7. Government of Mauritius, *Sessional Paper No. 2 of 1961*.

8. Menade, J.E., *Economic and Social Structure of Mauritius*, Frank Cass, London, 1961.
9. Government of Mauritius, *Sessional Paper No. 6 of 1966*; Government of Mauritius, *Sessional Paper No. 2 of 1963*.
10. CSO, op. cit.
11. Ministry of Agriculture, Press Conference, *L'Express*, 1 March 1984.
12. Government of Mauritius, *Report of the Commission of Inquiry of the Sugar Industry*, Government of Mauritius, 1984.
13. ibid.
14. *L'Express*, 18 April 1984 and 11 May 1985.
15. Ministry of Agriculture, *White Paper on Agricultural Diversification*, Government of Mauritius, 1983.
16. Ministry of Economic Planning and Development, *White Paper on Rice and Flour*, Government of Mauritius, 1982.
17. *Social Sciences in China*, vol.1, 1, 1980.
18. Ministry of Agriculture, Press Conference, op. cit.
19. Ministry of Agriculture and Natural Resources, *Draft Mauritius Sugar Authority Bill*, 1984, Government of Mauritius, 1984.
20. Ibid.

6. AGRICULTURAL POLICY AND THE LIMITS OF AGRICULTURAL DEVELOPMENT IN MOROCCO

A. Bouami & M. Raki

INTRODUCTION

The economic development of the Western world was based on agriculture. Radical changes in the rural areas had opened up the way to a new social structure, and to a massive migration of peasants to industrialising centres where their integration was made possible by the level of technical development at the time. Furthermore the outside world played a significant part with the inflow of precious metals. In short, the Western world developed as a result of a combination of several factors whose concomitant and complementary evolution constituted the very dynamics of this development.

The problematics of economic development in underdeveloped countries in general is rather complex; a maze of political, economic, technical, social and cultural contradictions get, as a rule, in the way of an authentic economic development. The growth of a number of economic aggregates may create illusions. Industries may be set up here and there and grand projects may be carried out without necessarily being able to generate a dynamic autonomous, self-centred development. In fact, food shortages are increasing, and the balance of payments and terms of trade are deteriorating.

It was believed for a few decades and more specifically in the 1950s and 1960s that industrialisation was the unique answer to the economic development of underdeveloped countries. However, the negative results of a number of experiences have recently led to the revival of agricultural development as a priority.

Like other underdeveloped countries, Morocco experienced several changes in the conception of economic development. In the early post-independence period (1956), the first government, which had progressive tendencies, intended to introduce a national industrialisation which was believed to guarantee economic development.

However, due to essentially political reasons, the plan was dismissed. A new government came to power, which gave precedence to agriculture in the sequence of plans which were implemented. The 1973-77 plan was an exception however as it was drafted with a new priority, the development of industry with a view to promoting the export of manufactured goods.

But all these projects of economic development were applied without achieving any major results, in that the main issue was and still is neglected: instead of promoting the development of the other sectors as planned, agriculture has become a burden for the state which is constantly backing it up with injections of sizeable amounts of funds.

What are the bases of agricultural policy in Morocco? What are the constraints on agricultural development in Morocco? These are two major questions to which this study attempts to respond.

NATURE OF THE STATE AND AGRICULTURAL POLICY

Since independence, agriculture has been regarded by the state as a priority sector whose development could generate the financial resources needed for the industrialisation of the country. This function assigned to agriculture was to be accomplished through the development of export crops. But not only did this orientation of agriculture turn into an obstacle to the generation of surplus which could be mobilised for industry but it also brought about agricultural distortions. Indeed, except for sugar crops and dairy products, food crops production has virtually stagnated since independence. Over the years, this stagnation, coupled with the demographic explosion, have contributed to the ever-increasing food shortages.

The Agricultural Development Priority: A "Constraining" Choice

Soon after political independence and a period of wait-and-see policy resulting from the struggles of hegemonic forces which emanated from the national liberation movement, the agricultural sector which had responded negatively to supposedly popular technical experiments was promoted as a priority with the adoption of the 1960-64 five-year plan.

Such a choice, it would seem, was made due to the significant role of agriculture in Moroccan economic life and the necessity of making the hydro-agricultural infrastructure inherited from the colonial era more profitable. In actual fact not only was this legacy a small one since it merely served a minority of colonial masters but it was also especially designed to produce the electric power needed for mining.

The 1956-60 Period—A Wait-and-See Phase

This period was marked by the launching of "operation labour", a programme meant to modernise the "traditional" sector. Initiated in 1957, "operation labour" was aimed at encouraging peasants to take up modern farming techniques. Underlying this objective was a purely economic preoccupation—that of increasing grain production so as to meet the needs of a fast-growing population.

It was believed that the introduction of the use of tractors on peasant farms combined with the use of selected seeds, fertilisers and a two- or three-year rotational system—growing winter crops (grains) next to spring root crops—would further grain production.

To be profitable, tractors have to be used on a large acreage. However, the peasant farms did not meet this requirement. As an alternative, the consolidation of small fragmented holdings into contiguous farmholdings ranging between 50 and 500 hectares was proposed.

In the beginning, peasants were so hostile to this consolidation of fragmented holdings that it led sometimes to riots. They opposed the removal of boundary stones and remained unconcerned with the activities carried out on their land by "work centres". In a nutshell, peasants did not lend their support to this experiment.

So the experiment covered only a small part of the area under cultivation, the size of which shrank every year, dropping to 94,000 hectares in 1961-62 from 167,000 hectares in 1957-58.

This farming experiment yielded good results in the first three years during which it generated an increased production of wheat—an increase of 6.3 quintals per hectare was recorded in 1957-58. However, this increase declined to 3.2 quintals in 1958-59 and further down to 2.6 quintals in 1959-60.

In a relative sense, the operation contributed to a relative increase in job opportunities. This positive aspect was the result of the introduction of new crops (pulses) and of the limited use of farm machinery. However, the experiment failed due to political, social and financial reasons.

Political reasons. The big landowners who were, as a matter of principle, hostile to any action undertaken by the small peasants began discrediting the experiment. They then started competing with the administration by hiring their tractors to the peasants at lower than government prices.

Social reasons. Those peasants whose farms were involved in the experiment did not support the project; as a result it was bound to fail from the very beginning. Their attitude was due to the fact that they did not trust an administration which was seeking to force the experiment on them.

Financial reasons. As of 1960, the experiment was no longer profitable. Indeed, payment expected from peasants for the work carried out on their farms was made under difficult conditions (1961 drought) and the number of tractors in use (1,000) dwindled as of 1960 until tractors were no longer used.

The 1960-64 Five-year Plan

After the failure of this new farming experiment, authorities emphasised the need to carry out a reform of structures within the framework of the 1960-64 five-year plan. These structures were regarded indeed as being the major obstacles to the modernisation of agriculture. The budget allocated for the implementation of this plan amounted to 201 billion francs. The state was supposed to contribute up to 40 percent of this amount and the private sector 25 percent in the form of self-funding and 35 percent in the form of loans. 118 billion francs were earmarked for dry farming, 11 billion francs were to go to arboriculture, 8.5 billion francs to livestock breeding while only 40 billions francs were allocated to water production. Investment with immediate results was given priority over investment with long-term results as can be seen from the following breakdown of funds allocated to dry farming.

Investment with immediate results	
Ploughing	13.5 billion francs
Seeds, fertilisers and	
small equipment	96.68 billion francs
Investment with long-term results	
Land clearing	7.05 million francs
Banks etc... or banquettes	64 million francs

The private sector did not, in the end, contribute its share of the budget, which amounted to over two-thirds of the planned investment.

The only investments made were for experiments which were to be funded by the state and in fact they exceeded the planned percentage, i.e. 40 percent. Thus public investment ran up to 86 billion francs against the 79 billion francs originally planned. These funds were allocated to big water projects.

The 1965-67 Three-year Plan

This plan was aimed at solving two basic issues, that of finding ways and means to develop agricultural production without disrupting the agrarian structures and that of finding financial resources.

The solution to the first issue was a dualistic development which was to give priority to the modern sector. The drafters of this plan saw the failure of the new farming techniques experiment as unavoidable. The traditional sector was indeed unprepared for the techniques of modern agriculture, especially as this traditional sector is made up of "small" peasants who most often are not real farmers. Their entreprises cannot therefore be qualified as farms. Hence the modernisation of smallholdings was not possible in so far as the final output would not be the major resource on which they would live throughout the year but rather a "secondary resource".[1]

Such a vision of the rural world, one may say, clearly reflects the advanced level of proletarianisation and impoverishment of small peasants. By the same token it points to the incapacity of planners to promote a coherent policy in favour of the whole rural world.

Behind this display of their incapacity there was in fact a plan, that of creating a "modern agricultural system run by nationals". This plan was to mobilise "comprehensive means of action which would guarantee its take off and its future success".

A problem then arose, that of finding the core element to "this modern agricultural system to be run by nationals". Reference is being made here, of course, to the allotment policy—whereby land is distributed to poor peasants—but it only covers poor peasants and involves only colonial land—250,000 hectares which was officially recuperated in 1963. In fact, from the sequence of events it appears that the official text on the control of landed property which was published in 1963 was to be the major element serving this policy, namely, transferring colonial property to a particular category of "nationals".

How was this type of agricultural development to be introduced? In other words, how could the capital formation of this new agrarian bourgeoisie be speeded up? Was it necessary to undertake "integrated operations consisting in financing the whole process of production or rather act indirectly by regulating methods of farming (prevent absentee landlordism, prohibit land-hire etc..)" and granting subsidies and loans? Had state intervention not reached its highest peak yet? Apparently not, for the major issue at stake was not of a technical nature . Indeed a political choice had been taken. What remained to be done therefore was to find the means to implement these policies. The only constraint on this option was financial.

Morocco did indeed undergo a very serious economic crisis in 1964-65 characterised by an outstanding deficit in the budget of the state and in the balance of trade. The authorities intended to fund particularly the internal organisation of big farmholdings but they lacked the required internal funds. So they were going to rely mostly on external resources. Thus, out of the 87 billion francs which were to be invested in agriculture, 50.5 billions were to be provided by IBRD and AID.

Public investment to be made in agriculture—87 billion francs —ran up to the level of investment made in the previous plan; 51.4 billion francs were earmarked for irrigation. Such a financial package covered the funding of large irrigated plots, small-scale and average water projects, as well as research on the Loukkos and Tafilalet projects. The *bour* areas which were fertile were to be allocated 15.7 billion francs whereas the remainder, that is, 16.4 billion francs, was to go to the so-called marginal areas. Thus over 80 percent of loans were made out to rich areas.

Grain production in the three-year plan was to be maintained at the same level as in the previous plan while meat production was expected to increase slightly from 180,000 tonnes in 1965 to 195,000 tonnes in 1967. The production of seed oil was supposed to double in this plan; however in spite of this new level of production the seed oil deficit would be maintained at its 1964 level, that is 48,000 tonnes.

Regarding export production, however, although no target was set for the quantities to be produced, provisions were made to maintain them at the same level and to increase them if possible to increase the volume.

One must also stress the higher place given to sugar production under this plan. While 20,000 tonnes of sugar were produced in 1964, sugar production was expected to reach 70,000 tonnes in 1967.

The 1968-72 Five-year Plan

Ideological context and policy choices of plan. In the post three-year plan period, the authorities spoke in a rather subdued manner. Indeed production was stagnant, and the debt situation had worsened as had the unemployment situation. The picture thus painted even implied a high level of rural exodus and "social threats rooted in a rapid urbanisation process".

However, in spite of the ringing of alarm bells, the underlying reasons mentioned above never went beyond the traditional grumbling of technocrats about the failure of private initiative, the shortage of foreign aid and the burden of the demographic explosion.

As a matter of fact in painting such a bleak picture the authorities wished to show that the situation arose from the "wrong options" which were taken in the early 1960s and which were still in force under the three-year plan. Apparently, profitability was not sufficiently stressed and the IBRD requirements confirmed it. This financial agency usually only funds projects which promise an adequate internal return, namely, projects with maximum capacity to diffuse capitalist relations of production.

However, as a rule, this bank contributes to the funding of

basically agricultural projects under one condition: the state must bring about a more balanced development, especially by taking steps toward reducing the regional and social disparities. The drafters of the second five-year plan took therefore a more moderate tone than with the first one: modern agriculture has to be developed, of course, but at the same time the traditional sector should not be neglected.

In fact the authorities had to take this into account as a result of the popular discontent which broke out in 1965. However, although there was a change in the rhetoric, as far as the implementation of programmes was concerned, the options made in the 1965-67 three-year plan were being reinforced.

In view of the difficulties faced by Moroccan agriculture, production grew only moderately: 2.1 percent per annum, in the first phase.

Indeed it was necessary, on the one hand, to pursue the policy of big dam construction projects which alone could ensure a quick growth in the long run and, on the other hand, to continue the intensification activities. These could only become successful with the establishment of a legal framework to protect economically viable farms.

The estimated budget was some 260 billion francs that is, over three times the investment made in the three-year plan. This investment was to be made almost entirely by the state. The breakdown of this investment was as follows:

Dams	74.6 billion francs
Agriculture	176.5 billion francs
National Promotion	15 billion francs
Total	266.1 billion francs

The state was to account for 241.6 billion francs of this investment, which represented 37 percent of public investment—655.2 billion francs—as a whole. The external resources involved totalled 173.3 billion francs, i.e, 26 percent of overall public investment.

The following table gives an idea of the "dam" orientation of this plan under which seven dams, including that of Arabat, considered as the most important, were to be constructed.

The immediate consequence of this choice was the acceleration of equipment in large irrigated plots as can be seen in the following table on public investment allocated to agriculture:

Irrigated areas	67.1 billion francs
Bour areas	57 billion francs
Stock breeding	13 billion francs
Farm production support activity	18.2 billion francs
Miscellaneous	21.2 billion francs
Total	176.5 billion francs

The irrigated areas including the construction of dams absorbed over 75 percent of agricultural investments.

Suggested reforms. The main objective was to create as many economically viable farms as possible. One of the incidental means used to that end was the distribution of land to poor peasants. The "agrarian reform" programme was cut down to half of what was anticipated in the three-year plan: it then covered 75,000 hectares. However, a major decision was made under this five-year plan; to recover all remaining colonial land. This decision became effective only by the end of the 1968-72 plan.

But how was it possible to create as many economically viable farms as possible without raising questions about the very structure of land ownership?

It was therefore essential to encourage owners of large estates to undertake further intensification activities. Their land base had to be safeguarded. The recuperation of the surplus produced by state investment was timidly mentioned; but the measure was never enforced. As for small and average landholdings, it was necessary to avoid their fragmentation by reforming death duties and by granting loans in order to extend these landholdings.

How could the modernisation of farms be accelerated? For one thing it was necessary to extend the Agricultural Loan scheme to more peasants as only 10 percent had so far been recipients. Furthermore, the state had to make available to the peasants "all the necessary supplies at the right times and at fair prices". The acquisition of these means of production was encouraged by granting subsidies and by the formation of producers' groups.

When examining this institutional framework one notes a big gap between the objectives and the means to attain them. How could agricultural production be boosted when the agrarian structures remained unchanged? In fact the development programme proposed under the plan aimed, first of all, at consolidating big farmholdings by providing them with the necessary financial resources. The introduction of the investment code in July 1969—a real charter of agrarian capitalism—did not confirm this trend. Of course, only in official speeches was the concern for the improvement of the situation of small-size and medium-size farms presented as a priority. But in fact this preoccupation boiled down to preventing their fragmentation rather than making them part of a dynamic development process.

Finally, this plan was more geared toward improving the social situation than finding a solution to the deadlock. One of the facts indicative of this new trend was that the rural exodus was no longer seen as a negative phenomenon and migration "abroad", particularly from the rural areas, was further encouraged.

The 1973-77 Five-year Plan

Under the 1968-72 plan, Morocco registered a slightly higher growth than in previous years. However, this modest growth was characterised by two elements: it was premised on a fragile base and it mainly benefited the rich.

Results in every sector of the national economy exceeded projections. The GDP recorded a 5.6 percent per annum increment when a 4.3 percent increase was projected. Farm production increased by 6 percent when only a 2.1 percent growth was anticipated. An exceptionally good harvest in 1968 and a good one in 1972 were responsible for such a performance. In other words, exceptional climatic conditions were the primary basis of such growth; second, the price paid to achieve such growth was a massive debt. In 1972 the public debt accumulated represented 25 percent of GDP as against 21 percent in 1968. The payment of debts absorbed therefore 8.6 percent of export revenues in 1968 and climbed to 9.4 percent in 1972.

In the face of such "prosperity" the state envisaged the implementation of a huge investment programme whose volume increased while being carried out, following the "exceptional" inflation of the price of phosphates. The volume of investment envisaged totalled 11.2 billion dirhams (1100.2 billion francs), i.e., double the amount recommended by the previous plan. As for priorities in directly productive sectors, the emphasis was put on industry which was to receive 37.2 percent of the total investment while only 15.8 percent was to be allocated to agriculture.

As for budgetary allocation, 26.2 percent of public investment was to go to agriculture while industry was to be allocated only 16.1 percent. Furthermore while the state was supposed to fund 80 percent of investments to be made in agriculture, in industry it was to account only for 40 percent of investments in this sector.

The outstanding feature in this plan was, however, the emphasis laid on the construction of dams, a decision which was made under the previous plan. This emphasis was even strengthened since 160,000 hectares of land were to be developed through the large-scale water programme as opposed to 120,000 hectares under the 1968-72 plan. Besides, allocations to irrigated areas exceeded two-thirds of the allocations to agriculture.

This plan is also characterised by its focus on improving peasants' incomes. This is not a new orientation, in fact, but, contrary to the past, a vast scheme of distribution of 400,000 hectares of colonial land was adopted. The decision was reached following the severe riots which shook the state apparatus in 1971 and 1972. But only 160,000 hectares were finally distributed, that is half the land coverage which had been distributed to the peasants between independence and the end of this plan.

The 1978-80 Three-year Plan

This was a "debate" rather than a plan of action. It was also an aborted plan in that it was supposed to cover five years originally. When drafting this "still-born" plan, the authorities had called upon technicians, researchers, producers. Many interesting studies were carried out, unfortunately under hasty conditions (studies on production potentials, funding capabilities, the impact of agricultural policy etc.).

Out of these the 1978-80 three-year plan only took the criticisms made of the results from previous plans into account. Morocco's agricultural situation worsened considerably; thus the agricultural trade balance was more in the red than it used to be, there even emerged rural unemployment and peasants were getting lower incomes.

Two basic reasons were put forward to explain this situation; government had concentrated all its efforts on irrigated areas, to the detriment of dry areas; and the pricing policy of agricultural products was responsible for stagnant production. It was necessary to cut off subsidies of imported farm products and to increase basic commodity prices sharply so as to stimulate national production and thus reduce the level of food dependency.

However, the orientation of this plan did not change. Dam construction was still privileged with allocations of 66 percent of planned investments (250 billion francs) were still geared toward irrigated areas. However, a programme for the "reconversion" of export crops, the early crop plan, was undertaken in response to the protectionist measures taken by the EEC.

The 1981-85 Five-year Plan

The policy choices made under this plan were the same as under the previous one. However, the 1981-85 plan allocated resources in favour of the dry zones. The investments in the agricultural sector were estimated at around 10.55 billion dirhams.

The major feature of this plan was its concern with the *bour* (dry) zones. Indeed they were allocated 51.86 percent of the overall investments whereas this percentage stood at 45 percent in the previous plans. The amount allocated to stock breeding also increased from 5.84 percent of total investments made under the 1978-80 three-year plan to 10.96 percent of the investments projected under the current plan.

But given the financial difficulties faced by Morocco and the successive years of drought that affected the country, the plan had serious problems meeting its objectives thus leading to the continuation of an agricultural policy in favour of irrigated areas and of big landowners.

CONTRADICTIONS IN AGRICULTURAL DEVELOPMENT

The State's Double Strategy:
A Compromise between the Bourgeoisie and the Big Landowners.
The state tried to develop a capitalist-oriented agriculture while maintaining a balance between the various social forces in the rural areas. The attempt made to implement agricultural development through structural reforms such as the operation labour project and the 1960-64 five-year plan, finally proved to be quite an adventurous enterprise. The state hastily returned to a purely technical and economistic view of agricultural development. This sheds light on the "parallelogram shape of social forces" of which the Moroccan state is composed.

Political independence offered the Moroccan political system an opportunity to take a number of political choices and notably those which could sanction its survival given the political strength of the existing social forces.

The 1956-59 period was an "experimental" phase during which the state was faced with two alternatives. It could either adopt a progressive (nationalistic) strategy and make a clean political, social and economic break with the types of alliance which had characterised the precolonial political system (e.g. the Makhzen alliance—the landed oligarchy). Or it could, on the contrary, perpetuate these types of alliance.

The 1956-59 period which had been qualified as being a wait-and-see phase was characterised by some sort of indecision. The combination of this "indecision" coming to an end and the implementation of the first alternative in 1959-60 by the Abdellah Ibrahim's government could not guarantee the survival of the Moroccan political system. This alternative was therefore quickly abandoned and replaced by the second which was characterised by its attempt to take into consideration the interests of the landed oligarchy. On the other hand, the integration of the Moroccan economy into the international division of labour required the development of a bourgeoisie by the Moroccan state. The state was therefore faced with a rather delicate task, that of developing a bourgeois class without harming the interests of the landed oligarchy, the mainstay of the precolonial and colonial alliance system.

So the state adopted a double strategy: the precolonial alliance between the landed oligarchy and the central power was reproduced in the post-independence period; and the development of a bourgeois class was necessary for the insertion of the Moroccan economy into the international division of labour.

From that time to the present the state has been trying to transform the landed oligarchy into an agrarian bourgeoisie and to consolidate the nascent agrarian and industrial bourgeois classes which emerged during the French protectorate (1912-56). The double strategy was not however, directed at two different objectives but at one (which consisted of instituting the total and exclusive domination of capitalism) through an open noncommital policy towards the two social classes, the bourgeoisie and the landed oligarchy.

This attempt to transform the landed oligarchy and to consolidate the status of the bourgeoisie was sought through the promulgation of various laws, namely: the 1963 law on the control of fixed assets operations, the principal instrument in the transfer of colonial property into the hands of the new agrarian bourgeoisie; the July 1966 law on agrarian reform which, however, did not have any practical effect until 1971-72; the agricultural investment code published in July 1969 and which was fundamentally oriented towards the strengthening of the agrarian bourgeoisie, with measures aimed at encouraging the development of economically viable farms (possibility of expropriation and the creation of a land market); reformulation of the 1972 agrarian reform law in order to redefine the land tenure structures; the distribution of land should be geared mainly toward small peasants resulting in the gradual elimination of the small *melk* and the *melk*isation[2] of consolidated land.

The Limits of Agricultural Development

This series of laws did not produce the expected results; agricultural production continued its very slow growth, hardly matching population growth. Apart from 1968 which was marked by very positive climatic conditions, the figures for domestic agricultural production remained unchanged between 1960 and 1967, with a limited growth between 1969 and 1975.

In the 1970s, agricultural production recorded sharp rises mainly due to climatic fluctuations.

After a period of stagnation between 1960 and 1974, agricultural exports started to decline consistently, mainly at the level of market gardening products.

On the other hand, food imports continued to grow.

The deterioration in the balance of food trade resulted in the decline of the rate of agricultural export earnings used to cover total imports. This rate fell from 52.6 percent in 1967 to 18.7 percent in 1980.

TABLE 6.1
Agricultural Gross Internal Product Growth Rate

Year	AGIP Growth Rate
1973	8.2%
1974	10.0%
1975	8.6%
1976	10.6%
1977	13.2%
1978	23.6%
1979	1.7%
1980	6.2%

Source: 1978-80 *Three-year Plan*, Vol.1, page 86,
Bank of Morocco, 1980 Financial Year, Annex 2.

TABLE 6.2
Evolution of the Quantities of Food Products Exported
('000 tonnes)

	1960-64	1965-69	1970-74	1975-79	1980
Mean of period	1,329	1,348	1,551	1,286	1,559
Index (basis 100 = 1960-64)	100	101	117	97	117

Source: "El Ayachi Alla, 'Dépendance et extraversion alimentaire du Maroc''
graduate thesis, INSEA, p. 99, Rabat, 1983.

TABLE 6.3
Global Evolution of the Volume of Imported Food Products
('000 tonnes).

	1960-64	1965-69	1970-74	1975-79	1980
Mean quantities imported during the above periods	887	1,087	1,169	2,053	2,431
Index (basis 100-1960-64)	100	122		231	274

Source: El Ayachi Alla, "Dépendance et extraversion alimentaire au Maroc",
graduate thesis, INSEA, p. 99, Rabat, 1983.

The failure of its agricultural policy pushed the state into making some changes. Until 1972, the state had considered the agricultural sector as a homogeneous entity capable of yielding agricultural surplus which could be exported with a view to financing the purchase of capital goods required for the industrialisation of the country. This surplus could also cover the costs of the importation of basic food products. In 1972, however, the state adopted a new attitude towards agriculture, ascribing it a dynamic role in which it is expected to reinforce the development of industry through the expansion of the internal market.

It was in this context that the state planned to distribute 400,000 hectares of land recuperated from colonial ownership. This objective was only attained halfway, for half of the expropriated land in 1972 was "illegally" bought, mainly by the new agrarian bourgeois class.

Today, state-owned land has considerably dwindled resulting from the absence of limitations on the size of big private farmholdings. The state is therefore pursuing the objectives of extending the internal market by involving small- and medium-scale producers in the modernisation of their own farms. As a result of this policy, the provisions of the agricultural investment code on irrigated areas are being extended to the *bour* areas by launching so-called integrated projects, particularly in the "favourable *bour* areas", and by increasing producer prices.

The orientation given to agriculture—which simply sanctioned the political choice taking into consideration the interests of the bourgeoisie and of the landed oligarchy—considerably limited agricultural and consequently economic development as specified below.

Financial limitation. Agricultural development was to be almost entirely funded by the state. This situation constituted a serious handicap both to the agricultural sector and to the pace of modernisation of this sector.

Socio-economic limitations. Agricultural development was still based on the growth of the so-called modern sector. It consisted of giving priority to export crops and encouraging products destined for the internal market, within the framework of an agricultural policy which did not take into consideration the enormous potentialities of small- and medium-scale farmers.

THE IMPACT OF STATE INTERVENTION
IN AGRICULTURE

Agriculture plays a significant part in the Moroccan economy. Its share in the GDP amounts to 13 percent, that is 10,887 million dirhams in 1981 out 76,445 million dirhams. It also accounts for 25 percent of export revenues and provides the population with half of the job opportunities.

From its inception therefore, the state was forced to take a great interest in the evolution of the rural world, a situation which has been rather disquieting for the central power.

The population is growing at such a tempo that agricultural problems keep worsening, hence the growing shortages. Prior to the drought period, that is between 1978 and 1980, meat was the only commodity for which the demand was satisfied by local supply. As for grains only 72 percent of the local demand was met whereas 54 percent of the demand for sugar, 50 percent of milk and 18 percent of oils were met. Since 1980 these shortage have been increasing with the drought which has been on for four years already.

We shall identify state intervention through its economic and social impacts.

The Economic Impact
In spite of the large amounts of credit allocated to agriculture, its development remained unsatisfactory, whereas, paradoxically, agricultural production was undergoing a relatively rapid modernisation process.

The paradox results from the fact that growth in the level of productive forces was not accompanied by a significant increase in production. The explanation of this fact has to be found in the uneven character of the modernisation process on the one hand and the failure of this process to bring about any radical modification in the production process on the other. "The modern sector" was drawing substantial benefits from modern technologies whereas the "traditional sector" underwent a "pseudo"-modernisation process consisting of truncated mechanisation, low levels of fertilisation etc.

a) Low level of development of the productive forces. In the meantime, agriculture in Morocco remained barely mechanised. In 1979 according to estimates, there were about 20,500 to 24,000 tractors, with one tractor to 366 hectares, whereas the optimal number of tractors required was one tractor to 100 hectares. Selected seeds were barely used; thus, cereal farming which involved selected seeds comprised 20 percent of soft wheat, 10 percent of hard wheat and 0.4 percent of barley.

Contrary to the mechanisation process, fertilisation developed remarkably. About three million hectares are now fertilised, that is, 40 percent of the total area under cultivation. Fertilisers were used 7.3 times more between 1956 and 1975, reaching 170,800 tonnes from 23,500 tonnes of various types of fertilisers (nitrogenous, phosphatic and potassic). The sharp increase was mainly due to the launching of the fertiliser project in 1966, which covered about 200,000 hectares annually till 1972.

However, the use of fertilisers was very inadequate. Indeed 22 fertiliser units were being used per hectare whereas the average standard recommended by the agricultural research head office was 132 units. According to this standard, the country needed one million tonnes of fertilisers whereas in 1975 it only used 170,800 tonnes.

These figures conceal the differential growth rates between the irrigated areas and the *bour* areas.

Uneven growth. Irrigated areas were highly mechanised. Thus, in 1979, irrigated plots which constituted 20 percent of the area of grain production utilised 44.3 percent of the state tractors while the unfavourable *bour* area (dry) with less than 400 mm annual rainfall covering 40 percent of the area of grain production utilised only 19 percent of state tractors.

With reference to fertilisation, the lion's share went to irrigated plots which absorbed 47 percent of the fertilisers supplied, while 41 percent went to the favourable *bour* and 21 percent to the unfavourable *bour* area. In 1974-75, export crops and industrial crops which were generally under irrigation absorbed 44 percent of the supply of fertilisers whereas food crops used 47 percent.

In the area of citrus fruits 357 fertiliser units were used per hectare whereas 313 fertiliser units per hectare were used for beets, 17 f.u. per hectare for grains and 10 f.u. per hectare for pulses.

An export-oriented growth. The export orientation of agricultural development growth could be seen through the value of agricultural exports in relation to the total value of crop production and through the relatively high amount of credits accorded to export crops. In 1975, 18 percent of the overall crop production was exported. Thus, almost one fifth of the plant production was absorbed by the external market while total agricultural exports represented 24 percent of the global volume of crop production. The market value of crops which could be exported was higher than that of products absorbed by the internal market. These export-oriented crops were concentrated in areas in which the quasi-totality of public investment was made.

Almost half of the crop production is harvested from a little over 10 percent of the cultivated area. Nearly all the crops geared exclusively to export—100 percent of greens and 83 percent of orchards—are concentrated in the irrigated areas. Half of the global volume of crops produced in these areas supplies external markets. The total market production—early vegetables and fruit production (citrus fruits, vine and others)—accounts for over 87 percent of the tonnes of food commodities produced within the irrigated areas.

TABLE 6.4
Distribution of Agricultural Production (by areas)

	Cultivated areas (hectares)	%	Production (tonnes)	%
Irrigated areas	740,000	12.2	4,200,000	46.2
Favourable *bour* areas	2,160,000	43	2,600,000	29
Unfavourable *bour* areas	2,269,000	44.8	2,220,000	24.8

The Social Impact

The social impact of state intervention in agriculture can be seen through two types of measures aimed at bringing about some social changes in the countryside. These changes are determined by the very nature of the mode of capital accumulation and they include the distribution of land to a section of the peasantry as well as the implanting of capitalist relations among the traditional ruling class, through the advantages offered by the 1969 agricultural investment code.

a) The agrarian reform. The distribution of land in independent Morocco started in 1956 with that of 3,410 hectares to 267 peasants in the Basse Moulouya and Ghard areas. This rate of land distribution was maintained (although it sometimes fell) until 1967. 1969 marked the beginning of a new phase (which was shortlived) in the distribution of farm or farm-oriented lands to farmers. 400,600 hectares of land were distributed to 20,729 farmers from independence until 1977.

Since 1967, that is one year after the 4 July 1966 agrarian reform law was promulgated, land distribution has been accelerated. However, the average price of land actually distributed is well below the estimates officially announced. For example, 400,000 hectares were to be distributed under the 1973-77 five-year plan but the acreage actually distributed under this five-year plan covered only 215,371 hectares, that is 53 percent of the originally planned distribution.

The lands distributed were basically those recuperated from the colonial estates. On the whole, these estates which covered an area of one million hectares in 1956 were distributed as follows: 500,000 hectares were bought mainly by the new agrarian bourgeoisie, predominantly made up of bureaucrats, 350,000 hectares were distributed to small farmers and 150,000 hectares were declared government property and administered by government enterprises (SODEA, SOGETA, COMAGRI etc.).

Consequently, the land reserves which enabled the state to distribute land according to the political and social situations were gradually exhausted. What was left was the land of the public sector, which technically, however, could not be distributed in the form of individual plots. The state did have control over plantations (citrus fruit, grape vines, rosaceae) through which it contributes 30 percent of the total citrus fruit exports and 60 percent of the grape vine production. The state-owned orchard estate was so huge that it was coveted by the agrarian bourgeoisie which claimed it, asserting that private sector management was more efficient than state management.

Besides, about 60 percent of selected grain seeds are produced on these state farms which consequently have conflicting interests with the agrarian bourgeoisie. Indeed, the latter is involved in this activity and the profits it accrues from it could drop due to state pricing policy and competition.

All these factors limited the policy of land distribution which was based on the July 1966 texts and was supported by the sole recuperation of colonial estates. Consequently, the state could only continue distributing land if it adopted the principle of limiting private property. Failing this, the state tried to promote the development of the forces of production within the framework of so-called integrated projects. These projects concerned the non-irrigated areas and constituted, on the whole, and may best be hypothesised as, a simple transfer of methods used in the hydro-agricultural development programmes.

The launching of these projects could only reinforce big land-ownership in the *bour* area with some fallout for the middle peasants, intensification of agricultural extension services, growth of inputs distribution networks, construction of roads and tracks etc.

b) Acceleration of capitalist relations among the traditional ruling class. The agricultural investment code promulgated in 1969 constituted the legal framework of state assistance to agriculture. As expressed in the AIC the state should discourage, as much as possible, private investment. To that effect it should give assistance in various forms: credit facilities, bonuses, subsidies, tax adjustments. Under this contract, the peasant on the other hand had to make a rational and judicious utilisation of the equipment acquired through government assistance. This assistance was to be given without any discrimination, to big and small farmers, given that the code stipulated that assistance should be given, for the acquisition of new agricultural equipment, "to owner-occupiers, tenant farmers and to corporate groups which were formed to jointly use agricultural equipment". In practice, this assistance was more profitable to the big farmholders than to the small farmers. In promulgating the AIC, the state's objective was to help reorganise the traditional ruling class into an agrarian capitalist class through the use of various productive techniques rationally, rather than providing the small and average farmers with financial resources to enable them to develop their farms rationally.

CONCLUSION

The development of agriculture in Morocco is facing a number of constraints due to the political grounds on which the state has based its conception of agricultural modernisation. Economic and social development plans have accorded top priority to agriculture, with considerable credit facilities granted to the sector. However, the fundamental issue still remains unresolved. The food deficit has worsened, for the state has implemented a policy of uneven development, giving more attention to the irrigated lands reserved for the cultivation of export crops. The deterioration of the terms of trade for agricultural exports and the narrowing of the international market which has worsened as a result of the broadening of the EEC to include Spain and Portugal raise doubts as to the adequacy of the state's policy choice in agriculture.

The agricultural sector is marked by a very low growth of production, land concentration which is detrimental to the maintenance of social peace and, particularly, to growth in other sectors because of diminishing demand, limited land reform etc. The state tries to implement a policy of rational land development without questioning the political power of the landed oligarchy.

The state tries to maintain a precarious balance between the landed oligarchy and the bourgeois class. This is why the technical aspect of agricultural development prevails over the political aspect.

NOTES

1. *Rapport Plan Triennal 1965-1967.*
2. Exclusive appropriation of land.

7. AGRICULTURE AND STATE POLICIES IN EGYPT

T. Abdelhakim

INTRODUCTION

Egypt is situated in north-east Africa in a temperate desert area. Its agriculture is completely dependent on the Nile river, the only water source in the country. Farmland covers three percent of the total one million km^2 area. With its 45 million inhabitants and a significant population growth of about 2.5 percent per annum this country has a very dense population (1,000 inhabitants per km^2 and as a result there is a very strong pressure on farmland.

Egypt has always been said to be a "great agricultural country" in view of the major role which was and still is being played by agriculture in the economy. Agriculture accounted for 28 percent of GNP in 1970-75 and 24.9 percent in 1980. Agricultural assets represented 48 percent of total assets in 1970 against 40 percent in 1979 and the rural population accounted for 60 percent of the total 1966 population and 56 percent of the 1976 population. The significance of agriculture as an economic activity is also due to its being the major source for satisfaction of food needs.

Agriculture has a very long established relationship with the state in Egypt. The state (or central power) has been in existence in Egypt since ancient times and it has always been involved in agricultural activity: 1) as the only holder of arable land in the country—the peasant was merely a farm hand (cultivator) with a usufructuary right; 2) as organiser and promoter of important water works (irrigation, building of dams, embankments etc.) a vital activity for Egypt's agriculture.

State involvement in agriculture never stopped all through the history of Egypt. Only the extent of its involvement varied with governments in power—and even with political regimes—and with the various historical eras.

Like most Third World countries Egypt was in the past self-sufficient and exported agricultural goods. Since the 1970s, however, the crisis has started, with changes in the economic structure and the

population explosion: shortage of farm products, especially food products, dependence on the outside, deficit in the agricultural trade
balance etc.

Table 7.1 shows changes in the imports of a number of food
products:

TABLE 7.1
Changes in the Imports of Some Food Products
('000 tons)

Year	1974	1976	1978
Wheat	2,550	3,358	2,930
Flour	258	404	960
Corn	388	458	730
Sugar	117	170	371

Food imports accounted for 16.4 percent of the total value of imports
in 1970 and for 21.2 percent of this value in 1977.

Compared with the overall value of exports, food imports were
estimated at 21 percent in 1970 and 79 percent in 1978. The deficit
in the agricultural trade balance started in 1974. From E£84.7 million
in 1974, it climbed to 115.1 million in 1975 and fell to 105.7 million
in 1978. It should be pointed out that the increase in the value of
agricultural imports was mostly responsible for this deficit, as
agricultural inputs accounted only for 10 percent of the total value
of agricultural imports in 1978.

These few figures give an idea of the dimensions of the
agricultural and food crisis in the country, especially after adding other
parameters such as the small size of arable lands, the poor potential
of irrigation waters, runaway population growth etc.

However, agriculture has not always been a crisis sector, in fact
it used to be the driving force of the economy in Egypt, and for a
long time it generated the surplus needed to start the process of
economic development in the 1950s and the 1960s.

Aware of the significance of agriculture, the state used its land
policy, its pricing policy etc. to manage the production, appropriation
and allocation of the agricultural surplus. With reference to state
agricultural policies a distinction can be made between *two periods*
and the orientation of each of these differs with the global political
option. We shall study each period by examining its objectives, its
means and the levels of state involvement.

FIRST PERIOD: STATE AGRICULTURAL POLICIES FROM 1952 to 1970

State involvement in agriculture in this period *was very strong* and a number of areas such as domestic and outside commercialisation of a major crop like cotton were almost completely under state control.

After political independence, the 1958 government adopted an economic and social development policy to industrialise the country and recuperate the national market.

The agricultural policies of that time had three obejectives: 1) to eradicate the holders of large estates as this class was the main ally of the monarchy and of colonisers; 2) to expand the internal market by increasing the buying power of the rural masses; 3) to increase and appropriate the agricultural surplus which was to be used to finance industrialisation, armaments etc.

In order to reach these objectives, the state often adopted very drastic policies and measures in various areas. The most dramatic among such policies and measures was obviously the Agrarian Reform issued three months after the 1952 change of government. It was also necessary to mobilise funds to increase agricultural output, to invest in order to intensify farming and to set up all embracing control structure for agricultural production and surplus.

Land Policies: The Agrarian Reform and the Creation of Service Cooperatives

Before the first Agrarian Reform in 1952 this was the structure of land ownership:

large estates (over 50 feddans). These covered 35 percent of the total farmland area and were in the hands of 0.4 percent of owners;

medium-size property (between 5 and 50 feddans). These covered 30 percent of tht total farmland area and were in the hands of 5.3 percent of owners;

smallholdings (under 5 feddans). These covered only 35 percent of the total farmland area and were in the hands of 94.3 percent of owners.

An outstanding feature in this structure was a very high concentration of land ownership, as 65 percent of the total farmland area was held by 5.7 percent of landlords. It should also be pointed out that most of the large estate owners were absentee landlords.

Not only did Agrarian Reform in Egypt change the structure of landed property but it also tried to organise the landlord-tenant farmer relationship, to protect the right of farm hands and to create cooperatives. Three Land Reform Acts were passed. The first one, Act 178 of 1952, set the ceiling of land ownership at 200 feddans per individual.

In 1961, Act 127 brought this ceiling down to 100 feddans per individual and the 1969 Act set it lower at 50 feddans per individual and at 100 feddans per household.

Land exceeding these dimensions was taken over by the state and redistributed to poor and "landless" peasants in such a way as to make the latter owners of 2 to 5 feddan smallholdings.

Large estate holders were to receive compensation for loss of their estate and peasants receiving land were to pay its price to the state over 30 years. Regarding the landlord-tenant relationship, the Agrarian Reform Act set the rent at seven times the land tax in cases of farm lease while in cases of sharecropping this rent was not to exceed half of the output (crops).

The same Act also regulated the establishment of cooperatives: these had to provide farmers with seeds, fertilisers, tractors etc. and had to buy their goods.

What was the impact of the Agrarian Reform on the land structure in the rural areas in Egypt? Table 7.2 shows changes in this structure.

It can be noted that smallholdings represented 35.5 percent of the total agricultural area before the first Land Reform and 57 percent of it in 1965 while in 1974-75 they covered only 49.7 of the agricultural area. However, large estate holders (more than 50 feddans) owned 6.1 percent of the farmland in 1965 and 16 percent in 1974-75 while the percentage of holders as a whole did not change: 0.2 percent. Holders of medium-size property (10-50 feddans) were in fact the greatest beneficiaries of the change in the land structure: the farmland area they held climbed from 10.9 percent of the total area before the 1952 Land Reform Act to 23 percent in 1974-75 while their percentage remained almost unchanged compared with that of holders as a whole: 2.4 percent before 1952 and 2.1 percent in 1974-75.

To sum up, let us say that the Agrarian Reform had a double impact on the land structure in Egypt. On the one hand, it reduced concentration of land ownership without making it disappear and, on the other hand, it greatly increased the fragmentation of holdings. The new power relations established in the rural areas in the post-Agrarian Reform period can be expressed in the following manner: in 1963-64, 85 percent of farmers with under 5 feddans holdings received only 50 percent of the loans made by farm cooperatives whereas the other half of these loans went to 15 percent of farmers with between 5 and 50 feddans holdings. As a matter of fact, state land policies in the 1950s and 1960s promoted household farming and medium-size property and one can see that the "wealthy peasant" (class) expanded. However, after experiencing a slight improvement in their living conditions, the latter were reduced again to poverty

TABLE 7.2
Changes in the Land Structure

Size of property (feddans)	After the 1952 Act		After the 1961 Act		After the 1969 Act		1974-75	
	% Holders	% Area	% Holders	% Area	% Holders	% Area	% Holders	% Area
Smallholding (-5 fedd)	94.3	46.4	94.6	52.1	95	57.1	94.9	19.7
Medium-size property								
5-10 fedd.	2.1	8.8	2.6	8.7	2.5	9.5	2.7	10.9
10-20 fedd.	1.6	10.7	1.6	10.7	1.3	8.2	2.1	2.3
20-50 fedd.	1	13.6	0.8	13.4	0.9	12.9		
Large Estates								
50-100 fedd.	0.2	7.2	0.2	7	0.2	6.1	0.24	16.3
100-200 fedd.	0.1	7.2	0.2	8.2	0.1	6.5	-	-
over 200 fedd.	0.1	5.9	-	-	-	-	-	-
TOTAL	100	100	100	100	100	100	100	100

with the 1970s crisis which affected first of all the poorest people in the rural areas.

Pricing Policies

In a market economy prices become one of the major means of action and of orientation in production and consumption. In the agricultural sector the state in Egypt has selected the pricing system as its major tool to control the level and flows of output as well as surplus production, allocation and appropriation. However, before explaining this process, it should be pointed out that prior to pricing agricultural goods the state used to control farming plans all over the country.

Indeed, farm cooperatives set up by the Agrarian Reform Act were meant to: 1) supply farmers with such inputs as fertilisers, seeds etc. at "reasonable" prices, i.e., at state prices; 2) implement and oversee the implementation of the farming plan (also called rotational cropping) in villages and in the region. This is a binding plan worked out by the Ministry of Agriculture. Farmers who do not abide by it are penalised by their own cooperatives. The penalty for such a breach is either to be refused the right to buy from the cooperative inputs that are necessary to production or to pay a fine; 3) buy from producers some of the major crops such as wheat, cotton, rice etc. at state prices either for export or for local consumption.

As of the 1960s, agricultural output increased in Egypt with the change-over from a basin check irrigation system—whereby farming was only possible once a year—to a continuous irrigation system—whereby land was under cultivation twice or three times a year. Thus approximately 6 million feddans of farmland area yielded as much as about 12 million feddans since the same plot was farmed twice a year.

At the central level, the Ministry of Agriculture worked out a rotational cropping system whereby it determined the exact area to be used for growing each of the crops in the various regions and villages.

Thus the state directly supervised the volume of output of each of the main crops.

In the 1950s and 1960s emphasis was on cotton which was the first export crop in Egypt. But from the 1970s on and with the cotton crisis on the world market (competition with the Indian and American productions on the one hand and with synthetic fibres on the other hand) the area on which cotton was grown was reduced.

The other main crops included: wheat, corn, rice, barley, broad beans, lentils and sugar cane. They were traditional crops (or what is called field crops) and in 1960 they covered 92 percent of the farmed area whereas vegetable crops and fruits covered only 5.9 percent of it.

The *pricing policy* was implemented in three ways: 1) the whole production had to be delivered to the state at the state price; 2) part of the output was delivered to the state at state price whereas the other part was sold at the market price; 3) producers' prices were regulated by the law of supply and demand.

Crops which had to be delivered to the state included cotton, rice, onions, wheat, groundnuts, broad beans and lentils. The delivery quota varied from crop to crop: 100 percent for cotton production, 66 percent for rice production, 57 percent for onion production, 27.6 percent for wheat. The state purchased these products through cooperatives to which farmers had to sell their goods.

The state made substantial profits with this commercialisation system. Indeed these profits were made as a result of the *difference between state prices paid* to producers and *sales prices* on the domestic or outside maket; furthermore this difference has to be multiplied by the volume of production which this system applies to. Research conducted on pricing policy in agriculture and on the "hidden" tax (or indirect taxation) on agricultural income, attempted to estimate the hidden tax percentage for each export crop—cotton, rice, onions and groundnuts. It was found that the hidden tax on the production unit amounted to the difference between the purchase price set by the state and the sales price (on the domestic or foreign market) whereas the hidden tax percentage expressed the percentage of hidden tax levied in relation to the total income made on each crop. This percentage depended therefore on two factors: 1) the difference between the purchase and the sale prices; 2) the quota of each crop which has to be delivered.

Table 7.3 indicates the hidden tax and its percentage for cotton, rice, onions and groundnuts.

Regarding cotton, from 1964-65 to 1970-71 the payment made by producers as hidden tax ranged between E£10.8 and 23 million; this amount later increased to E£36 million and 49.5 million to 1971-72-73. It then reached respectively E£195.7 and 110.7 million in 1973-74 and 1974-75 because quotations for cotton soared on the world market. The percentage of hidden tax on cotton ranged from 22.4 percent of the global income made on this crop to 51.6 percent, in the second half of the 1960s, and reached 83 percent in the first half of the 1970s.

As for rice, the hidden tax ranged between E£22.8 and 58.1 million up to the early years of the past decade. The percentage of this tax ranged between 37 and 60.3 percent of the total revenue made on rice, later reaching 60.3 percent and 50.4 percent in 1974 and 1975 respectively.

The hidden tax levied on onion production varied between E£3.5

TABLE 7.3
Hidden Tax on Agricultural Products

YEAR	COTTON		RICE		ONIONS		GROUNDNUTS	
	Hidden tax (in million E £)	Hidden tax Rate (%)	Hidden tax (in million E £)	Hidden tax Rate (%)	Hidden tax (in million E £)	Hidden tax Rate (%)	Hidden tax (in million E £)	Hidden tax Rate (%)
1964-65	19.1	38.9						
65-66	15.9	168.2						
66-67	12.9	51.7						
67-68	14.8	38.4	36.2	45.1	4.6	76.7		
68-69	16.2	32.2	37.7	43.2	6.2	36.1		
69-70	10.8	22.4	22.8	37.0	5.0	73.0	0.51	23.1
70-71	23.0	25.1	25.1	44.6	3.5	51.5	0.44	-4.0
71-72	36.0	31.5	25.7	44.3	5.3	54.1	0.64	35.6
72-73	49.5	41.5	35.2	52.3	10.2	61.4	0.72	12.1
73-74	195.7	82.8	58.1	60.3	7.5	45.7	1.21	4.6
74-75	110.7	83.2	47.0	50.4	9.5	51.1	0.82	27.4

and 10.2 million for the 1968-1975 period while its percentage stood at 45.7 percent and then 68 percent during the same period. One can see that this percentage exceeds that levied on cotton and rice in the same period.

Regarding groundnuts, the hidden tax ranged between E£0.44 and 1.2 million in the 1970-75 period while the hidden tax percentage during the same period varied between 24 and 40.8 percent of the total revenue from this crop.

Obviously then, agriculture recorded a substantial surplus which was channelled into other economic sectors. However, a number of elements were subsidised in the agricultural sector; these included inputs, intermediate products used by this sector and investments. The real question to be raised is the following: does this agricultural surplus exceed the subsidies to and investment in this sector? If so what is the value of the net agricultural surplus?

The value of this agricultural surplus was estimated at E£21.8 million in 1960-61, E£44.7 million in 1967-68 and E£66.8 million in 1969-70. Another study estimated the agricultural surplus (after deduction of subsidies to inputs and means of production) at E£1,200 million in 1974-75 and at E£500 million in 1982.

One may then conclude that, through its pricing policy (purchase and sales prices), the state determined the exact amount of surplus to be extracted and appropriated this surplus. But among the farmers who are those paying the most for this surplus extraction?

A study of the structure of property to the distribution of major crops shows that 46 percent of the overall area under crops which had to be delivered to the state can be classified under smallholdings (under 5 feddans); 46 percent of the area sown to wheat, 38 percent of that under cotton and 33.8 percent of that under cultivation of onions belong in this same category: smallholdings covering less than 5 feddans.

Small producers accounted for between 62 percent and 68 percent of all the producers of broad beans, onions, rice and groundnuts; they also formed 80 percent of cotton and wheat growers. Small farmers were then undoubtedly those who were producing the bulk of major and export crops which were subjected to state pricing policy.

Investment Policies

In the 1950s and 1960s the only investments made in the agricultural sector were *public investments*.

Investment policies had a double aim: vertical and horizontal expansion. Vertical expansion required a change-over from a basin check irrigation system—whereby land could only be farmed once a year—to a perennial irrigation system which could double the farmland area. Another factor was necessary to reach this objective: control of the Nile and maintenance of the waters at the same level all year long.

The construction of this Aswan high dam was the greatest water development project ever carried out in the country. It revolutionised the irrigation system and as a result of its energy production it was possible to bring electric power to the whole of Egypt's rural areas. For ten years, i.e. during its construction, the bulk of investments made in agriculture were to finance this project.

Horizontal expansion means an increase of the farmland area, which can be reached either by improving on poor quality soils or by encroaching upon the desert—that is by changing part of it into arable land. This land development process required huge investments which also consumed a great chunk of the investments in the 1950s and 1960s in the agricultural sector.

Between 1953 and 1959 the area developed and farmed was estimated at 79,000 feddans in the valley area and some desert areas. During the first five-year plan (1960-65) 536,000 feddans of land were developed and farmed, 453,000 of which were in the valley area and 33,000 in the desert; whereas during the following plan (1965-70) only 276,000 feddan, were improved upon, against only 21,000 between 1971 and 1975. Therefore between 1952 and 1975 the whole area which was developed covered 912,000 feddans 775,000 of which were farmland. Of this area 600,000 were actually farmed with different levels of output. These lands were managed and distributed to small farmers in keeping with the Agrarian Reform Act: peasant households were settled on 5 feddan plots on which they grew mostly fruits and vegetables. The production was commercialised for the local market or for export by state companies mostly.

Technological Policies

The technological issue was a rather comprehensive issue as it embraced several areas: the use of inputs (inorganic fertilisers, high quality stock, seeds etc.), scientific research, production techniques, irrigation techniques, the mechanisation of farm works etc. All of these matters cannot be studied in this paper, so we are just going to deal with the mechanisation issue.

In the 1950s, the use of farm machinery developed more or less depending on the type of machinery. Table 7.4 shows the development of mechanisation in agriculture. In spite of this development, it was found that the energy used on farming one feddan could be broken down as follows: 18 percent originated from human sources, 44 percent from animal sources, 38 percent from mechanical sources.

There were few tractors available compared with the farmland

TABLE 7.4
The Development of Mechanisation in Agriculture

Year	1960	1961	1970	1975
Tractors	9,972	12,837	17,566	20,809
Irrigation Machinery	15,170	29,387	13,026	12,750
Reapers	2,951	9,178	1,829	2,876
Sprinklers	-	-	12,087	12,310

area. In Egypt the ratio was the following: 1 tractor to 280 feddans (against 38 feddans in Italy, 36 in England and 27 in France) and after adding a 200 percent agricultural expansion one finds that there was one tractor to 560 feddans.

SECOND PERIOD: AGRICULTURAL POLICIES FROM THE 1970s TO THE 1980s

State involvement in agriculture in this period was less strong in view of the liberalisation of Egypt's economic life. An open-door policy was declared in the economic sector as of 1974-75. Indeed an appeal was made for the investment of foreign capital in the ''development process'' and local private capital was granted facilities.

But ten years after adopting this policy it appears that it had a harmful economic and social impact: runaway inflation, growing economic and social disparities and inequalities, de-structuration and imbalance of production and consumption. Growing dependency is also one of the obvious consequences of this policy which has been increasingly criticised in a number of recent economic studies.

Land Policies

One of the first decisions made was to do away with the 50 feddans per individual ceiling set on farm property. New owners who are mostly purchasing land in the new farmland area developed by the state finally hold 200- or 300- feddan plots.

A number of expropriated landowners have also been given back their land by the state either partly or in full depending on cases. There are no statistics yet as to the exact area covered by these lands which were given back.

Rampant urbanisation is also a threat to Egypt's farmland. Cities continuously expand over neighbouring farmland expecially as smallholders find it more profitable to sell their land instead of farming it. There are many protective laws on farmland which ban any construction but social and economic systems are above laws.

Pricing Policies

The pricing system described in the first part still applies but with slight changes. The dissatisfaction of small producers (growing the largest part of such major crops as wheat, cotton, corn, broad beans etc., led the state to increase its producer price for those crops. But at the same time it increased the land rent.

The issue of subsidies to farm products to keep the consumer price relatively low often crops up. It should be made clear that the bulk of subsidies is consumed by imported agricultural products. In 1973, for instance, imported wheat accounted for 91.8 percent of the subsidised product and in 1979, for 97 percent. The same holds for lentils: in 1973, 50 percent of the subsidised quota was imported and this figure climbed to 93 percent in 1979.

With reference to the consumer price, in view of the fact that such produce as fruit and vegetables were not marketed by the state, they were sold to consumers at administered prices so as to limit traders' profit margins. With the "liberalisation" of the maket the sale prices of these goods varied with market conditions: artificial shortages, black market, monopolisation etc. Quotas were increased for other basic commodities like rice, wheat, sugar etc. which were sold on the free market.

Investment Policies

Public investment in agriculture fell considerably from 13 percent of overall agricultural investment in 1970-73 down to 8 percent between 1971 and 1978.

Investment in this sector was allocated in such a way as to promote horizontal expansion projects—new farmland development projects. Investment in these projects represented 68 percent of global investment in agriculture up to 1975 and 45 percent from 1978 to 1982.

Since the 1970s state policy has been encouraging the investment of either domestic or foreign private capital in the purchase of the new farmland that it has developed and improved without actually farming it.

Investors purchase large estates—at low prices and with very good terms of payment—on which they build "modern" farmsteads equipped to produce speculative crops mostly geared toward exports—e.g. flowers, medicinal plants, citrus fruits etc.

Technological Policies

As a result of previous policies, technological gaps have started to appear between productive well-equipped farmsteads and small "traditional" farms.

The mechanisation issue, which cropped up a long time ago and which remained unsolved on account of the severe fragmentation of farmland, needs an even more urgent solution today in view of the lack of farm labour.

As a result of the rural exodus and emigration to Arab oil-producing countries, two phenomena which have accelerated in the past years, there is hardly any labour force left in the countryside.

Research was carried out on the use of small tractors meant for small size areas. However, there are costs of production and productivity problems for which no solution has been found yet.

Traditional irrigation techniques are seriously questioned for two reasons: a heavy waste of water is recorded with these techniques; irrigation and drainage ditches cover a large area of farmlands: 20 percent.

Debate is open on the introduction of new irrigation techniques especially "spray irrigation" or "drip irrigation" and research is being carried out on costs and adjustment but no final conclusions have been reached yet. Attempts are being made once again at introducing these new techniques in new areas which are neither developed yet nor equipped. On the old farmlands, however, that is the valley area, conditions are still the same.

BIBLIOGRAPHY

Abdalla, A., *Distribution du revenu en Egypte entre 1970-1980*, Doctoral thesis, University of Cairo, 1963.
Annuaire Statistique, 1952-1978.
Korayen, K., *Distribution du revenu dans la campagne Egyptienne*, Cairo, 1980.
Metwal, S., "La sécurité alimentaire et la structure agraire dans une perspective d'indépendance," paper presented to the Sixth Congress of Egyptian Economists, Cairo, March 1981.
Nassar, S. and Mustafa, M., "La participation de l'agriculture égyptienne dans l'accumulation du capital dans les années 80", paper presented to the Fifth Congress of Egyptian Economists, Cairo, March 1980.
Shabana, Z.M., "Perspectives de solution du problème alimentaire en Egypte", paper presented to the Fifth Congress of Egyptian Economists, Cairo, March 1980.

8. THE STATE AND FOOD POLICY IN COLONIAL ZIMBABWE 1965-80

Thomas D. Shopo

We were in a position to produce eagerly sought-after raw materials in the mineral field so, with a little bit of ingenuity, we were soon manufacturing a large range of industrial products to meet local needs which were previously satisfied by imports, and also to offer large quantities in exchange for those products which were not able to be produced. The strength of our position soon revealed itself because we became masters in our own house through the good fortunes of sanctions which had so effectively removed the villains who had previously manipulated our development and had dictated our progress through the fraudulent device of international finance... Our strategy and policy was designed to place a maximum labor force in employment in the realisation that producing food for all was the first priority and the most effective weapon to counter discontent. Britain had relied heavily on internal discontent developing to disrupt our economic activities, and had banked on an internal revolution erupting to give her the excuse to step in on the pretext of saving the European.

W.H. Nicolle, (Secretary for Internal Affairs in the Smith Regime), writing in *Rhodesian Journal of Economics*, Vol. 5.4, 1971, p. 2.

INTRODUCTION

Diagnoses of Africa's present food and agricultural crisis are increasingly being centred on the role of the state. The World Bank's so-called "Berg Report" of 1981 attributed the present food crisis to inward-looking policies, biased against exports, that have included overvalued exchange rates, quantitative trade restrictions, excessive government spending and the lack of "incentives" to agricultural producers.[1]

The solution proposed by the World Bank has been in the form of a package of neo-liberal strategies—that to correct the distortions and inefficiencies created by protectionist policies, the "state" in Africa should take a back seat, cut back on parastatal activity, and hand over control to private capital investment. The basic underlying assumption of these policy prescriptions has been that African countries possess comparative advantages in growing certain types of crops on which they should concentrate their productive energies; further, that there exist "realistic prices" which will be reflected in the international and free markets. Implicit has been the view that there is no conflict between the proposed solutions to the proposed problems, and the pressing need to ameliorate the conditions of the rural poor. There has thus been little attempt at specifying the nature of the "state", which is perceived in all this in purely technocratic terms, reducing the problem to one of ignorance and the implementation of "wrong policies".[2] And those proponents of the "World Bank solution" with a penchant for the historical have not only seen structural continuities from colonial times to the postcolonial era, with regards to state interference with the free interplay of market forces, but have gone so far as to draw analogies between apartheid states, like colonial Zimbabwe, and South Africa, and post-independence Africa. One such leading light writes:

> In the independent states of black Africa as in South Africa, economic life was extensively politicised; the apparatus of state power was valued for the control it conferred on the distribution of income and wealth; economic status and gains depended on political influences, political office or sectional economic support.[3]

The Lagos Plan of Action, in contrast to the Berg Report, called for the state to become more involved as an "autonomous actor"[4] in the process of development. There was however little in the way of specifying what type of state was more likely to become an "autonomous actor". This has left the door wide open for neo-classical interpretations based on what has been perceived as the historical "success stories" of settlers in Kenya and colonial Zimbabwe to support the call for more state involvement. It has thus been suggested that the "state" should be seen as an arena solely for political action—so that the key to agricultural progress and food self-sufficiency is for farmers to gain power and to use it to shift relative prices, government taxes and subsidies in their favour.[5]

This "urban bias" model, however, attempts rather clumsily to sidestep the awkward reality of intra-agricultural class differentiation, by divorcing the political objectives of various rural classes from the material base of their interests in agricultural production and land ownership. The real central political issues implied by championing the peasants, which are about intra-agricultural choices of modes of production and land ownership and not about rural-urban struggles, are not dealt with. Further it has not yet become clear how the whole notion of "individual rationality" which, it is assumed, will prevail once an "agricultural faction" has gained a foothold in government, will result in an improvement which will trickle down to other producers.[6]

In this chapter, without getting too embroiled in conceptual controversies on the state, it is hoped to demonstrate in the case of colonial Zimbabwe between 1965 and 1980, that the whole issue of food and agricultural policy cannot be resolved at the level of prescriptions of either more or less state intervention, but that the parameters of state activity are ultimately determined by the specific economic form in which unpaid surplus labour is pumped out of the direct producers.

Contrary to the Berg Report's observation that in those African countries with less state intervention, certain gains in agricultural productivity have been made, it has been pointed out that, whatever the social costs involved, the substantial growth experienced in colonial Zimbabwe between 1965 and 1980 was largely attributable to active state involvement in the economy.[7] But given the efforts presently being undertaken to effect a transition to socialism, and to dismantle the form of the state imposed on Zimbabwe by the Lancaster House constitution, a more critical assessment of the settler colonial state and its policy instruments has to be undertaken. This is particularly so with regard to the agrarian question, where certain "general views" need to be confronted. To cite one Zimbabwean economist:

> the general view in Zimbabwe is that commercial farmers are seen not as a failure but a success element in agriculture. Whereas the latifundia and other feudal forms of ownership distorted the agrarian structure in a negative manner, capitalist commercial farmers in Zimbabwe are seen to contain some positive elements from productivity and efficiency considerations that ought to be preserved.[8]

Certainly, the aggregated national statistics for agricultural and food production are impressive. Taking the years 1961 and 1965 as the base period, Appendix 8.1 shows that while there was a cumulative decrease in total agricultural production per capita of 25 percent for the years 1961-80, food production per capita declined by a mere one percent. And a preliminary application of the FAO balance sheet methodology using data from the commercial sector showed that the country's per capita availability of food was more than 2,600 calories per day in 1978, above the FAO optimum of 2,390 calories per day.[9]

It has, however, to be noted that statistics in Zimbabwe have until recently placed more emphasis on the "commercial" aspect of agriculture. Statistics for the period do not trace what happens to marketed food items right down the food chain. And beneath the national level, the problems of inadequate food and nutrition are glaringly problems of poverty and inadequate command over sufficient resources to exchange for sustaining food.

The food problem in Zimbabwe has therefore to be seen not at the technical level of agricultural production, but within the framework of the reproduction and circulation of aggregate social capital, where the state has been a reproduction of the capital relation itself.

Historically, the most outstanding characteristic of the agrarian structure of colonial Zimbabwe was its dualism, in which the state supported the white commercial sector (both family farms and large company estates), which possessed the most fertile land, with access to national and international markets, credit, technology, extension services, credit, manufactured inputs and consumption goods. The "traditional" or communal sector was assigned unproductive land, producing in the main for family consumption and local markets. Its parameters for agricultural growth had since the first decade of the century been closely defined by the state. Within such limits, the marginal improvements which had taken place with regards to marketing and pricing by 1965 had little effect on the pervasive historical patterns of poverty within the sector.

Important too was the inter-related character of the agricultural sector with other sectors of the economy. By 1980, it is estimated that one out of every two employed Zimbabweans worked in the country's food system and those closely allied with food industries such as tobacco and cotton.[10]

The first part of the chapter will review trends in the structure of the colonial Zimbabwean economy from 1965 to 1980, with particular emphasis on agriculture; the second part will consider the evolution of state product and marketing policies, and credit policies, and the last part will attempt to draw together the fragmentary historical data that are available to bear on developments within the communal sector from 1965 to 1980, and inferences that can be drawn with regards to the pauperisation of the masses.

THE GROWTH AND STRUCTURE OF THE COLONIAL ZIMBABWEAN ECONOMY 1965-80—AUTONOMOUS DEVELOPMENT OR DEPENDENT REPRODUCTION?

Analytically, developments in the Rhodesian economy between 1965 and 1980 can best be viewed by distinguishing between the two major departments into which the total production of society can be divided: Department I—the production of the means of production—capital goods, and Department II—the production of the means of consumption or consumer goods.[11]

The latter can further be broken down into (a) the necessary means of consumption or those which enter the consumption of the working class and (b) the luxury means of consumption (consumed in the case of Zimbabwe largely by the settlers) which enter into the consumption of the capitalist class.

In the reproduction schemas of Volume II of *Capital*, Marx showed the conditions of proportionality which must obtain between the departments for reproduction to prosper. The concept of expanded reproduction points to the tremendous self-expansion drive of capital which constantly moves on to a greater and higher level of accumulation. Finally, the movement of capital must be seen as a combination of three different "functional" forms of circuits: capital in the form of money that buys the elements of production (money capital), capital in the form of the industrial labour process that combines productively the elements of production (productive capital) and capital in the form of industrial products which are sold as commodities containing surplus values (commodity capital).

In order to avoid analysing the colonial Zimbabwean economy in terms of the contradictions of advanced capitalism, it is however necessary to draw out the fundamental differences between Marx's reproduction schema, and that of "dependent reproduction" which characterised societies like colonial Zimbabwe.[12]

(a) The element of foreign trade is an integral aspect of the dependent reproduction schema. Thus, owing to the moulding of the dependent economy by imperialism and the size of the internal market there is a marked degree of dependence on foreign trade for realisation. In the case of Zimbabwe, at the time of UDI in 1965, almost the whole of mining output was exported, while 80 percent of the motor assembly industry's inputs was imported.[13] The prosperity of agriculture was heavily dependent on export earnings, with the agrarian sector divided primarily between the export-oriented enterprises and those producing for internal consumption.

(b) The division of labour between the departments of social capital was complicated by the ossification of precapitalist relations, through the development of communal forms of rights to landed property in the so called "tribal" subsistence agriculture as distinct from "commercial" agriculture. And in the racially ordered consumption regime, the settler state had through a plethora of native labour regulations, in particular the Master and Servants Ordinance of 1905 (which was repealed in 1980), regulated consumption for wage earners in the agricultural and mining industries, whereby food rations were an integral part of unit labour costs.

(c) The absence of an internal capital goods industry meant that the export sector fulfilled that role (by earning the currency which bought the machinery). This means that increases in productive investment/gross capital formation as happened in Zimbabwe's manufacturing industry in the first ten years after UDI (see Appendix 8.2) may have led to further imports but not an internal expansion of Department I which is the main dynamic element in advanced capitalism. From 1967 to 1976, a quarter of gross fixed capital formation for the country as a whole was accounted for by the manufacturing industry. Thereafter it declined steeply to become negative by 1979. Given such structural arrangements in the economy, despite the growth that took place (through import substitution), there was not an integrated expansion of the means of subsistence and the means of production in the process of expanded reproduction. Though there was in existence by 1965 a well developed consumption exchange regime, the fact that the settler state employed a deliberately less developed internal market for food production meant that regardless of any gains made in agricultural productivity, the logic of the mode of accumulation forced the majority of African households to become progressively pauperised as they were forced to purchase food commodities when retail market prices rose and consumed the products of their household production when consumer prices were relatively low.[14]

The consumption of the working class was therefore not a dynamic element in the type of capitalism that characterised colonial Zimbabwe, but it was rather the exchange relations established with international capital which were to determine (to a greater or lesser extent) the conditions of capital accumulation. State intervention at the macro level after 1965 was therefore mainly directed (through sanctions-busting) at maintaining these conditions, and at the same time promoting through import substitution, the viability of Department II (b) (luxury and durable consumer goods), in order to stem the tide of white emigration.

The immediate economic impact of the imposition of international economic sanctions on Rhodesia after 1965 was to reduce the real rate of economic growth. But this was only temporary, as the economy "took off" in 1967 and grew rapidly at an average real rate of growth of 7.5 percent until 1974; the relatively high growth rates can be attributed to utilisation of unused capacity in the country following the break-up of the Central African Federation, and also state import substitution policies.[15]

In the years 1965-74, it should be noted that the real growth in GDP during the decade was unequally distributed between individuals and social groups. It has been estimated that total white personal incomes grew from $369.1 million in 1965 to $925.6 million in 1974, whilst the European population rose by 31 percent in the period. Therefore, white incomes per head increased by 40 percent, to a level of $3,062. By contrast, while black personal incomes also rose in per capita terms during the decade by 41 percent, the population rose more sharply than that of whites (35.9 percent) and African per capita income in 1974 was $72.7.[16] But beginning in 1974, the economy received powerful exogenous shocks, in the form of the world economic recession that had been triggered off by huge increases in the price of oil. The rise in import prices relative to export prices brought about a deterioration in the country's terms of trade, which declined from 85.1 in 1973 to a record low of 74.3 in 1976 (1964 = 100).

The state was able to facilitate swift manipulation of such factors as import quotas, resource allocation and prices in order to blunt the short-term effects of the recession. For example, the rate of inflation as measured by the "European" consumer index increased by an average of 2.9 percent a year between 1965 and 1973; for 1974 and 1975, the average annual increase was 7.6 percent. Between 1975 and 1976, the economy declined at a rate of 1.4 percent and further declines of 7.4 percent and 3.1 percent were recorded in 1977 and 1978 respectively.[17] The gross domestic product between 1975 and 1978 fell in real terms by 12.1 percent and in per capita terms the decline was over 20 percent, meaning that living standards were cut by a fifth.[18] Much of the real improvement in the standard of living, which in any case had been unequally distributed between racial groups and classes, was wiped out. Further, the total number of workers in formal wage employment, which had reached a peak of 1,055,000 in 1975, declined continuously between 1976 and 1979: it fell by 65,000 or 6.2 percent of the peak level in 1975.[19]

In considering the economy of colonial Zimbabwe between 1965 and 1980, it is therefore possible to identify two periods: the decade 1965-74, when the economy grew, and the half decade following when it went headlong into decline. Though exogenous "shocks" such as the world recession were partly responsible for the decline, more fundamental were the structural incongruencies of the economy. In the words of one observer:

> In crude terms the economy was like an automobile stuck in low gear, slowly moving forward at full engine acceleration. The economy during the first decade of sanctions moved forward (i.e. expanded) and had the general confidence of the European community in Rhodesia. The fuel to drive the engine consisted partly of import substitution, use of excess industrial capacity and increased productivity. By 1975, the first two elements were just about exhausted and there were limits as to what the latter element could continue to achieve. New fuel had to consist of exporting secondary products, (e.g. consumer goods) increasing mineral exploitation and producing capital goods, all of which increased the need for foreign capital[20]

Within the broad trends in the economy outlined above, the share of agriculture in GDP fluctuated around a generally declining trend.

The most significant change in the country's agricultural commodity structure (as a result of sanctions) was the shift from tobacco as the major cash crop to grains, fibre crops, and the expansion of the livestock industry. Between 1965 and 1979, tobacco declined from 56 percent to less than 25 percent of agricultural sales, while livestock sales increased from 22 percent to 30 percent. An equally significant structural change which emerges from Table 8.1 is the growth in export crops, such as coffee and cotton, which as the Chairman of the Agricultural Marketing Authority wrote, "gave a greater diversification to the mix of export commodities and were easier to mask (in the sanctions context) than tobacco".[21]

Impressive gains were also made for wheat, sorghum and crops directly consumed by high-income groups or crops (e.g. sorghum) consumed by livestock and poultry which in turn are sold to high-income groups. Maize, the staple food crop, underwent substantial declines in levels of output after 1975. In the 1974-75 season, 1,337 million tons had been delivered to the Grain Marketing Board; this fell to 877,000 tons in the 1978-79 season, which represented a decline of some 34 percent. This was despite the fact that the area devoted to "commercial" maize production had progressively increased from 175,800 hectares in 1965 to 298,700 in 1972.[22]

TABLE 8.1
Trends in Pattern of Output
(Value of Sales)

	1965		1979	
	Value in $Millions	Percent of Total	Value in $Millions	Percent of Total
Tobacco	70.0	56.2	92.0	24.7
Maize	9.7	7.8	30.0	8.3
Sugar	12.0	9.7	33.5	9.0
Wheat	0.14	*	18.1	4.8
Cotton	0.96	*	54.9	14.7
Soya-Beans	0.02	*	11.9	3.2
Coffee	0.02	*	10.6	2.8
Cattle Slaughterings	19.5	15.7	86.7	23.2
Pigs	2.9	2.3	5.4	1.4
Milk	5.3	4.3	21.5	5.8
Others	3.8	5.1	7.7	2.1
Total	124.34	100	373.1	100

* Less than 2.
Source: CSO.

There were also important changes in the structure of ownership of land during the 15 years of UDI. The trend towards land consolidation, first begun as a way of maintaining economic viability, was given added impetus during the war of liberation. Farming companies registered each year steadily grew in number; 139 were registered in 1974, with a nominal capital of $43 million.[23] This increasing incorporation of farming enterprises represented a departure from the "familial" pattern of ownership, and a move towards the penetration of agri-culture by agribusiness. By 1980, transnational corporations dominated the production of sugar, tea, maize, cotton and forestry products.[24]

The history of agricultural enterprise during the UDI years was however hardly one of success. Taking net farm incomes; in general terms, the years 1970-74 saw a steady improvement in farm incomes with net farm income at constant prices rising from $33 million to $101 million in 1974, the year of the oil and commodity price boom. From 1974 the position deteriorated; there was an overall output increase of 22 percent but the effect on net farm income was swamped

TABLE 8.2

Net Farm Income for the Large-Scale Agricultural Sector ($ Million)

Harvest Year	Total Output	Total Input	Gross Income Before Transfers	Drought Relief Tobacco Quotas[1]	Transfers Levies Interest Depreciation	Net Income After Transfers	Net Income at Constant Prices
1970	184.0	136.3	47.7	3.8	18.1	33.4	33.4
1971	226.0	152.2	73.8	-0.3	21.2	52.3	50.8
1972	250.9	165.6	85.3	-0.3	23.4	61.6	57.4
1973	626.9	182.5	80.4	9.3	25.3	64.6	58.0
1974	368.9	219.5	149.4	-0.2	27.8	121.4	101.4
1975	384.9	250.4	134.5	-	30.6	103.9	80.9
1976	415.1	273.4	141.7	11.6	34.6	118.7	85.8
1977	404.5	300.2	104.3	6.9	38.5	72.7	49.2
1978	430.1	320.3	109.8	-	41.7	68.1	43.3
1979	451.1	350.1	101.0	3.0	43.0	61.0	35.1

1. Tobacco farmers were entitled to repurchase tobacco quotas.

by a 59 percent increase in total inputs together with an annual inflation rate of 10 percent[25] (see Table 8.2).

Turning to indebtedness, from 1965 to 1972, it is estimated that agricultural indebtedness rose by 98 percent, and that only about one in seven white commercial farmers earned enough to pay income tax.[26] This was largely attributed to the droughts of 1968 and 1972, but downright inefficiency on the part of commercial farmers cannot be discounted, as the Secretary of the Agricultural Assistance Board reported concerning the drought of 1968:

> Weighed down by accumulated debt from a succession of adverse climatic seasons, surrounded by broken down and often obsolete machinery, and with the usual sources of credit no longer available to them, many farmers were faced with the bleak prospect of having to leave the land and seek other employment for which they probably had neither the aptitude nor training.[27]

Between 50 and 60 percent of gross farm indebtedness was represented by short-term borrowings. The short-term credit position for the agricultural sector over the period 1965-69 is set out in Table 8.3.

TABLE 8.3
Short Term Credit to the Agricultural Sector ($ Million)

March	Commercial Banks	Agricultural Finance Company	Cooperatives & Companies	Total
1965	28	9	11	48
1970	31	13	15	59
1975	57	23	39	119
1976	64	30	33	127
1977	67	39	33	139
1978	65	46	44	155
1979	63	41	33	137

Source: CSO.

Given the very high liquidity in the money market, the sharp reduction in short-term credit, due to the combination of factors,such as the war, but more importantly the weak financial state of the agricultural industry, agriculture was not considered a profitable avenue for investment.

Another indicator of the "failure" of commercial agriculture during UDI are the estimates for gross capital formation in agriculture which after rising up to 1975, then show a general pattern of decline (see Table 8.4).

TABLE 8.4
Gross Fixed Capital Formation in Agriculture
($ Million)

Category	1970	1975	1976	1977	1978
Buildings	3	4	6	5	4
Civil engineering works	10	24	20	16	17
Transport equipment	6	10	11	8	6
Plant/Machinery	4	15	14	16	15
TOTAL	23	53	51	45	42

Source: CSO, National Accounts.

It should be noted that the figures for transport equipment and plant/machinery which have a high import content probably represent replacement factors as distinct from new investment since the period under review was one of acute foreign exchange shortages, in which priority would have more likely been given to replacement rather than growth.

STATE PRODUCT PRICING AND MARKETING
AND CREDIT POLICIES 1965-80

The so-called "success story" of Rhodesia was not one of market liberalism as the state even before UDI had been actively and pervasively interventionist. Until 1939, the most conspicuous role of the state in the economy had been in agriculture, in such areas as providing loans, pricing, regulation of labour supplies etc. The Second World War led to special measures to mobilise resources and regulate the economy, and many of these were to become permanent. The Federation in the 1950s required heavy public investment in energy infrastructure and basic industries. But by far the most important period for the growth of state involvement in the economy were the years after UDI.

State control of the economy grew both in terms of the instruments available to policy-makers and in terms of the comprehensiveness of coverage and the incidence of usage. It has been estimated that the Gross Public Debt grew from $409 million in 1963 to over $600 million in 1971.[28] The imposition of economic sanctions prompted the development of internally organised strategies to offset the costs of sanctions and distribute those costs in accordance with a combination of political and economic considerations in a society under siege.

The extent of state involvement was heavily entrenched in the agricultural sector. Two broad objectives can be identified in the product pricing and marketing policies which were to be implemented. First, there was the need to keep a balance between domestic and world market prices of agricultural commodities; the domestic price was regarded as the level desirable for political and social reasons, i.e. maintaining the high living standards of settler farmers; secondly, there was the desire to avoid sharp changes in retail food prices on the home market so as to minimise inflation. But whatever trade-offs were made between these two broad objectives a paramount consideration was ensuring the reproduction of the state through contradictory strategies of maximising export realisations for agricultural produce and simultaneously entrenching inequality by maintaining the living standards of the settlers at levels sufficiently high to discourage their emigration. And as farmers, increasingly, came to be regarded, as not only food producers, but "protectors" of the country against the terrorist "menace", their subsidisation through easy credit and writing off of taxes etc., entered into the "efficiency considerations" necessary for the reproduction of the settler political economy. Very conservative estimates show that from absorbing 15 percent of the state budget in 1965-66, the war effort came to absorb a third of the budget by 1979-80. And if one takes into account the fact that none of the statutory bodies set up to assist the agricultural sector was able to show a profit for the entire sanctions period, it becomes increasingly apparent that in the trade-offs involved in evolving policy instruments for marketing, pricing and credit, the objective of achieving food self-sufficiency was purely subsidiary to one of reproducing the "apartheid" state.

State intervention in agricultural pricing and marketing has a long history in Zimbabwe, and right from the start, had been motivated by attempts to insulate the internal market from international markets. Following the collapse of prices for agricultural products during the Great Depression the state was under pressure to avoid measures which would have increased surpluses. The Maize Control Act of 1934 stipulated that all African-grown maize had to pass through approved agents of the Maize Marketing Board[29].

Following the outbreak of World War II and the consequent food shortages, Africans were encouraged to produce for the market, and in 1940 the Grain Marketing Act established Board agents in the producing areas who were required to accept delivery on the basis of a cash payment related to the prescribed grade prices. The range of controlled products was extended in 1950 to cover groundnuts and sorghum.[30] Between 1949 and 1953, 22 percent of the country's marketed maize was imported. And though maize then accounted for 50 percent of all African sales, it only represented 38 percent of production from settler farmers who were then cashing in on the tobacco bonanza. To further encourage maize production in remote areas by African producers regional pricing was introduced in 1957 in the form of a Transport Equalisation Fund administered by the Minister for Native Affairs.

But as a result of the growing "black market" in maize, the Grain Marketing (Transitional Provisions) Act, 1965, brought about a reversion to uniform pricing, as it was considered that:

> Apart from the question of whether it is equitable to require producers with naturally low transport costs to subsidise those not so favourably placed there is the fact that by artificially depressing producer prices in the primary marketing sector at points close to consumption centres, producers and traders are induced to circumvent official channels.[31]

The Grain Marketing (Transitional Provisions) Act, 1965, established a Maize Trading Fund to which moneys appropriated by Parliament relevant to maize marketing and to which losses arising from the maize trading account would be transferred.

After the imposition of sanctions, the need to segment markets, and centralise pricing in favour of white farmers was made more imperative. To assist agricultural diversification away from tobacco the state announced in November 1966 the payment of a subsidy to commercial farmers on their purchases in bulk of both nitrogenous fertilisers and diesel fuels.

And to protect commercial farmers from oversupplies emanating from "inefficient producers", the Grain Marketing Act of 1966 was passed. It aimed at controlling "illegal" sales of maize by Africans in urban areas. The country was divided into two areas for the purpose of marketing maize. Growers in Area A (Commercial Area) could only sell their produce within these areas to the Grain Marketing Board whereas an Area B producer (Communal Area) could trade without restriction in his area, but was compelled to sell to the Grain Marketing Board if his product was to be moved into an A Area. Using such a segmented market and a uniform pricing policy was aimed at achieving maximum protection for the commercial producer. The local selling price for maize fixed in 1961 was to remain static until 1975-76 and the producer price was calculated as the net realisation on local and export sales and was paid as an initial price, estimated in advance of the Grain Marketing Board's intake year which started on 1 May. This price was to be supplemented by additional payouts as actual realisation permitted. Export realisation was expected to become an important determinant to the producer (see Appendices 8.4 and 8.5). In 1969-70 the Rhodesia National Farmers Union concluded a maize price agreement which guaranteed a minimum producer price, and rebate prices for livestock feeding purposes.

In 1967, all pricing and marketing for "Controlled Products" were centralised under the Agricultural Marketing Authority, which assumed overall control of marketing boards, and had the task of recommending, in consultation with producer associations and the state, producer prices for beef, milk, maize, cotton, sorghum and groundnuts. The Authority became fully operational in 1971, and in 1972 a unified policy for all marketing boards was announced, the main features of which were that all the capital required for the acquisition of fixed assets would with minor exceptions be provided by the state; all existing state capital provided for the acquisition of fixed assets and all future capital provided by the government would become irredeemable; and no capital contributions would be required from the net trading surpluses. All this was meant to ensure that the producer received the best price possible.[32]

State pricing policy thus had far-reaching effects on the whole agricultural sector after 1965. Before UDI, regulated products had contributed only 35 percent of the total value of agricultural output.

By 1973, they contributed about 77 percent. This was initially because of the diversification away from tobacco which had been sold under a free auction system, and had contributed 34 percent to the total agricultural output. The change in its marketing therefore swung the balance heavily toward regulation. In 1973, only 23 percent of the country's agricultural output derived its prices from the "free" interplay of supply and demand, i.e. sugar, poultry, vegetables, pigs, tea and fruit.[33]

Estate crops like tea and sugar were the main beneficiaries of state pricing policy. In the case of sugar, after 1965, the sugar industry accepted a determined price for the internal market as a compensation for the "loss of markets" as a result of sanctions. The need for the state to regulate the marketing and pricing of these "estate"-type crops was obviated by the fact that, by 1965, transnational corporations in Zimbabwe had developed highly sophisticated marketing and selling operations.[34]

Price control for the regulated products assumed varying forms. It usually extended to the producer price only, or applied to both the producer and marketing prices, as in the case of maize.

Statutory bodies in their role as residual buyers were required by legislation to accept all produce that farmers delivered to them. Commercial farmers were therefore certain of having a buyer for all their produce at pre-announced prices. The state also had channels through which it could distribute subsidies when necessary and could in addition encourage production by means of the pricing system. Thus, a fixed price system was operated for beef cattle with prices being announced before the beginning of the calendar year. And in order to encourage the expansion of certain crops (e.g. sorghum in 1972) pre-harvest prices were announced. In times of drought and falling production, long-term guaranteed prices were instituted (for beef see Appendix 8.6).

There were a number of principles underpinning this method of pricing. The first was that one product was not to subsidise another. This was a result of pressures from local commodity associations which in negotiations with the Agricultural Marketing Authority insisted that the full net marketing realisations had to be returned to the grower of the product. The second was that, for those products which were in deficit within the country, the guideline was to be import parity or incentive levels to encourage local production, so that if prices were kept below world prices for long periods, producers had a strong case for arguing that their product be exported in order to earn the prices prevailing on the world market. Third, that, for those products which had an exportable surplus, local prices were to be related to export parity. This was done for fear of penalising consumers and to keep the inflation rate at low levels.

These principles however proved very difficult to put into practice, especially given the declining terms of trade for agricultural products after 1974. But even internally, regarding those products in which the country was self-sufficient, the emphasis on maximising the benefits to the white commercial producers incurred many losses in revenue for the marketing boards and the state.

The Grain Marketing Board, for instance, established whether a product was in deficit or not by preparing harvest estimates prior to the marketing year. But throughout the period this proved very difficult, especially in the case of maize produced in African areas, as was noted in the Grain Marketing Board's report for 1965:

> The problem of estimating African marketable surpluses of controlled products remains as intractable as ever and estimates still have to be based on indications rather than any form of measurement.[35]

In the 1965-66 season, based on its estimates, the Board announced that it would be unable to meet local demand from locally produced supplies, and yellow maize had to be imported. But by the end of the marketing season the Board found itself with a surplus, because of late deliveries, and there was congestion at its delivery points. As there was no government subsidy payable for locally produced maize, the losses incurred through the importation of yellow maize were met by the government—and the local selling price remained unchanged. In 1974-75, sales to the Grain Marketing Board again exceeded estimates by 29 percent,[36] and the Agricultural Marketing Authority reported:

> The subsequent marketing season of 1974/75 saw the Authority's marketing agencies handling the largest sales ever made by Africans of those products for which the Authority is responsible. Recent years and particularly 1972 and 1973, have been notable for the increased value of sales made by Africans but 1974 topped these by a considerable margin.[37]

The balance between the domestic price and export prices proved difficult to maintain. For products where the aim was to achieve self-sufficiency, and for which there was not expected to be an exportable surplus, the principle of paying producer prices at import parity rates was not adhered to. Towards the end of 1968 world prices for soya beans, wheat and groundnuts rose steeply, and from then until 1973 remained substantially above the local selling prices, despite increases in the latter in 1972 and 1973.

In the case of wheat and sorghum, incentive prices were offered to increase their production. But when world prices rose, the local selling price was increased and supplementary payments were made without a subsidy. But, as wheat production increased, a system was devised which allowed for the adjustment of selling prices and the regulation of bread prices. For dairy products, prices were above export parity despite the fact that supplies were in excess of local demand.

Further revisions in the pricing system were made, following the rise in world commodity prices in 1974, which had been triggered off by the increase in fuel prices. The Agricultural Marketing Authority was able to pay record producer prices for most products for the 1974-75 marketing year. But as far as commercial farmers were concerned, these prices were offset by the rise in input costs. The fall back in world market prices began early in 1975, and the announced producer prices for the 1975-76 season had to be revised downwards.[38] Those paid for beef however maintained an upward trend, and for 1974 achieved a 16.2 percent increase over the 1973 prices, and those for 1975 included a further 5 percent premium over 1974.

The maize price agreement which had guaranteed government support for a minimum producer price for commercial farmers was not reviewed and the state resorted to alternative means of assisting maize producers by increasing the local selling price which until the end of the 1975 season had remained static for 13 years. Rebates on maize used for stockfeeding were, however, retained.

There were further declines in commodity prices during the second quarter of 1977, then there was a slight recovery in 1978. The state, therefore, adopted a cautious approach to the guaranteed pre-planting price of maize to be sown during the latter part of 1977 in order to restrain production.

With the intensification of the liberation war, the distribution and marketing network was severely disrupted, and with the increasing concentration of people in urban centres, the Agricultural Marketing Authority was experiencing difficulties in meeting demand by 1978. Further, possibly as a result of the deadlock between the government and the commercial farmers, domestic self-sufficiency in maize was in serious jeopardy. But the winter crop for wheat exceeded domestic requirements by over 40 percent.[39]

In 1979, the government decided to carry out a review of the whole pricing regime that had been in operation since 1965. Traditional pricing procedures were abandoned in favour of a short period ''of intensive discussions covering all regulated farm products''.

For the beef industry, negotiations were completed in late 1978 for the introduction of a new price schedule. The price increases granted to beef producers in 1978-79 were the largest in the history of the industry. Maize producer prices were left unaltered but there were substantial increases in the pre-planting prices. A quota system was introduced for wheat and, in the absence of profitable export opportunities, the state curbed future plantings.

FARM CREDIT POLICIES

Agricultural Credit

While the manipulation of prices and markets by the state for the benefit of commercial farmers was an important element of agricultural policy, the greatest form of subsidy to them was in the form of loans. By 1974, Rhodesia could boast of having the cheapest agricultural credit in the world.

The history of agricultural credit in colonial Zimbabwe goes back to the setting up of the Land and Agricultural Bank of Southern Rhodesia in 1924, for the purpose of granting loans to white farmers. In 1947, the promulgation of the Land Bank Act resulted in a number of other state sources of agricultural credit being consolidated under the Land Bank. In 1963, the state again set up numerous funds to assist farmers.

After the Unilateral Declaration of Independence, in an effort to assist diversification away from tobacco, the Agricultural Assistance Board was set up. By 1968, partly as a consequence of the drought, the Board found itself committed to capital expenditure in excess of the provisions in its capital budget. It also found itself taking over many farms which were no longer able to operate. By June 1969, 99 farms were vested in the Board, and the Board itself was having great difficulties in staying above water, because many of the farms it took over had accumulated substantial unpaid arrears of short-term loans. The Board reported for the 1969-70 season:

> In many cases it became clear that only a succession of reasonably good seasons would enable the desired financial rehabilitation to be achieved... It became evident that a number of the assisted farmers had a tendency to be of the "fair weather" variety and in a few instances all confidence had been lost.[40]

In 1969, a statutory authority was set up to amalgamate all sources of agricultural credit provided by the state. The Agricultural Finance

Corporation, which began operations in 1971, absorbed the Farm Irrigation Fund, the Agricultural Diversification Scheme, the Tenant Farming Scheme, the Coffee Scheme and the Landowners Development Loan Scheme.

The Agricultural Finance Corporation was to place greater emphasis on the granting of medium- and long-term loans at low interest rates—6 percent per annum. Table 8.5 is a partial schedule of the purposes for which loans were granted from 1971-72 to 1979-80.

With the rapidly increasing input costs, crop production costs escalated almost fivefold between 1971 and 1980. As a partial remedy to this, the Agricultural Finance Corporation in 1973 made arrangements to ensure that assistance for crop production would be paid to marketing boards or tobacco floors to which crops if grown would have been sold. This meant that lending institutions such as the corporation had to recover part of their investment in crop production through the operation of stop orders registered against crop proceeds, thus restoring the creditworthiness of most farmers for the following season's requirements.

TABLE 8.5

Purpose for which Loans Were Granted by A.F.C. ($'000)*

Year	Purchase of Land	Erection of Buildings & Dip Tanks	Erection of Fencings & Security Items	Purchase of Implements	Crop Production Costs	Payment Bonds	Purchase of Live stock
1971-72	3,001	160	91	912	13,480	444	1,139
1972-73	3,272	195	34	866	13,792	340	1,244
1973-74	4,143	592	108	1,333	15,082	781	1,627
1974-75	6,302	991	87	2,642	23,317	498	3,159
1975-76	3,380	703	102	2,295	31,194	219	2,166
1976-77	4,694	576	197	2,849	38,447	1,915	2,596
1977-78	3,339	655	7,420	2,979	44,052	2,039	2,117
1978-79	3,725	1,234	418	3,779	54,577	1,588	1,872
1979-80	2,715	961	138	3,659	63,040	1,904	988

* Excludes Special Development Loans.
Source: Agricultural Finance Corporation.

After the opening of the north-eastern front by ZANLA guerrillas in 1973, "security" considerations became more important than the purely productive assistance being granted to commercial farmers, as the AFC reported at the close of the 1973-74 season:

...the Board, on being satisfied that the occupation of a particular farm was necessary in the interests of security of the area, relaxed its normal requirements regarding the applicant's own capital contribution.[41]

By 1978-79, the terms of trade for agriculture had deteriorated by nearly 9 percent of what they had been in 1975-76, and the Agricultural Finance Corporation accounts show that for 1978-79, its bad debt account was in a parlous state: no less than $1.75 million had to be provided from the year's income and expenditure account against anticipated losses in respect of loans guaranteed by the government.[42]

SELF-SUFFICIENCY OR PAUPERISATION OF HOUSEHOLDS?

In the foregoing sections of the paper we have seen that in colonial Zimbabwe there was a very marked disjunction between the internal market for food and the international market for commodities. While this is always present in the capitalist mode of production the logic of accumulation and expanded reproduction in labour-reserve economies like colonial Zimbabwe meant that because a well developed consumption-exchange regime was imposed on a deliberately less developed internal market for food production, peasant and worker households become pauperised: they were forced to purchase food commodities when retail market prices rose and consume the products of their household production when consumer prices were relatively low.

As a consequence of what can be termed the "statistical mind" of settler colonialism it is not possible to get precise quantitative data to demonstrate the fact of impoverishment, so that this part of the paper relies very heavily on analytical and conceptual deductions.

Pauperisation here is defined as:

a condition under which households cannot continue to reproduce their means of subsistence at a constant level when a particular level of consumption has been established on both nutritional and social grounds.[43]

From which it follows that "the replacement of direct food consumption by purchased food may alleviate pauperisation rather than aggravate conditions for subsistence farmers", as "purchased food consumption permits the total flow of annual consumption to increase even over the year.[44] And more importantly, under ideal conditions the development of an internal if not international market for food production and consumption should enable households to consume food that is produced under conditions where the average productivity of labour (because of large-scale production) and of land (due to better ecological conditions) is rising.[45]

All this should mean that the "development of an internal market for food should accompany the growth of commodity production, if households are not to be pauperised".[46]

But, as pointed out earlier in this paper, the peculiar form of dependent reproduction and the division of labour in the social capital characteristic of the settler political economy meant that local food markets were too little integrated, and the development of an internal market in food has not matched the growth of internationally determined specialisation in household commodity production.

The form in which official statistics for agricultural production in the so-called communal areas exist (Appendix 8.8) make it almost impossible to deduce either empirically or theoretically the distinction between directly produced forms of consumption and purchased forms of consumption, the assumption being that less marketed output means more directly consumed output.

This has been referred to as the "doctrine of fungibility":

> Fungibility is the extent to which a rise or fall in the availability of a "resource" can be treated as if it were a change in cash funds and thus converted swiftly, conveniently and without loss into a change in whatever activity maximised benefit.[47]

In Zimbabwe, both radical scholarship and the state subscribed to this doctrine of fungibility. In a seminal collection of readings, Palmer and Parsons brought together the concepts of peasantisation and underdevelopment to show the "historical contradictions in capitalist development first stimulating the growth of a peasantry to supply its foodstuffs and then proceeding to break up that peasantry in order to obtain its labour".[48] The widespread poverty of much of rural Central and Southern Africa was linked to the colonial undermining of rural self-sufficiency and the process of peasantisation. Underdevelopment, it was claimed, was not merely a matter of increasing economic distortion, dependency and subordination for the mass of the people, but also a matter of absolute impoverishment against previous standards of living. The problem of scale is not touched on, the assumption being that the production of occasional surpluses during the early decades of the century represented the same process of peasantisation as the full-time cultivation of an export crop like tobacco and cotton. Thus, for Zimbabwe, the cataclysmic decline of the Zimbabwean peasantry is chronicled: 1914 for Phimister, the slump of 1921-23 for Arrighi and the Great Depression for Palmer.[49] Despite these crises it appears amazing that African cultivators remained sufficiently resilient to respond to "market forces" once more in the 1950s and 1960s.

No Rhodesian Native Commissioner would have seriously questioned the goal of food self-sufficiency, and the so-called "two pyramid" policy in the 1930s under which the Land Apportionment Act was passed was predicated upon making "natives" self-sufficient in the rural areas. After the Second World War, the commercialisation of African agriculture was in fact encouraged and the Native Land Husbandry Act envisaged the ending of rural-urban migration by making the peasant self-sufficient.

And after 1965 there was a drive by the settler state for the achievement of a final apportionment of land on a racial basis by the enactment of the Land Tenure Act of 1969. The so-called development of "African Tribal Areas" showed an overwhelming concern with conservation, the underlying object being to ensure adequate food supplies for a rapidly increasing population within the lot allotted to Africans.[50]

SOME CONCLUDING REMARKS

The experience of Zimbabwe with regards to state involvement in agricultural production and the food question clearly demonstrates that while state "involvement" in itself cannot be a panacea, this should not lead one to propose the disengagement of the state. The peculiar conjunctural conditions for intensified state involvement after 1965, the "nature of the state" itself and the "apartheid" society state policy was geared to reproduce have to be borne in mind.

Zimbabwe inherited a state apparatus developed by and for the settlers. The whole problem of dismantling that state apparatus, or even utilising some of the policy instruments to effect a transition to socialism are likely to prove intractable. While the argument might be put forward that the policy instruments inherited could be utilised for the interests of the workers and peasants, this could in the long run be sidetracked by mere tinkering with race-discriminatory credit policies, leaving the much more fundamental question of land redistribution untackled.

With regards to the latter, an important legacy of the Unilateral

Declaration of Independence era is the heavy indebtedness of the commercial farmers, which has resulted in the overcapitalisation of land. And given that the "commercial" definition of "utilised land" is not presently applied in the real physical sense of utilisation but also incorporates the "market value", the acquisition of land by the state for redistribution could be seriously hampered.

And finally, for the purposes of effective agricultural planning, the policy instruments for pricing and marketing need to be reconsidered. Recently, accolades have been showered on the farmers for delivering 900,000 tons of maize instead of the Agricultural Marketing Authority's harvest estimate of 560,000. Meanwhile, the state has already committed itself to the importation of yellow maize from the United States of America.

NOTES

1. The World Bank, *Accelerated Development in Sub-Saharan Africa: An Agenda for Action*, Washington, World Bank, 1981. See also FAO, *Regional Food Plan For Africa*, Rome, UN, 1981; USDA, *Food Problems and Prospects in Sub-Saharan Africa*, Washington, USDA, 1981.
2. Mkandawire, T., "State Policies and Agriculture in Africa", CODESRIA Project Proposal, 1983 (mimeo.), p. 6.
3. Rimmer, D., "Domestic Food Supplies in British Tropical Africa", in Rotberg, R.I. (ed.), *Imperialism, Colonialism and Hunger in East and Central Africa*, London, D.C. Heath and Co., 1983, p. 159.
4. Organisation of African Unity, *The Lagos Plan of Action*, Addis Ababa, OAU, 1981.
5. Bates, R.H., *Essays on the Political Economy of Rural Africa*, London, Cambridge University Press, 1983, especially pp. 107-33.
6. For an extensive review, see Berry, S., "The Agrarian Crisis in Africa, A Review", Boston University (mimeo.), 1983.
7. Riddell, R., "Market Focus versus State Intervention", *Financial Gazette*, 17 June 1983, p. 4.
8. Mumbengegwi, C., "Some Observations on the Problems and Prospects of Socialist Agricultural Transformation in Zimbabwe", *Economic Quarterly*, Vol. 18, No. 14, 1983, p. 27.
9. Moyo, E.E., "A Preliminary Application of the F.A.O. Balance Sheet Methodology to the Zimbabwe Food Situation". Economics and Markets Branch, Ministry of Agriculture (mimeo.), 1980.
10. Agricultural Marketing Authority, *Economic Review of the Agricultural Industry of Zimbabwe 1980*, Harare, AMA, 1980, p. 3.
11. See Marx, K., *Capital*, Vol. II, London, Lawrence and Wishart, 1977, Part III, "The Reproduction and Circulation of the Aggregate Social Capital".
12. On the schema for dependent reproduction see Munck, R., *Politics and Dependency in The Third World: The Case of Latin America*, London, Zed Books, 1984, pp. 32-34.
13. Curtin, T. and Murray, D., *Economic Sanctions and Rhodesia*, London, Institute

of Economic Affairs, pp. 13-14.

14. For empirical demonstration of this in another context see Cowen, M.P., "The Commercialization of Food Production in Kenya after 1945", in Rotberg, *Imperialism, Colonialism and Hunger*, op.cit.

15. Government of Zimbabwe, *Annual Economic Review of Zimbabwe*, Harare, Government Printer, Ministry of Economic Planning and Development, August 1981 (Cmd. R.Z. 11-1981), p. 1.

16. Clarke, D.G., "Structural Trends Affecting Conditions of Labour for African Workers in Rhodesia", *Rhodesia Journal of Economics*, Vol. 10, 2, 1976, p. 68.

17. Government of Zimbabwe, *Annual Economic Review*, 1981, p. 11.

18. Reserve Bank of Zimbabwe, *Quarterly Economic and Statistical Review*, Vol. 1, No. 1, 1980, p. 11.

19. Government of Zimbabwe, *Annual Economic Review*, 1981, p.11.

20. Strack, H.S., *Sanctions, The Case of Rhodesia*, Syracuse University Press, 1978, p. 6.

21. Agricultural Marketing Authority, *Economic Review of the Agricultural Industry of Zimbabwe*, p. 6.

22. Ibid., p. 96.

23. McKenzie, J.A., "White Farmers and Government Agricultural Policy 1965-1974", undated mimeo.

24. Sachikonye, L.M., "The Agrarian Question in Zimbabwe: Transnational Corporations and Labour in Agrarian Capitalism", ZIDS Seminar, mimeo, 1983.

25. Agricultural Marketing Authority, *Economic Review of the Agricultural Industry*, p. 8.

26. Strack, H.S., op. cit., p. 95.

27. *Report of the Agricultural Assistance Board for the Year Ended September 1968*, p. 5.

28. Clarke, D.G., "The Growth and Economic Impact of the Public Sector of Rhodesia", *Rhodesian Journal of Economics*, Vol. 6, 3, 1972, pp. 48-49.

29. See Keyter, C.F., *Maize Control in Southern Rhodesia 1931-1941*, Salisbury, Central African Historical Association, Local Series No. 34.

30. Dunlop, H., "Efficiency Criteria in Primary Marketing. An Analysis of African Marketing Policies in Rhodesia", *Rhodesian Journal of Economics*, Vol. 4, 3, 1970, pp. 10-19.

31. *Report of the Grain Marketing Board of Rhodesia for The Year Ended 30 April 1965*, p.17.

32. Agricultural Marketing Authority, *Reports and Accounts for The Year Ended June 1972*.

33. Hunt, A.F., "Agricultural Pricing in Rhodesia", *Rhodesian Journal of Economics*, Vol. 7, 1973, p. 189.

34. Ibid., p. 193.

35. *Report of the Grain Marketing Board of Rhodesia for The Year Ended 30 April 1965*, p. 8.

36. *Report of the G Marketing Board for the Year Ended 1975*.

37. *Agricultural Marketing Authority, Reports and Accounts For the Year Ended 1975*.

38. Ibid. p. 9.

39. Agricultural Marketing Authority, *Reports and Accounts For the Year Ended 1979*, p. 8.

40. *Report of the Agricultural Assistance Board for The Year Ended 1970*, p. 2.

41. *Report of the Agricultural Finance Corporation for 1973-74, p. 4.*

42. *Report of the Agricultural Finance Corportion for 1973-74*

43. Cowen, M.P., "The Commercialisation of Food Production in Kenya after 1945", p. 204.

44. Ibid.
45. Ibid.
46. Ibid.
47. Ibid, p. 209.
48. Palmer, R., and Parsons, N., "*The Roots of Rural Poverty in Central and Southern Africa*, London, Heinemann, 1977, p. 2.
49. Phimister, I. "Peasant Production and Underdevelopment" in *The Roots of Rural Poverty in Central and Southern Africa*, pp. 255-67; *The Agricultural History of Southern Rhodesia*, in ibid. pp. 221-25; Arrighi, G., *The Political Economy of Rhodesia*, The Hague, Mouton, 1967.
50. Dunlop, H., "Development in Rhodesian Tribal Areas", *Rhodesian Journal of Economics*, Vol. 8., 1974, p. 182.

APPENDIX 8.1

Classification of Sub-Saharan Africa
According to Agricultural Production Performance, 1961-80

Total Agricultural Production	Increase	105	Food Production[1] Stagnant	95	105 Decrease	95
Strong Increase 130	Swaziland	(152,142)				
Increase 105 130	Burundi Cameroun Ivory Coast Malawi Rwanda	(111,111) (127,120) (116,122) (127,119) (128,127)				
Stagnant 95 105	Botswana Sudan	(102,112) (99,112)	Central African R. Gabon Sierra Leone Zambia	(99,101) (101,101) (99,99) (100,101)	Kenya Liberia	(99,91) (104,90)
Decrease 80 95			Chad Congo Ethiopia Ghana Guinea Lesotho Mali Mozambique Niger Nigeria Somalia Tanzania Upper Volta Zaire			(82,82) (90,90) (85,84) (83,83) (92,89) (87,94) (91,88) (81,85) (82,82) (87,88) (85,85) (90,93) (95,93) (90,91)
Strong Decrease			Zimbabwe (75,99)	Angola Gambia Mauritania Namibia Senegal Uganda		(70,94) (79,79) (74,74) (77,77) (71,70) (76,90)

1. The figures in brackets give the index for total agricultural production per capita followed by the index for food production per capita. They are based on 5-yearly averages for the period 1976-80, in comparison with the base period of 1961-65 = 100.

Source: FAO, *Production Yearbook, 1980*, Rome, 1980.

APPENDIX 8.2

Rhodesian Economic Indicators 1965-75

Indicators	1965		1975	% Increase
Population-African	4,260,000		6,110,000	43.4
Population-Non-African	230,600		308,900	33.9
Ratio-Non-African-African	1:18.47		1:19.78	
Total net European migration		-23,430		
Total net European migration				
(1966-70)		17,550		
Total net European migration				
(1971-75)		22,430		
Tourists from other countries	208,725		244,404	17.1
Total African school enrollment				
(year)	654,241		868,689	32.8
African employees	656,000		945.000	44.1
Non-African employees	89,700		119,700	33.41
European Consumer Price Index				
(1964/100)	101.7		149.22	
Sales of principal crops &				
Livestock (m)	124.2		305.4	46.7
European-owned cattle (m.)	1.63		2.78	70.6
Mineral output (m)	63.9		169.8	165.7
Index of volume of manufacturing;				
Production (1964/100)	108.7		204.3	87.9
Building plans approved				
(m. 1966-75)	17.7		82.5	365.0
Electrical energy consumed				
(m.kw. hrs)	2,028.9		6,136.4	202.4
Rhodesia Railways-net tons				
hauled (m. tons 1968-75)	9,886		12.800	29,5
Retail Trade Value Index	100.0		266.5	166.5
Imports (m.1965-72)	239.6		274.2	14.4
Exports & re-exports ($m. 1965-72)	309.2		328.5	6.2
Terms of trade (1964=100)	100.8		79.9	-20.7
Gross national income ($m., 1965 prices,				
1965-74)	722.2		1,352.6	87.3
Gross fixed capital formation ($m.)	100.0		400.0	300.0
Gross public debt ($m., central & local				
governments)	557.0		896.1	60.9
Rhodesian Stock Exchange Index	118.3		375.1	217.1

Source: Rhodesia, Central Statistical Office, *Monthly Digest of Statistics*, March 1973 and April 1976.

Notes: $m. = million Rhodesian dollars. 1972 was the last year that import and export statistics were made public by the Rhodesian government.

APPENDIX 8.3
Zimbabwe: Consumer Price Index, 1964-80 (1964=100)

Year	Higher income all items	% change	Lower income all items	% change	Average rate of price increase%
1964	100.0	-	100.0	-	-
1965	101.7	1.7	102.5	2.5	2.1
1966	104.3	2.6	105.7	3.1	2.9
1967	106.4	2.0	108.2	1.4	1.7
1968	108.7	2.2	109.7	2.3	2.3
1969	111.7	2.8	110.1	0.4	1.6
1970	115.6	3.5	112.4	2.1	2.8
1971	119.1	3.0	115.8	3.0	3.0
1972	124.3	4.4	119.1	2.8	3.6
1973	128.8	3.6	122.8	3.1	3.4
1974	138.5	7.5	130.9	6.6	7.1
1975	149.2	7.7	144.0	10.0	8.9
1976	162.6	9.0	159.9	11.0	10.0
1977	178.2	9.6	176.4	10.3	10.0
1978	190.0	6.6	193.6	9.8	8.2
1979	211.5	11.3	220.1	13.7	12.5
1980	231.0	8.4	232.0	5.4	6.9

Source: CSO, *Monthly Digest of Statistics*, April 1983.

APPENDIX 8.4

Maize: Producer and Selling Prices (Per Tonne)
Producer Price delivered in bulk or bulk equivalent
Selling Price[3]

Marketing Year	Pre-Plant-ing Price	Prescribed Producer Price by Grade					Ex GBM Depot
		A	B	C	D	Sup-plementary	
1959-60[1]		39.68	39.13	38.03	-	-	4.17
1960-61[1]		34.72	34.17	33.62	-	-	4.17
1961-62		26.07	25.52	24.72	-	-	4.17
1962-63		29.57	29.02	27.92	-	-	4.17
1963-64[2]		31.92	31.37	30.26	-	8.93	4.17
1964-65		35.79	31.37	30.26	-	3.30	4.17
1965-66		34.04	33.48	32.37	-	3.66	4.17
1966-67		28.76	28.20	27.65	12.77	2.45	4.17
1967-68		29.13	28.57	28.02	12.58	-	4.17
1968-69		32.53	31.97	31.42	14.05	0.92	4.17
1969-70		30.97	30.42	29.87	14.44	1.97	4.17
1970-71		32.97	32.41	31.86	21.38	5.46	4.17
1971-72		30.05	29.49	28.94	19.46	2.46	4.17
1972-73		25.88	23.33	24.78	18.19	4.50	4.19
1973-74		36.37	35.82	35.27	24.95	1.80	4.19
1974-75		40.11	39.56	39.01	28.68	3.40	43.24
1975-76		35.00	36.45	35.90	25.57	11.25	51.54
1976-77	42.00	44.00	43.45	42.90	32.57	4.00	51.54
1977-78	46.00	52.00	51.45	50.90	40.57	-	51.54
1978-79	48.00	53.00	52.45	51.90	41.57	-	57.07
1979-80	56.00	60.50	59.90	59.25	52.55	-	63.00
1980-81	75.00	85.00[4]	84.15	83.25	73.80	-	89.00
1982	120.00						

Notes
1. Free bags issued to producers.
2. Introduction of bulk delivery incentive 1st May 1963.
3. Up to 1973-74 prices relate to 91 kg. bag. 1974-75 onwards relate to bulk price per tonne.
4. In addition bonus payments of $5/tonne for increased plantings of 15% in area and EDB of $10/tonne up to 30 April and $5/tonne up to 31 May.

APPENDIX 8.5
Maize: Marketing Year, Supply and Disappearance
('000 tonnes)

Intake Year	Opening Stock	Supply			Disappearance			Closing Stock[1]
		GMB Purchases	Imports	Total	Domestic	Export	Shrinkage	
1959-1960	187	241	-	241	214	87	-	127
1960-1961	127	252	-	252	238	33	-	108
1961-1962	108	495	-	495	161	348	2	94
1962-1963	94	335	-	335	163	261	-	5
1963-1964	5	237	35	272	172	85	-	20
1964-1965	20	252	24	276	253	13	1	29
1965-1966	29	337	45	382	347	4	1	59
1966-1967	59	525	15	540	193	808	-	97
1967-1968	97	876	-	876	193	681	-	99
1968-1969	99	424	-	424	315	182	1	25
1969-1970	25	961	-	961	227	673	1	85
1970-1971	85	628	-	628	410	243	1	59
1971-1972	59	1,112	-	1,112	310	717	2	142
1972-1973	142	1,400	-	1,400	232	891	7	412
1973-1974	412	550	-	550	445	356	3	158
1974-1975	158	1,337	169[2]	1,506	371	879	6	406
1975-1976	408	1,007	-	1,007	387	758	7	263
1976-1977	263	959	-	959	393	297	2	530
1977-1978	530	941	-	941	504	420	9	538
1978-1979	538	877	-	877	545	554	6	310
1979-1980	310	512	149[2]	661	635	265	7	64
1980-1981[e]	64	815	95	910	723	86	3	162

Notes

e = Estimates

1. Figures prior to and including 1963-64 are derived.
2. Includes external purchases to meet export commitments. True export figures should be deflated (1974-75 all external purchases 1979-80 65 000: External purchases).

APPENDIX 8.6

Cold Storage Commission Producer Prices, Local Wholesale Sale Prices, Net Export Realisation

Year	Producer prices c/kg			Local av. wholesale price c/kg[1]	Export realisation c/kg[2]
	Basic price excl. condemned	Additional payouts	Final payment excl. condemned		
1965	30.18	-	30.18		37
1966	33.99	-	33.99	30.80	37
1967	35.22	0.27	35.49	32.14	36
1968	35.04	1.16	36.22	33.84	39
1969	35.44	0.50	35.94	33.57	39
1970	35.23	0.43	35.76	33.92	36
1971	36.16	0.60	36.76	33.97	35
1972	38.38	2.00	40.38	34.97	38
1973	41.17	7.64	48.81	37.41	48
1974	49.62	7.20	56.82	41.47	66
1975	54.87	4.09	58.96	44.82	57
1976	53.79	3.21	57.00	47.42	48
1977	54.65	3.26	57.91	47.66	40
1978	54.03	3.23	57.26	51.42	38
1979	70.46	-	70.46	59.39	56
1980	81.11	-	81.11	63.01	53

Notes
1. Sides forequarters only. Gross wholesale price before allowing for cash discount.
2. Realisation ex-slaughter plant and includes from 1970 onwards the realisation on the local disposal of export residue (bone and fat).

APPENDIX 8.7
Agricultural Output

Year	Communal Lands			Commercial Areas		
	Sales through Marketing Agencies	Production for Rural Household Consumption	Total	Gross (at current prices)	Index of Gross Output (at 1964 price)	Index of Producer Prices
$ million	$ million	$ million	$ million	$ million	1964 = 100	1964 = 100
1964	6.4	26.2	32.6	143.1	100.0	100.0
1965	6.1	25.4	31.5	145.0	93.2	108.3
1966	7.2	40.4	47.6	150.6	113.4	90.8
1967	8.3	57.6	65.9	153.1	117.7	88.5
1968	5.6	38.4	44.0	141.5	97.4	98.3
1969	10.7	52.7	63.4	179.2	129.6	93.2
1970	8.9	44.4	53.3	175.7	122.3	98.8
1971	13.0	58.9	71.9	217.6	154.9	97.4
1972	20.1	62.7	82.8	241.6	168.0	97.4
1973	18.4	44.6	63.0	252.9	140.1	125.6
1974	26.3	84.9	111.2	356.0	174.7	142.2
1975	26.6	68.6	95.2	371.3	169.4	153.4
1976	28.2	80.2	108.4	401.1	182.7	152.6
1977	22.0	84.0	106.0	392.5	168.8	163.7
1978	22.5	51.8	74.3	418.2	167.1	178.0
1979	16.9	85.5	102.3	438.5	164.8	190.3
1980	28.9	117.1	146.0	591.6	185.2	231.4
1981	79.5	185.1	264.6	775.6	271.6	232.5

9. STATE POLICIES ON AGRICULTURE AND FOOD PRODUCTION IN MALI

Kary Dembele

Mali is a landlocked country situated in the Sudano-Sahelian zone. Its vegetation consists of savanna woodland and grassland thinning out progressively toward the Sahara. In relief, the land forms a series of lateritic plateaux rising to heights ranging between 300 and 400 m. In the desert areas there is practically no vegetation cover at all. But a modest amount of rain, between 300 and 500 mm, falls annually in the semi-desert Sahelian zone, which stretches over some 200,000 km² in the Niger Bend and the Central Delta regions. Here the vegetation is sparse, with thorny scrub most in evidence. The zone is mainly suitable for animal husbandry.

The remainder of the territory falls into two zones: the Central or Sudan zone, with annual rainfall totalling 600 to 1,000 mm, and is covered by dry grassland; the Southern or Guinean zone, where annual rainfall averages 1,300 mm, is an area of humid grassland. These two regions are suitable for both animal husbandry and crop farming, with millet, rice, maize, cotton and groundnuts thriving in the southern parts. Almost all the country's arable land is in this region. Land use distribution is as follows : fallow, 22 percent; crop farming, 5 percent; forests, 3 percent; pasture, 70 percent.

The dominant feature of the climate is the annual swing from the dry season (November to May) to the rainy season (June to October) during which humid winds blow in from the Atlantic Ocean.

Two great waterways run through the country: the Senegal River flows into the Atlantic, while the Niger, after a long northward trajectory, turns south and pours its waters into the Gulf of Guinea.

Not much is known about the population of Mali. General census data published in December 1980 indicate that in 1976 the country had a resident population of 6,394,918. Given a growth rate of 2.6 percent, Mali's population would have totalled over 7 million in 1980. Of this total 83.2 percent live in the rural areas. Population density ranges from 0.7 inhabitants per km² in the northern region of Gao to 15 inhabitants per km² in the region of Segou. Average life expectancy in Mali is about 38 years. Nearly half the population (44.3 percent) is less than 15 years old. These statistics say a great deal about conditions of demographic reproduction in Mali. They are particularly eloquent as to the efficiency and coverage of the country's health system. An incidental note : according to the latest census, out of every 1,000 Malians, 30 suffer from various disabilities, the severest being blindness, leprosy, sleeping sickness, tuberculosis and river blindness. There is also malaria, responsible for a substantial drop in the productivity of a large part of the rural population. This state of affairs is not new. At least in part, it can be traced to the country's colonial heritage. In the wake of independence, the state pledged to change the condition of the peasantry as inherited from the colonial process, by means of a new agrarian policy. To achieve this aim it decided on the implementation of a five-year development plan assigning a significant status to agriculture. In the event, implementational problems cropped up before the end of the plan period, and the original plan was replaced by a series of policies, each presented, at least in rhetoric, as peasant-oriented.

In the implementation of these successive policies, three phases may be distinguished. The first phase covers the period from independence until 1968; the second phase covers the decade thereafter; the third phase begins in the late 1970s.

FROM AGRICULTURE-ORIENTED INVESTMENT PLANNING TO PEASANT REGIMENTATION: 1961-68

The Abortive Plan

According to the provisions of the first five-year plan, the lion's share of investment resources was to be allocated to the agriculture and transport sectors. As shown in Table 9.1, these two sectors were expected to absorb almost two-thirds of total investment expenditure. This gives an idea of the government's original intentions, to wit: to work towards improved integration of the rural and urban milieux, to provide the country with outlets to the outside world, and to stimulate agricultural capacity so as to enable that sector to help improve peasant living standards as well as to provide the country's urban population with ample supplies of cereal.

In the breakdown of investment allocations for agriculture, it was planned to give pride of place to land improvement as well, to a lesser extent, as to the supply of production resources to peasants.

TABLE 9.1

Productive Investment forecast under the 1961-66 Five-Year Plan

Agriculture	Industry	Hydro-electric dams	Geographical prospection	Transport etc.	Total
38	21	6	12	22	100

So much for good intentions. Coming down to realities, we find that the stated objectives were scarcely ever attained. Worse still, the plan, which drew three-quarters of its funding from foreign sources, resulted, even in that portion of it that was in fact implemented, in a consolidation of options and orientations inherited from colonial days. As Table 9.2 makes clear, the Office du Niger, which was nationalised in 1960, swallowed up the bulk of agricultural investment, yet accomplished no fundamental change at all in the settler-type agricultural system under which extension supervision was provided for some 57,500 peasants on an area of about 55,000 hectares.

TABLE 9.2

Actual Investment as a Percentage of Projected Investment

	Land improvement	Extension staffing	Equipment	Forestry	Animal husbandry	Office du Niger	Total
Projected Investment	57	16	18	3	6		100
Actual investment	19	?	17	13	2	49	100
Actual investment as % of projected investment	16		46	38	97		48

The initial plan, in effect, was agriculture-oriented. But in short order, the new state turned the first five-year plan upside down in the process of implementation. A lot of plan funds were spent on such unproductive projects as the construction of a National Assembly building and a presidential palace. Thus, the funds provided for equipping the agricultural sector and the rural areas were diverted away from the peasant population.

The Establishment of Extension Structures

In marked contrast to the lack of investment in agricultural equipment, a different type of investment did occur : an over-abundance of administrative structures were set up for the purpose of providing extension services for peasants. Technical extension services for peasants under the USRDA, the ruling party's umbrella consisted of four structures. At the bottom was the Basic Sector (BS), operative at the village level and among village associations ; at the district level there was the Rural Expansion Zone (ZER) as well as the Rural Development Service (SDR) at the level of the administrative area; at the regional level there were the Economic Services; finally, at the national level came the Rural Action Services.

The basic sector (BS) covered about 2,000 to 3,000 inhabitants, in other words 150 to 200 family farms, who cultivated 1,000 to 1,500 hectares of land. The objective of the basic sector was to maintain a *close relationship* between peasants at the village level and technical agents who actually supervised family farms. The BS was in charge of hydraulic development. It was oriented toward the intensification of industrial crops.

The BS was led by an instructor, assisted by "rural group leaders" who were not paid. These "rural group leaders" who were trained in what were called here "seasonal schools" were peasants who were supposedly open to progress and to modern agriculture. Their duty consisted in *propagating progress in rural areas*.

At the village level, cooperatives or pre-cooperatives were set up to allow peasants to acquire equipment either individually, or collectively. To this end they could borrow from agricultural credit institutions. These pre-cooperatives were allegedly intended to facilitate the acquisition of inputs by peasants and to supply them with basic commodities (salt, sugar, oil etc.). Thus they could avoid falling victims of speculators and other middlemen, remnants of the cash crop economy in the colonial period.

At the district level existed the rural expansion zone (ZER) defined as being a technical, economic and administrative unit. In practice, however, there was a confusion between 7 these functions. Furthermore, the political role prevailed over all these functions. The ZER covered a district consisting of about four BSs and approximately 10,000 to 15,000 people.

The Cooperative Training and Agricultural Modernisation Centre (CCEMA) was set up at the district level. It undertook various activities: it bought and sold wares as well as agricultural goods on behalf of GRAs;[1] it transported into the district wares needed by peasants ; it undertook farming operations—crop processing and technology; it also conducted farming experiments and showed peasants how to use new techniques and equipment. The CCEMA also produced and distributed selected seeds to peasants and organised workshops and discussions on ways and means to modernise agriculture. Finally, it was assigned a social mission—teaching peasants how to read and write, introducing them to first aid, organising collective discussions and radio listening sessions as well as sports and folkloric events.

SDRs were set up at the level of the administrative area and headed by rural development leaders, agronomists who also acted as technical advisers to "circle commanders" for matters related to rural development programmes. ZER leaders were supervised by the head of the SDR.[2]

The cooperative was represented at the level of the administrative area by SMDRs[3] whose terms of reference were defined by Law no. 60-8 of 9 June 1960. SMDRs ran GRPSMs,[4] basic cooperative structures, through GRAs and their primary federations.

SMDRs also managed primary funds which had to be set up in each administrative area for agricultural credit. They only dealt with the cooperative aspect in rural areas. They did not handle technical activities in rural development programmes nor did they equip the rural areas in any way.

Consequently SMDRs, as relays of SIPs[5] which existed during the colonial period, inherited the colonial mentality, therefore remaining bureaucratic parasites living off the exploitation of peasants.

At the regional level, there was a governor who was appointed by the Head of State. He was in charge of the implementation of regional economic and social development programmes. He ran such departments as rural development, livestock breeding, forestry, trade, tax etc. Each region had its own "farm" which undertook applied research, experiments, demonstrations of rural modernisation and selected seed growing, and to each farm was attached an agricultural learning centre to train agricultural instructors and later "advanced" farmers.

At the national level there was a Division of Rural Development which was placed directly under the control of the Minister of Planning.

This division was in charge of rural extension, popularisation of agricultural equipment and strains. It also managed the Central Agricultural Credit Fund which extended loans to rural groups (GRSMs) and rural cooperatives.

At the national level, there was also a National Cooperation Centre which undertook research on cooperation laws, advertised, trained senior staff, monitored the activities of SMDRs and finally oversaw the organisation and marketing of products purchased by SMDRs,

GRAs and GRPSMs from peasants.

The ruling party (USRDA) tried to use village "tons", one of the traditional farming associations, for every activity involved in the mobilisation of the rural labour force and in the supervision of seeds. These village "tons" were traditional farming associations of 10- to 30-year-old youth. They rented out their labour force on Mondays and Fridays (sacred days on which extended families did not undertake any heavy agricultural task) to farmers who had limited means (insufficient manpower, some farmers may suffer from malaria, onchocercosis etc.).

This service was usually paid for in kind (goats, millet etc.). At any rate such an income was used to finance the annual feast which was organised by the youth, usually in May, that is, prior to the beginning of the rainy season.

The USRDA *compelled* farmers to cultivate the so-called collective farms (*fama foro*: the *fama's* farm that is the king's, the emperor's, the political power's, the state farms) on Mondays and Fridays, instead of cultivating their own farms as they usually did on such days, after working on extended family farms, as we have indicated earlier. As a result, the farmers were torn between work for "tons", or work on collective or individual farms. If a farmer chose to work for "tons" on Mondays and Fridays, he *was taxed with being unpatriotic*, he was even sent to prison. However, if he devoted Mondays and Fridays to work on collective farms, he did not receive any income, for this income was in most cases embezzled by the instructor, the head of the ZER.

The situation gradually changed over. Each local political leader—heads of districts having replaced the former heads of townships—exploited "tons" as sources of a cheap wage-earning labour force and even as labour without pay.

Consequently, "ton" members, who were often young, preferred to leave "tons" or chose to migrate to the cities in an attempt to escape the political constraints of a party which gradually proved unable to solve either the problems of the Malian peasants or the country's economic problems in general.

Under such circumstances, the rural populations were inevitably reluctant to give support to the cooperatives which, additionally, lacked technical means as well as loans. These cooperatives, which were mismanaged—even the most basic accounting system was non-existent—and lacked transportation as well as other logistical means, were run and monopolised by a social group which had nothing to do with the peasantry. This social group was made up of traditional chiefs, and civil servants running the village and technical assistants.

"Collective farms" which, supported by an age-old custom, were to constitute the cornerstone of Malian socialism, were forcefully imposed. As a result, they yielded on the average, *30 percent less* than farms on which no technical assistance was provided. This clearly shows how reluctantly peasants participated in a farming system which they did not choose. Furthermore, the economic incentives which were to change the behaviour of these small-scale farmers were not granted. *Prices of agricultural produce, which were frozen* until 1965 so as theoretically to extract accumulated funds in order to finance industrial development, led to *lower production and to the development of black markets* which were in the hands of private traders.[6]

It is now easy to understand why peasants, disappointed and even hankering after the colonial period, made the following statement on the impoverishment they experienced after independence: "Life is said to be pleasant on this earth but *lady mouse* is not aware of this for, when it comes out of its burrow, there is a club threatening its head and when it wants to enter it again there is still another club threatening its head." In the first case, it was the club brandished by colonial rule and its cash crop economy. In the second case, it was the club held by what Jacquemot referred to as the new class of profiteers. Although these peasants lived in abject poverty and suffered from hunger, they could not help but express their sense of humour when, after watching the lifestyle of those ruling the country—who were sometimes known as the "newly rich"—they ironically added the following: "together we hunted and killed the big game but now a minority of those hunters wrap themselves around with the game skin to scare us."

Subsequently, producers opted for passive resistance—they refused to market their surplus production which they concealed either in the village or in their families, sometimes even in neighbouring villages etc.—and for indifference vis-à-vis the USRDA etc.

Sometimes they decided to fight openly, but these struggles hardly led to meaningful social changes. They were no more than peasant uprisings which often shook both the party and the state apparatus in various localities. However, these struggles finally sped up a situation which culminated in 1968 in two events, the constitution was left in abeyance and changes were made in the ruling state apparatus.

EMERGENCE AND IMPLEMENTATION OF "DEVELOPMENT UNITS" TO OVERCOME THE CRISIS

Three-Year Economic Recovery Plan

After what was known as the November 1968 events, Mali worked out a three-year plan for financial and economic recovery. The main target of this plan was to increase production and thus re-establish two basic economic balances, the balance of trade and the state budget.

According to the government, this plan led to the following results:

1) A 4.75 percent average yearly growth rate of the GDP in the 1969-72 period, against the projected 5.4 percent growth rate. In previous years it had registered a 2.35 percent growth rate. During the three-year financial and economic recovery plan, the per capita GDP had therefore recorded a 2.25 percent increase per annum.

2) The deficit in the balance of trade fell sharply from MF 8.8 billion in 1968 to MF 3.8 billion in 1971. It increased again in 1972 to MF 8.6 billion. It seemed that the new increase was due to the import of capital goods. (Though the nature of these capital goods was nor officially specified: it was just stated that capital goods accounted for 34 percent of imports in 1972 compared to only 20 percent in 1969.) Generally speaking Mali's external trade is not very diversified either in terms of its products or its partners. Cotton, live animals, their by-products and dried fish alone accounted for 80 percent of Mali's total exports in 1971. About 65 percent of Mali's exports were absorbed by France. (But it was not mentioned that this situation is quite different from what it was previously. Indeed about 75 percent of Mali's trade was effected with East European countries.) African inter-regional or intra-regional trade climbed only from 12 percent in 1960 to 18 percent in 1972. It is interesting to note that to avoid speaking of Mali's external dependency, the following statement was made: "the Malian economy remains (therefore) strongly affected by the fluctuations in the world price of a few basic commodities and by the changing market of its main commercial partners."

3) The recovery plan provided for MF 77.5 billion of investments, about 15 percent of which were to be funded locally. Actual expenditure amounted to MF 58 billion, 46 billion of which were financed by external resources, that is 75 percent of the total. It was officially recognised that the economic and financial recovery plan was not quite a success; however, its failure was attributed to the drought. During the implementation of the economic and financial recovery plan, Mali experienced three years of drought, mostly as a result of a very scant, uneven distribution of rainfall, the low level of the Senegal and Niger Rivers and a sharp fall in the level of ground water. Consequently, wells almost dried out and it became a crucial problem to supply populations and livestock with water.

The sharp decline in the level of rainfall had dire consequences on the country's economy especially on the rural sector. Cereal production (mainly millet and rice) fell short of projections by 400,000 tons in 1972-73. Cereal stocks being estimated at 600,000 tons, this figure was at least 300,000 tons short of national food requirements which amounted to about 900,000 tons.

Cash crops such as cotton and groundnuts were also affected and at best, the level of production achieved in 1972-73, almost equalled that of preceding years.

In terms of stockbreeding, overall loss amounted to at least 30 percent of total livestock. This could be explained by the fact that the country's 6th region lost about 80 percent of its cows which represented one-third of the country's overall cow population.

Other sectors such as fishing and health were also seriously affected by drought. A number of diseases broke out while others caused by malnutrition and food deficiencies (especially in refugee camps) became widespread.

Interestingly, drought made it very hard for peasants and stock breeders to reimburse their agricultural loans and to equip themselves properly; hence an increase in rural migration to the cities.

Political authorities also contended that state assistance to the populations increased government expenses (but no assessment of these expenses was made). Another consequence of the drought was an increase in the budgetary and external trade deficits. These are the main official conclusions on the results achieved by the Three-Year Recovery Plan which admittedly was a failure. Another Five-Year Plan (1974-78) was then drawn up.

The Five-Year Plan: Priority to Food Production

Among the priorities defined for the various sectors of the economy, the Five-Year Plan (1974-78) gave priority to increased food production. Agriculture, forestry, livestock breeding and fishing were the main beneficiaries of the plan's investments. Emphasis was also laid on the secondary sector which was to be reinvigorated and whose share in gross domestic production was expected to increase from 15.6 percent to 20.8 percent.

The main objectives assigned to the plan were:

> to satisfy the basic needs of the whole population especially in terms of cereals and water. The Plan stated that "the cereal requirements must be met by national production under the implementation of this Five-Year Plan".
> to rebuild and expand the combination of livestock and agriculture;
> to enhance national agricultural products by developing agro-industry;
> to pursue the construction of access roads in order to put an end to the country's isolation;

to consolidate economic independence through reduced deficit in the balance of payments, increased productivity of State-owned companies and enterprises, the participation of more nationals in development operations and in the economy;

to meet social needs including food balance, health and education (right to education);

to carry out the construction of a number of dams (Selingue, Manantali) in order to control water resources and to develop Mali's hydro-agricultural potentialities.

The plan provided for MF 395.2 billion in investments, 34 percent of which were allocated to rural economy and 15 percent to the social sector.

The Five-Year-Plan was revised at least twice. Initially, planned investments amounted to MF 395.2 billion. Then, they were increased to over MF 900 billion. But, due to lack of external funds for certain agricultural projects, the plan was revised again and its provisions reduced. On 30 September 1978 it could be broken down as in Table 9.3.

TABLE 9.3

Structure of the Five-Year Plan Investments (MF billion)

	Projection	Amount invested	Amount invested as % of projection
Rural Economy	224.3	150.9	67%
Secondary Sector	159.8	123.6	77%
Communication & Housing	121.4	94.7	78%
Social	38.2	33.9	89%
Unclassified	-	-	-
Total	543.3	407.6	75%

In this 1974-78 plan domestic efforts accounted for 6.4 percent of total financing whereas the remaining 93.6 percent were made up by external funds—48.8 percent in grants and 51.2 percent in loans.

Analysis of results achieved under the plan shows that the growth rate of industrial agriculture was significantly higher than that of food agriculture, yet the latter was theoretically allocated a greater share of investments.

In fact, when one considers food crops for human consumption, one notes that traditional crops and those consumed, especially in

TABLE 9.4
Evolution of Production (MF billion)

Year	Food Agriculture	Ind. Agriculture	Livestock breeding	Fishing	Total
1972	27.3	9.3	30.6	8.0	75.5
1978	35.4	14.1	39.2	9.4	98.1
Growth rate	+ 4.4%	+ 6.6%	+ 4.2%	+2.7%	+ 4.5%

urban areas, evolve very differently. Thus, among cereals, the growth rate of wheat, maize and rice is much higher than that of millet and sorghum.

TABLE 9.5
Evolution of Output as of 1974

	1974-75	1975-76	1976-77	1977-78	1978-79
Millet-sorghum	100	106	113	119	115
Maize-wheat	100	148	168	232	276
Paddy	100	109	120	143	167

Basis of Agricultural Policy under the Three- and Five-Year Plans
The two plans described focused on the creation of state-owned enterprises for specific sectoral actions—rural development actions.

Generally speaking, as of November 1968, various tentative policies, especially agrarian policies, were adopted.

"Collective farms" were quickly abolished as they represented the most obvious aspect of the unpopular agrarian policy applied by the USRDA and peasant cooperatives were left barely managing (primary federations).

Shortly afterward, other bodies were set up to supervise the rural areas. These were ODRs[7] which were set up by late 1969 to oversee single crop farming—cotton in areas under the supervision of the CMDT, groundnuts in areas covered by the OACV[8] which later became known as ODIPAC.[9]

Such units were designed: 1) to generate production by populations and to develop the country, as intended by national and regional objectives; 2) to generate income for the state and the local communities by providing the former with tax revenues mainly under the IAS [10] system, and especially with custom duties on exports. These custom duties were estimated at MF 3.4 billion in 1976, 2.8 billion of which were levied on crops produced within the framework of the above units; 3) to ease peasants' access to inputs, the SCAER[11] being considered inaccessible by peasants, the SCAER was dissolved in 1982; 4) to guarantee the peasant a relatively reasonable market at a fixed, albeit low, price; 5) to make selected seeds available to peasants; 6) to increase the number of farm agents and establish a closer relationship between the latter and the peasants; 7) to bring about the financial autonomy of these units, thus enabling them to get funds directly from donor or lending countries.

These were the objectives assigned to some thirty state companies and enterprises.

It should be pointed out outright that these units were established haphazardly as financial resources became available. As a result Mali's agricultural sector became more dependent on external funds, on international capital.

But, in fact, these units did not meet expectations. Indeed, almost all of them were financed from the national budget (at least insofar as the salaries of the Malian staff working for these units were concerned) and therefore constituted a burden on this budget and on the national economy.

They had been urged to do their best to be *self-financing* but none of them was in a position to do so. Instead, they burdened the national budget. One easily understands then why these units were unable to inject new dynamism into rural development, especially as mismanagement was another feature of these units to reckon with—a non-existent or ill-organised book-keeping system—and this was conducive to embezzlement.

Although these units were supposed to be the best channel to request and have access to outside funds as well as the ideal means to train and assist rural producers, they gave the new class of profiteers an opportunity to become rich.[12] With reference to the increase in production, the units, as one can note, encouraged industrial crops as well as rice even though this cereal, which required more water and care, was considered a *luxury food* in the old Mali (contrary to millet which is the traditional staple food).

As far as producers were concerned, these units contributed, in fact, to marginalise those of them who did not get the opportunity to be supervised, who were more or less left without any technical assistance or credit. Among the others, only the more privileged ones performed fairly well. For instance, under the Segou Rice Unit each peasant (33 percent own less than one ha of land) produced officially about 1,500 kg of rice/ha. However, he had to pay the equivalent of 100 kg as ground rent as well as 10 percent for threshing and he was forced to sell 400 kg of his harvest to the Mali Agricultural

Products Marketing Board (ODAM); peasants always disputed the way in which these amounts were calculated and assessed.

Eventually, the peasant ended up with 850 kg. This was not enough to feed his family (seven members on the average). Hence he was forced to buy grain back. He thus found himself caught in endless indebtedness. If he wished to increase his production, he had to hire oxen and required agricultural equipment from richer peasants or civil servants—the rent was set at MF 20,000 per hectare for a plough. However, 70 percent of peasants do not own any. On the whole, these units, no doubt, helped to increase the income of a number of peasants. They also contributed to higher incomes in the areas under their supervision and they partly boosted GDP. On the other hand, these expansion activities did bring about in their wake the explosion of traditional community structures. Emphasis was placed on cash crops (even though at some point most of them tried to integrate food crops).

Peasants, whether rich or poor, started getting interested in money since the emphasis was placed on cash crops (cotton, tobacco, ground-nuts) and since they were urged to produce surplus food or to produce so as to provide urban populations with food. Individualism as well as a new money-making mentality have made their way to the rural areas and tend to prevail. Furthermore, the power of the old men in these areas gradually gave in to the power of money now being wielded by rich peasants; hence a new model of reference. *New patterns of behaviour* were also being introduced into villages by extension workers and developers—they owned radio-cassette recorders, listened to the same type of music as the urban youth, also owned motorcycles purchased after selling cotton grown in one farming season, and used these to consolidate their position.

The credit policy applied to the agricultural sector by the various units widened social gaps as this policy was based on a banking motto, "never lend to those who do not meet repayment conditions". These units also contributed to the *emergence of a new category of peasants officially called "pilot or model peasants"*.

Finally, as a result of these development activities and of the increasingly prevalent market-oriented mode of production in the rural areas, some types of craftmanship declined. For example, traditional blacksmith groups disappeared or converted to something else, e.g. conversion through the "CMDT blacksmith Action" etc. Some of the village weavers migrated because of poverty. Around the same time, another class of craftsmen emerged (radio-, bicycle-, motorcycle-, watch repair-men, tailors etc.).

THE 1981-85 PLAN: TOWARD A GRASSROOT INTEGRATED PARTICIPATORY ENDOGENOUS DEVELOPMENT

Having failed with cooperatives and RDVs, the government decided to come up with a new agrarian policy, "a grassroot integrated participatory endogenous development" also called "grassroot self-centred development".

Thus one can read the following statement in a document entitled, *Five-Year (1981-1985) Economic and Social Development Plan, Proposed Guidelines*: "The low productivity of rural grassroot communities being responsible for a poor development, development should be based on grassroot populations rather than on large-scale national actions."[13]

Indeed, the original feature in this Five-Year Plan was that development was to be based on decentralised local actions for maximum local self-sufficiency. These actions should, therefore, be known as *self-centred actions* for their conception and implementation will vary with existing community structures, e.g. village "tons" or other similar structures. However, one should endeavour to make these structures as homogeneous as possible. They were also described as local actions[14] because they were mainly undertaken at the *level of villages* or groups of villages, for their own populations.

These local actions were specific in the sense that they were determined by the base rather than being imposed from the top. Furthermore they were implemented mostly by the people concerned and very rarely by outside companies and services. Finally, they benefited rural populations directly rather than through intermediaries.

Such was the new rural development policy proposed by the Five-Year Plan. Apparently its implementation had started here and there. The above-mentioned document stated further that:

Of course, projects will be designed by senior technicians. These will help villagers to identify, plan, implement and maintain their projects. They will provide villages with advice on how to choose the most productive income-generating activities. As a matter of fact these activities should generate enough income to finance their own development...

Depending on regions, projects which will be designed and carried out by rural communities may consist of the following:

—development of agricultural hydraulics and irrigation
—building of access roads, and rural tasks
—water works for human and animal consumption

—re-afforestation actions and control of the Sahelian phenomenon
—small local goods processing units, production and maintenance of equipment
—actions aimed at livestock and poultry development
—vegetable and fruit production and development of hand-picked products
—cultural and social actions, especially practical actions such as teaching populations how to read and write, community development etc...

Thus, in this policy, contempt was expressed for the peasant who was supposed to be unable to design (technicians rather than peasants themselves were to undertake this task) or to participate in designing projects from their earliest stage. This is in fact what an agrarian policy pretending to be based on traditional rural communities really looks like.

A number of organisations were set up to have this implemented by rural development communities. These organisations included local and regional development committees which were to work closely with village "tons". As regards the setting up of development structures, the plan stated that: "It is up to the development process to create its own structures which must be light, that is *non-bureaucratic, pragmatic and efficient*'.[15]

This is a short insight into the new rural policy in force after the dissolution of the USRDA.

With reference to local or regional development committees, it is stated in circular No. 17 of the Ministry of Planning that "'in almost every administrative division, local and regional development committees *lack information* on national projects implemented in their various localities". Does this reflect shortcomings in the institutions in charge of implementing the 1981-85 Plan? Or is it indicative of the type of relationships existing between them and government?

The village "tons" seem to spearhead the new rural development policy. However, as we have already pointed out, the USRDA wished to use these village "tons" as political bodies (people called them Mali "ton") to supervise the rural youth. Now, attempts are still being made to use tons as a *technical assistance body* at the service of the rural sector. The experiment carried out by the USRDA in this area as in so many other areas was short-lived, as indicated earlier.

The new rural and economic development programme is designed to meet the quantitative food requirements of the Malian population as a whole; to guarantee and steadily increase producers' incomes via a water scheme policy, a pricing policy which would work as an incentive to producers, and improved marketing channels for agricultural produce; to ensure a steady increase in farms' productivity by reinforcing technical supervision via rural development units, offices or companies etc. to ensure surplus production of agricultural goods for export; to ensure continuous education and training for producers etc.

What can be made of this agrarian policy programme? Does it differ from previous rural world-oriented programmes?

One is first surprised by the lack of links between the various structures which are to promote rural development : on the one hand there are rural development units, on the other hand the old cooperatives set up by the USRDA, which are either no longer operational or dying despite the desire to redynamise them especially through the assistance of the Federal Republic of Germany. There are, finally, village "tons" which are given legal status; it should be said that the latter are copied from regulations especially made for "modern" cooperatives or pre-cooperatives. Yet attempts are being made at applying them, as they are, to non-existent or dying community structures.

Will village "tons" develop into modern cooperatives? Will they eventually replace these cooperatives or the rural development units? This question is yet to be answered.

It should be added that in Mali, these traditional communities are presented as being not only homogeneous but also superior to both "capitalism and socialism"—two *concepts* that are only valid in Europe and America. Government resorts to these traditional community structures to try and find above all a development alternative, mid-way between capitalism and socialism.

But in fact, even in areas where "tons" seem to have succeeded, e.g. in areas supervised by the CMDT, their role mainly consists in collecting the farmer's cotton production and delivering it to the CMDT. They act as collaterals for the recovery of loans granted by the CMDT to peasants. The price of cotton is determined by the CMDT and the government rather than by "tons". "Tons" settle old and new conflicts among peasants themselves or between peasants and the CMDT. From a brief study of "tons" in CMDT-supervised areas it appears they are used to enhance the image of the CMDT which is one example or form of agri-business in the country. In CMDT-supervised areas a shift from a traditional basic association to a modern cooperative does not seem to be considered.

In any case, village "tons" are presented both in theory and in practice as the best form of participation by peasants in what is known in Mali as an endogenous participatory integrated development.

AGRARIAN POLICIES: A FAILURE

It is necessary after this discussion to raise a question as to the results of these successive plans and programmes. Instead of improving peasants' living and working conditions, the agrarian policies which have been applied for over twenty years plunged them into misery. For the past ten years Malian peasants have experienced nothing but chronic death and hunger as well as malnutrition. Drought, which seems to be a sort of Deus ex machina of all that affects peasants, was held responsible for all these evils.

Thus, *L'Essor* of 10 April 1981 described the cereal deficit that seems to have plunged peasants into the above situation as touching. In an article which informed the reader of some of the conclusions drawn and recommendations made by a team of donors on mission in Mali from 5 to 15 December 1981, it was reported that "Mali's production of cereals (millet, sorghum, maize, paddy...) *dropped very sharply* in 1980 compared to the level reached in 1979". Table 9.6 indicates the magnitude of this decline:

TABLE 9.6
Decline in Cereal Production (1979-81)

1979-80	1980-81		Differ. in%
Millet, sorghum, maize	1,007,500 T	603,200	T 40.2
Paddy	219,900 T	141,700	T 35.5
Total	1,227,400 T	744,900	T 39.3

Source: L'Essor, 10 April 1981.

To be sure, this shortfall did further worsen peasants' living conditions—we are particularly interested in this—but it should also be underlined that low-level civil servants and other city dwellers lived in greater poverty. In spite of a FM 5,500 monthly minimum wage, which has slightly increased, and a FM 30,000 average salary, these civil servants have to spend anything between FM 17,000 and FM 22,000 to buy a bag of rice, FM 110 to buy bread, FM 1,200 for two pounds of meat   (to these 1980 figures, must be added an increase of a least 45 percent). What should be done then to survive? There is only one solution even in cities; it involves attempts at participating as much as possible in embezzlement, looting, bribery and extortion.

In one fails here, there are still beggary, theft and prostitution left which are gradually penetrating and developing in Mali's rural areas. This cereal deficit led therefore to an acute food shortage and plunged the peasants into such poverty that human survival in rural areas was not even guaranteed.

Poor rainfall as well as its uneven distribution are usually made responsible for this cereal shortage and food crisis. To be sure, Mali has been affected by persistent drought; but what has been done to control or even eradicate it? What happened to the food aid granted to Mali up to this point? Did it really reach final destination, that is, did it really get to the peasants? and how?

To the first question, we can answer that since 1970, we have continuously argued that the future of Mali is mainly linked to the construction of big and small dams in areas where this is feasible. Such construction will help supply water for irrigation and by the same token increase agricultural production. It will then be possible to achieve food self-sufficiency, which is already dangerously jeopardised by the lead taken over food crops by export crops. The Malian agricultural sector needs huge investments if it is to come out of its stagnation, let alone its regression. The bulk of these investments should be earmarked for the construction of large and small dams. They should also be used to supply peasants with heavy equipment such as tractors, threshing-machines wherever possible, and with light equipment—ploughs, carts—in dry farming areas.

The dams which have been built have still not produced the expected results. It is too early to take stock of the situation but their construction is slow and requires funds which are not available to the state and which are often difficult to obtain from international aid.

Unless this prerequisite is met, Mali's cereal deficits and food shortages will go on and Mali will keep begging for international assistance. This aid originates from Western Europe and America which use it as a weapon, as a Damocles' sword hanging over the state. This is why we wrote in one of our unpublished papers that "Food aid or food cooperation is a current issue, however, should one speak of food aid or food as a weapon?"

NOTES

1. GRA: Rural group, an association which includes many villages.
2. SDR: Rural Development Society.
3. SMDR: Mutual Rural Development Company.
4. GRPS: Rural Production and Mutual Solidarity Grouping.
5. SIP: Local Provident Society.
6. Jacquemot, P. "Le Proto-Etat Africain: Quelques réflections autour de l'histoire contemporaine du Mali", *Revue du Tiers-Monde*, XXIV, No 93, January-March 1983.
7. ODR: Rural Development Unit.

8. OACV: Groundnut and food crop production unit. This unit is based on the former BDPA (Groundnut crop and development production board).
9. ODIPAC: former OACV, Integrated food crop and groundnut production development unit.
10. IAS: Business and service tax.
11. SCAER: Agricultural Credit and Rural Equipment Company.
12. Having accumulated wealth at the expense of these units some of these profiteers or technocrats set up their own business in the private sector. Some simply fled to seek refuge in neighbouring countries thus avoiding their malpractice being uncovered. But it was unveiled shortly afterwards as inspectors had already started unexpected inspection of the various units.
13. This draft plan will be amended; however, its philosophy will remained unchanged.
14. The same document states that self-centred development actions can also be set up at regional or even national level.
15. *Five-Year Plan Orientation Project, p. 5.*

APPENDIX 9.1
1974-78 Five-Year Plan: Balance Sheet
Funds Received And Actual Investment (in million FM)

Sectors and Sub-Sectors	Initial Projections	74-78	Updated Projections after 78	Total	RESULTS Funds received	RESULTS Funds used
Sector of Rural Economy						
Agricultural resources	97,041.0	106,809	43,133	149,942	102,300.3	54,740.6
Forest resources	5,060.0	2,690	222	2,912	1,865.5	1,262.7
Animal resources	24,480.0	55,796	1,163	56,959	35,354.0	18,351.4
Pre-investment	6,430.0	14,449	–	14,449	24,644.2	17,195.0
TOTAL	133,011.0	179,744	44,518	224,262	164,164.0	91,549.7
Secondary Sector						
Mining resources	13,149.7	16,219	4,630	20,849	24,419.0	13,348.9
Water and Energy	63,913.1	83,260	38,984	102,224	102,867.3	54,866.7
Industries	33,514.4	30,824	5,882	36,706	24,216	15,294.6
TOTAL	110,577.2	110,303	49,476	159,779	151,502.5	83,510.2
Public Equipment Sector						
Road Equipment	58,820.2	35,383	17,920	53,303	47,185.0	25,600.2
Railroad Equipment	8,599.0	11,449	6,320	17,769	18,616.9	10,243.7
River Equipment	3,181.1	3,438	360	3,798	3,130.0	2,898.6
Aeronautic Equipment	8,289.2	14,226	1,311	15,537	6,436.7	5,164.9
Post & Telecommunication	9,429.5	10,043	1,530	11,573	9,547.6	5,642.0
Tourism Equipment	2,753.2	2,417	–	2,417	2,317.0	2,317.0
Equipment & City Planting	22,809.0	7,729	8,908	16,637	19,998.5	10,274.1
TOTAL	113,881.2	84,685	36,349	121,034	107,239.7	62,140.5
Social Sector						
Training	24,090.7	16,307	9,344	25,651	21,788.8	11,644.8
Information	3,334.7	1,794	–	1,794	2,189.4	216.1
Health	10,260.5	8,101	2,675	10,776	7,808	4,760.5
TOTAL	37,685.9	26,202	12,019	38,221	31,786.3	18,621.4
TOTAL PLAN	395,155.3	400,934	142,362	543,296	454,692.5	253,821.8

APPENDIX 9.2

Funding of Projects Covered by the 1981-85 Five-Year Plan All Sectors
(in million FM 1980)

Sectors	1981	1982	1983	1984	1985	TOTAL 1981-85 Plan	After 1985	Overall Total	DOM.	Funding 1981-85 EXT.
Rural Economy	34,842	48,981	67,209	73,725	77,826	302,583	9,409	311,982	59,938	242,645
Secondary	54,127	82,602	74,068	52,115	29,451	292,664	150,839	458,494	23,610	269,054
Infrastructures	32,996	55,495	55,285	64,491	48,002	256,269	65,313	321,582	33,324	222,945
Social and Administrative	4,994	19,035	17,604	20,504	22,961	85,098	55,529	140,627	18,451	66,647
TOTAL	127,259	206,113	214,166	210,835	178,241	936,614	291,081	1,227,697	138,323	801,291

10. STATE POLICIES ON AGRICULTURE AND FOOD PRODUCTION IN BURKINA FASO 1960-83

T. Thiombiano

Burkina Faso, with its 7,000,000 inhabitants, is situated right in the West African heartland. It is a continental territory of 274,000 km² which shares frontiers with Mali to the north, Ghana and Togo to the south, Benin to the south-east, Niger to the east and Ivory Coast to the west.

Being a landlocked country—the closest port to Ouagadougou is Lomé, 950 km away—the bulk of the incoming and outgoing passenger and freight traffic transits through the Ivory Coast along the 1,100 km rail link.

CONDITIONS AND FEATURES OF AGRICULTURAL PRODUCTION

Natural Potentials
Since the long drought of the early 1970s, the country has been classified with Sahelian countries because its upper reaches touch the fringes of the Sahara. Strictly speaking, however, Burkina has an essentially Sudanese-type climate, with two main seasons, a rainy season from May-June to October-November and a dry season from December to the end of April.

Rainfall distribution over the country is uneven, ranging from 500 mm a year in the northern zone to 1,400 mm a year in the southern zone. Coupled with disparities in soil fertility, this results in four agro-climatic zones : the South-West, the Central Plateau, the Eastern Savanna, and the Sahel. Needless to say, boundaries between these zones are not rigidly fixed. They shift and change in accordance with the interplay of four factors: rainfall, soil quality and vegetation, population density, and the degree of internal accessibility. Quite clearly, these features correspond to regional variations in economic activities and their degrees of profitability. The zones are not naturally homogeneous. Any analysis based on them, therefore, can only provide an approximate idea of the geographical distribution of the country's agricultural and pastoral potential.

The South-West. Because of the area's natural endowments, most of the principal crops are grown here, and yields per hectare are relatively high. Here possibilities of extensive and intensive farming are also good, because there are still tracts of uncultivated land, while numerous rivers, streams and shoals provide ample water for irrigation. Not surprisingly, foreign investors have brought into this part of the country, where less than a third of the population lives, modern production factors : animal-drawn ploughs, tractors, sprayers, fertilisers and pesticides.

In this zone, which covers 30 percent of the country, government efforts were initially concentrated on industrial crops : cotton and sugar cane.

The Central Zone. This zone includes the whole of the Mossi Plateau, and covers the entire area between the 600 mm and 900 mm isohyets, except for patches in the eastern (Koupéla) and west-central regions (Koudougou). It covers 45 percent of the country and includes the northern reaches of the Dédougou, Koudougou and Ouagadougou Regional Development Units, as well as the whole of the Yatenga and Kaya Units. Population density in this zone is very high, reaching 60 people per km² in places. This central part of the country may be considered the Sudanese zone. The poor eroded cover consists mainly of a thin layer of sandy topsoil, acidic as a rule. The rainy season lasts about four months. Perennial streams are rare, the only significant one being the White Volta; in addition, there are a few lakes such as the Bam, providing water for irrigated farming and attracting settlements on their fringes.

Farms are usually small, and the main crops are cereals. All industries in the area are concentrated in Ouagadougou which, with a population of about 400,000, is also the main consumer centre. The town of Ouagadougou gets its vegetable supplies from unorganised peasants carrying on market gardening along the dam reservoirs, as well as from cooperatives in Kongoussi, Guiédougou and Lanfiéra in the northern section of the Black Volta area. From November to February, supplies of sweet potatoes come to Ouagadougou from Kombissiri.

In this zone investors are not particularly keen to venture into these Regional Development Units, because agricultural yields are low and profits from them correspondingly small, while there is no obvious bonanza to be expected from the area's mineral resources, or from its potential in animal husbandry. In 1970-72, the bulk of peasant income came from cereal farming which accounted for 69.9 percent of the region's income. The second main source of wealth is animal husbandry, accounting for 21.1 percent of income. Even though the zone is near the capital, in 1978, the rate of monetarisation of rural income was lower there than in any of the other three zones.

The Eastern Zone. This zone covers the south-eastern part of the country, with a small portion, the province of Gnagna, belonging much more to the Sahel. Counting in a part of the Koupéla Regional Development Unit, the Eastern Zone might be said to cover 20 percent of the national territory.

This zone, reputed to be the country's second bread basket after the South-West, is very sparsely populated, with a population density of less than 10 people per km² in places. Rainfall is slightly higher here than in the Central Zone. In Tapoa province, in fact, climatic and soil conditions are similar to those in the South-West Zone.

The zone has enormous agricultural, mining and water power potential, an index of great economic possibilities. The Kompienga Dam, estimated to cost about 36,000 million F CFA, will on completion supply electricity to a section of the country. The project is expected to help put a definite end to the region's isolation. There is already a phosphate mining enterprise at Kodjari which meets part of the country's fertiliser needs. In the 1980-81 farming season, the various Rural Development Units consumed 72,275 tonnes of this fertiliser, with the Eastern Zone alone using 81 percent of the total. On paper, then, this zone is as abundantly endowed in resources as the eastern and southern areas. The major drawback, however, is that the road network here is appalling, so, even though in its sheer immensity the zone offers great agro-pastoral opportunities, it has absolutely no industry. With regard to farming, the northern districts produce groundnuts and onions, the central districts soya beans, the districts in the far south grow yams, and the far eastern districts, where the soil is rich, are suitable for planting sweet sorghum and sugar cane. The main food crops grown are sorghum, pearl millet and, as an occasional off-season supplement, maize. Being close to the Sahel region, this zone offers quite lucrative opportunities for animal husbandry.

The Sahel Zone. This zone, bounded by the 600 mm mean isohyet, overlaps to a great extent with the Dori Regional Development Unit. It covers 5 percent of the country's area. Most of the zone's inhabitants are nomads, and animal husbandry is the dominant economic activity here.

In terms of relative value, this zone has lost some of its animal husbandry potential to the western and eastern regions as a result of the critical drought affecting the entire Sahel. For nomads here, farming is still a minor occupation, their main food items being milk or its derivatives. But, in reaction to the loss of their cattle herds, nomadic Peuls are apparently taking up farming in increasing numbers. In 1970-72, cereal production in the zone was only 3.6 percent of the country's total production ; by 1979-81, however, stimulated by cash and food shortages occasioned by livestock losses, the figure had risen to 4.6 percent.

Agricultural Production

Agriculture in Burkina Faso, as elsewhere, is subject to the impact of industrial development on market economies. It is this impact which resulted in the emergence of agro-industrial enterprises in the country in the 1950s. The emergent linkage of agriculture and industry—hardly a novel phenomenon—put a definite end to the homogeneity of the agricultural sector.[1] Due to this new orientation, agriculture in Burkina Faso developed in two directions: production for subsistence: this accounts for the survival of the "traditional" sector in agricultural production: and production for the international market.

Agricultural production for subsistence. This is the kind of agriculture in which the majority of the rural population is involved. As a rule, it is not very productive; those practising it use few modern production tools and hardly any surplus is produced. There is very little investment involved. Productivity is so low that the peasants have to plant their subsistence crops over a large part of the land they cultivate.

TABLE 10.1
Percentage of Land Areas Planted with Cereals
from 1978-79 to 1981-82

Farm Season	%
1978-79	88.7
1979-80	88
1980-81	90
1981-82	90.9

Because their yields do not increase, Burkinabe peasants are constantly forced to expand farm hectarage in an attempt to meet sub-

TABLE 10.2
Average Per Capita Consumption of Cereals and Leguminous Crops in the Rural Areas

Year	Rainfall in mm		Average per capita rural consumption (in kg)		Yearly average as related to 215 kg target
	Amount	Difference from year to year	Cereals	Leguminous crops	Cereals
1975	866	−69	227	21	+12
1976	797	48	195	19	−20
1977	845	−107	193	20	−22
1978	738	74	201	21	−14
1979	812	−83	220	21	−15
1980	729		196	21	−19
1981	764	35	230	21	+15
1982	752	−12	212	21	− 3

sistence needs. These needs have been minimally estimated at 215 kg per year. The question is whether, after all these efforts, the minimum target is attained.

Judging by Table 10.2, the following observations are in order: there is a permanent cereal shortfall although a relative improvement can be registered; leguminous crop production has stagnated; in the absence of an effort to modernise agriculture, climatic hazards have had and continue to have an impact on food production.

The assumption here is that urban food needs, in this case mainly those of the capital city, are still met from imports alone. The urban population, which was 6.4 percent of the total in 1975, had risen to 9.67 percent in 1982. According to a CEDRES survey, average annual per capita cereal consumption among urbanites was 135 kg in 1982.[2] Cereal requirements for the period were calculated at 84,700 tonnes while cereal imports for the same period amounted to 93,500 tonnes.

Generally, the extent to which food requirements are met varies from region to region. Urban dwellers for example have access to alternative sources of food: pasta and noodles, tubers like sweet potatoes, ordinary potatoes and yams, meat etc. In the Western Zone, fonio is an extra source of food, in addition to tubers. By contrast, there is very little variety in food production in the Central Zone, and cereals remain the predominant staple.

In this traditional farming system, the market is unregulated, and peasants are independent of the world capitalist system, in the sense that they are free to grow whatever they choose, in whatever quantities they need. Another aspect of this freedom is that under prevailing atomistic production conditions, producers can sell their produce in a situation of almost perfect competition. Isolated and unorganised, they have no way of obtaining credits for modernising their farms.

In practical terms, these peasants are also dependent on the world capitalist market : they consume industrial products, and they patiently look forward to obtaining from the state banks the loans they need imperatively for their development. Furthermore, the concept of traditional farming remains susceptible to change. After all, every type of agricultural production is likely to evolve eventually into an industrial farming system necessarily integrated into the world raw materials market. Thus for peasants in the Comoé basin, maize is increasingly becoming more than just a subsistence crop, because they know that the new Burkina milling complex needs it to keep functioning. The overwhelming majority of these peasants are poor, the poorest of them all being those in the Central Zone, whose main occupation is agriculture. Only a minority is engaged in other work of secondary importance, capable of generating considerable income. Skilled craftsmen and farmer-herdspeople in the Eastern Zone belong to this category. Such people may earn incomes comparable to those of peasants producing cash crops.

Industrial farming. In the particular situation we are dealing with, we have identified three cases, from the type of control used over production.

Direct control over agricultural production. When an agro-industrial corporation decides to produce the raw materials that it processes, we have a case of vertical integration. In Burkina the best known example is the SOSU-CO Corporation. To achieve vertical integration, such a corporation, usually a transnational, needs state help to acquire a piece of the national territory. The land thus ceded becomes the corporation's domain. As a rule, the rural districts thus handed over cover good farm land on which peasants formerly grew subsistence crops. (In the case of SOSU-CO, the corporation acquired a domain of 10,000 hectares.) When the transnational corporation integrates primary production directly into its chain of processing and marketing activities, it automatically pushes peasants into selling their labour power: they become regional labourers, short-term contract workers, or permanent full-time employees working for the factory. The peasant, formerly a free, independent producer, becomes a worker wholly dependent on foreign capital. Individual peasants continue to own plots on which they grow subsistence crops. But these plots are on infertile, marginal land. Such a disparity in soil fertility leads to a disparity in income on the produce market.

From the transnational corporation's point of view this integrated system presents clear advantages. There is one great disadvantage, however: the likelihood of labour unrest. To avoid this problem, some companies prefer a partial integration system.

Quasi-integrated agricultural production through indirect control of production. In such a system peasants continue owning their own land, allotted to them according to informal clan norms. In principle peasants act of their own free will when they decide to grow cotton, for instance. However, such a decision binds them, in practice, to sell their cotton crop to a single company, in this instance the textile corporation SOFITEX, an offshoot of the French Textile Development Company (CFDT). What forces peasants to sell their cotton to SOFITEX is that it is the corporation which, acting through the Regional Development Unit, supplies them with the necessary inputs, cotton seed and such up-to-date production factors as tractors, each of which was at one point estimated to cost 4,500,000 F CFA. To deal with this situation, individual peasants sometimes choose to cultivate a single large farm measuring about twenty hectares, on part of which they grow cash crops (cotton) while sowing cereal (sorghum, pearl millet, maize) on the remainder. Alternatively, they may work three or four separate fields simultaneously, growing a specific crop on each, while carefully separating cash crops from subsistence cereals.

The truth is that such peasants, even if they own simple ploughs, are thoroughly integrated into the capitalist world market. Certainly, it is the state which fixes cotton prices every year. But until recently all such prices fixed by the state were pegged to cotton prices on the major world markets. In recent years, however, because of the *state's monopolistic* control over the domestic market, even when world prices for cotton rise, the increase is not fully passed on to the peasant producer, or is simply not passed on at all.

For example, Nana-Sinkame[3] says the oil price rise of the 1970s affected cotton prices for two reasons: prices of synthetic fibres increased substantially; and interest in natural fibres picked up again, with the result that for a whole decade demand for cotton was higher than available supplies.

The question arises: did price increases on the world market produce positive effects on the national cotton market?

It appears from Table 10.3 that neither the demand nor the

price—especially—have been affected positively on the domestic Burkinabe market.

TABLE 10.3
Movements of Prices on the World Cotton Fibre Market
& on the Domestic Raw Cotton Market in Burkina

Year	Medium fibre (cents per pound)	Long fibre (cents per pound)	Raw Cotton, Burkina F CFA/kg 1st grade	2nd grade
1977	77.33	149.20	55	45
1978	71.95	138.60	55	45
1979	77.14	153.21	55	45
1980	93.73	153.49	55	45
1981	83.97	151.99	55	45
1982	72.51	124.87	62	45
1983	84.10	139.96	62	45

Could that be because the peasant was paying the heavy cost of state intervention? Was it mostly because peasants were not organised? Or are both factors simultaneously at work ? The last explanation—a combination of the peasants' lack of organisation and the state's heavy exaction—is the most reasonable. As a matter of fact this peasant who could be called a capitalist peasant because he owns modern production tools, produces for the market and employs wage labour, is himself a victim of market forces. Part of the surplus values he extracts from his labourers is in turn grabbed by the SOFITEX corporation.

The type of agricultural system outlined above is practised only in the Western Zone, especially in the Hauts-Bassins and Black Volta Regional Development Units, which specialise in cotton farming.

Although this type of agricultural system, as we have just seen, presents a number of disadvantages, it also offers more advantages to peasants formerly engaged in traditional subsistence farming. For in such a system the peasant is free to decide, every new planting season, whether to sow cotton or not depending on the abundance of the previous cereal harvest, and/or on the profitability of the previous year's cotton sales. Peasants in the Comoé basin, pushed onto marginal land as they are, have no such choice. They are caught up in a system in which they have absolutely no resources with which to turn the mechanisms of the market process to their own advantage. There is a third option: contract farming.

Quasi-integrated agricultural production through contract farming.
The first difference between contract farming and the system just

described is that here the peasants do not own the land they till. All they have is the right of usufruct, i.e. the right to the yield they produce from the land. They are under contract to produce specific crops. The moment they refuse to do so, they face expulsion. This kind of system operates on the irrigated perimeters of the Tougan sector in the Northern Zone of the Black Volta basin, where the Guiédougou and Lanfiéra cooperatives are situated. It is also found in the North-Central Regional Development Unit (Kaya), site of the Lake Bam cooperative.

The peasants, organised in cooperatives, are allotted land by the state. Inputs (fertiliser, tools) are supplied by the Volta Union of Farm and Market Gardening Cooperatives. On the basis of demand on the French, Dutch and other markets, the Union fixes production quotas and prices. The main export crop is French beans, grown here mainly for the foreign market.

According to the Burkinabe economist Ouali, "in 1980-81 the European market absorbed 80 percent of total production. In 1983-84 it took 100 percent of high grade bean exports and 85 percent of lower grade beans".[4] This past farming season, earnings totalled 487 million F CFA from 2,621 tonnes of produce. This means a unit price, free on board, of 186 F CFA per kg, which is a mere 24.8 percent of the retail price charged in France. French wholesalers charge an estimated 41 percent of the retail price on the Paris market. Following Ouali's estimate, we arrive at a retail price of 750 F CFA per kg on the French market for 1983-84. Under these conditions a French retailer earns a commission of 307 F CFA per kilo, while back in Koungoussi or Lanfiéra the producer is paid a paltry 54 F CFA per kg. The retail trader's commission is 5.6 times the producer's price. Nor is that all. As we have seen, the seeds the peasants plant are imported. They therefore have to pay for them from the same 54 F CFA per kg. Not surprisingly, peasants in the Lake Bam area who plant only French beans frequently wind up a farming season in the red.

Granted, there are here, as in every organisation, a few peasants who in spite of all the problems, manage to do a thriving business. In this particular case those who grow several different off-season crops are the most successful. But these are exceptions, and their presence cannot obscure the fact that the overall situation is negative.

Such is the situation of agricultural production in Burkina Faso, with its implications for future trends in social differentiation.

The logical consequence of the underlying policy is that cereal imports have to be increased to make up the domestic production shortfall. Every year these cereal imports cost 5,000 million F CFA, but total export earnings from cotton and French beans only average 3,900 million F CFA per year. Total exports, mostly raw materials, cover barely 40 percent of imports. And since 1978, the percentage of Gross Domestic Product represented by farm produce has stagnated at 24 percent.

The State and the Burkinabe Agricultural System

State intervention in Burkina from 1960 to 1983 posed absolutely no challenge to the basic operation of market forces. To be sure, the situation was a far cry from the classical 19th century theories of perfectly competitive markets. But the point is that imperfect as the market here was, it did not abolish the profit motive, the keystone of capitalism. At present, the Burkinabe state intervenes, through its planning prerogatives, to define industrial and agricultural policies. It determines guidelines for educational and training systems. It is also active in the field of public health, providing considerable support for first aid and preventive measures in such areas as vaccination and community clean-up campaigns.

The state is a shareholder in major industrial concerns (Faso Fani, the Sugar Corporation etc.). It manages companies operating in the distribution of mass consumer items, a task beyond the capabilities of a perfectly competitive market system. One might mention other enterprises : the National Cereals Bureau, Faso Yaar, the National Water Resources Corporation, the Burkinabe National Electricity Corporation etc. And the state, operating through the National Bureau of Dams and Development, builds dams and dykes in different regions of the country.

Nowadays, in discussions of the role of the state the major issue at stake is whether it is up to the public authorities or private initiative to see to the satisfaction of communal needs. According to classical economists and their converts, state intervention is justified only in situations with which private initiative cannot cope. It just so happens that in the immediate post-independence period, private enterprise in Africa was incapable of coping with the situation anywhere. Here capitalism was born at a time when the nation had no capitalists of its own. More and more, those companies dealing with the collection of primary produce and distribution of manufactured goods withdrew from the hinterland to the capital. And, to hedge against all type of risks, such companies as SCOA, CFAO and CFCI (French Trading and Industrial Company) tended to form mergers to limit competition, to stave off the threat of nationalisation, and to cut costs.

The country had only two banks: the Banque de l' Afrique Occidentale (BAO) and the Banque Nationale de Paris (BNP). Credit policies were highly selective: as the old economist's saying goes, "only the rich get loans". Yet on the eve of independence no national in this country had accumulated enough wealth to really qualify as a national bourgeois. The reasons for this lack of primitive accumulation include the following: there were no large landed estates; there were no major export cash crop plantations ; and the nation's productive forces had remained blocked in the absence of any industry.

The reason for this state of affairs was that the colonial powers intended the country to serve only as a labour reserve for plantations in the Ivory Coast and to provide recruits for the colonial army.

At the time when the Burkinabe petty bourgeoisie assumed state power in 1960, it had no economic base whatsover. The country as a whole had no real national bourgeoisie. There was therefore only one way left to ease the transition from colonialism to neocolonialism: the alliance of junior clerks, school teachers and a sprinkling of petty traders and hawkers.

Naturally, when this greenhorn bourgeoisie found itself in power, its first concern was to build up an economic base.

Building up the State Economic Base
Three elements indicate this: the development strategy followed; the development funding methods adopted; and pricing policies.

Agricultural development strategy. The overall objectives assigned to economic and social development in Burkina Faso have been defined in several official documents. Between 1960 and 1983, there were four development plans: the 1963-67, 1967-70, 1972-76 and 1977-81 plans.

— The 1963-67 Plan: drafted under the First Republic this plan was never implemented.

— The 1967-70 Plan: this called for a total financial package of 42,000 million F CFA. It was followed by an interim plan in 1971. During this period when the country's first real plan was being implemented, the absolute value of investments in agriculture up to 1971 was 5,500 million F CFA. We would point out, however, that because the country was dependent on external funding, it was only able to implement 64 percent of this.

— The 1972-76 Plan: estimated expenditure under this plan was expected to total 62,000 million F CFA, of which 14,000 million (22.5 percent) were earmarked for agriculture.

— The 1977-81 Plan: in all the plans, implemented or not, the intention to develop agriculture as an indispensable base for the country's economic take-off was constantly expressed. Among the objectives of the Third Plan one can note: the improvement of living conditions for the people; the reduction of underemployment; and the achievement of self-sufficiency in food through increased cereal production. This was the first plan expressing clearly the need to achieve food self-sufficiency in the context of the cereal production crisis affecting the country since the great drought of the early 1970s. However, these intentions were not followed by concrete actions.

Judging by the sectorial distribution of planned investment, agriculture did not enjoy the much-touted priority.

TABLE 10.4

Sectorial Distribution of Planned Investment

Sector	Investment percentage
Agriculture	21
Infrastructure	39
Industry	30
Social Services	10
Total	100

Putting aside the gaps between planned and actual investment occasioned by the fact that these plans depended on external funding, we may observe that the state always intended to develop agriculture through the accomplishment of the following tasks: the satisfaction of the population's growing needs—through intensified food production based on greater control over the rural environment coupled with more efficient water resource control; increased earnings for all Burkinabe citizens thanks to a steady shift from a subsistence economy to an exchange economy—according to official doctrine, this would come about through intensified production and the development of commercial circuits; increased production for export—without jeopardising the satisfaction of food needs; the development of the country's infrastructural facilities with an emphasis on roads and communication in general; water supply, the development of labour-intensive activities and the establishment of new schools adapted to the educational needs of rural children, the protection of the national heritages through the wide spread introduction of soil conservation methods and the intensification of the anti-desertification campaign.

Regional Development Units. As far as the authorities are concerned, the main structure for achieving this objective is the system of Regional Development Units (RDU).

Regional Development Units were started in 1966. Gradually the units came to cover the entire country, gaining a special status as institutions facilitating the devolution of responsibility for agricultural development to the district authorities. Regional Development Units have gone through three important phases.

At their inception they were primarily involved in the technical and economic aspects of rural development. They paid scant attention to social issues. This was the phase when these structures were dominated by foreign technical aid expects.

As from 1968 and especially from the mid-1970s, the military government, having adopted integrated rural development as the basis of its strategy for the rural sector, gave the Regional Development Units total control over inter-sectorial assignments in the regions, related to the new integrated policy. These assignments comprised agricultural production, infrastructural development, peasant literacy campaigns, the supply of inputs and credits, and marketing. All these assignments were interesting and important. Ultimately, they put the Regional Development Units in a position to operate as real catalysts of rural development. Alas, the new institutions proved incapable of coping with their overwhelming assignments. In short order the authorities, faced with the patent inefficiency of the Regional Development Units, drew up a set of new guidelines. To be sure, at that time there was a shortage of officers with the requisite political and technical training for the tasks assigned the Regional Development Units. These tasks were therefore revised, and from that point on these units were restricted to duties directly related to agricultural production. In 1974, however, the government again reversed its stand and restored the monopoly over intersectorial assignments to the Regional Development Units. The hope this time was that the units would operate profitably enough to become financially self-supporting, thus saving the state from having to pay all operating costs.

This vacillation was clear evidence that the Regional Development Units had been launched without proper design and planning. Occasionally they were even expected to double as civil engineering outfits. The result was that they kept duplicating efforts in other agencies.

Regional Development Units were initially designed to promote agricultural change.

As a rule, each Regional Development Unit covers four to six geographic sectors ; each sector is divided into five to seven sub-sectors. Each sub-sector, in turn, is divided into six to eight extension zones. Each such extension zone is defined to cover several villages.

The idea behind this way of structuring the Regional Development Unit is: to help farmers adopt improved production methods; to ensure uninterrupted interaction and information exchange between farmers and official research and planning agencies; and to encourage farmers to pool their resources in order to achieve results unattainable in isolation; in other words, to get them to create village associations or cooperatives.

To ensure the attainment of objectives assigned to the Regional Development Units, an Extension and Popularisation Service was set up with personnel whose duties were commensurate with their training and qualifications.

Extension officers in RDUs are primary school leavers who have failed to make headway in the formal educational system. At 20 to 35 years of age, such personnel are recruited for training as Popularisers. This training lasts a year at the most, and the personnel thus trained do not enjoy the status of civil servants. At times, when funds are low, they go several months on end without pay.

In keeping with their aims, Regional Development Units maintain services in each of the following areas; agricultural production, animal husbandry, equipment, community development, economic statistics and planning, credit, finance and administration. Funds for agriculture are channelled through these structures.

Agricultural financing. Until 1980, the National Development Bank (BND) channelled agricultural credits through the Regional Development Units. In practice, the Regional Development Units had a hard time recovering credits advanced to peasants. To make matters worse, the National Development Bank had absolutely no mechanism for checking the uses to which the funds it provided (in theory) for agricultural investment were actually put. All these problems led to the creation of an agency specialised in the handling of agricultural credit: the National Agricultural Credit Fund (Caisse National de Crédit Agricole), in August 1980, with a capital of 1,300 million F CFA. The state holds 54 percent of the shares, with the French Central Fund for Economic Corporation (CCCE), the BCEAO and the Banque Ouest-Africaine de Developpement (BOAD) sharing the remainder.

Credit allocated to agriculture. The bulk of bank credits are medium- and long-term credits. In any modern economy credit is a key factor; it helps money to circulate, and it enables those economic agents engaged in production to do their work, thus satisfying demand.

We are in no position to draw up a list of the percentage of total bank credits allocated to agriculture. Available data, however, indicate that this percentage was relatively low. Table 10.5 presents the pattern of its evolution:

An infinitesimal portion of credit goes to the agricultural sector. Part of the explanation lies in the nature of the existing banks and in the fact that the state has always practised an extremely liberal policy towards these banks. A second reason is the role of these banks in the development of the national economy. It is practically certain that

TABLE 10.5
Evolution of Bank Credits Allocated to Agriculture

Year	% of bank credits allocated to agriculture
1969	1.3
1974	0.7
1978	0.9
1981	2.4

in 1981 the percentage of credits to agriculture rose to 2.4 percent mainly because of the establishment of the National Agricultural Credit Fund (CNCA), with the resulting injection of credits into the agricultural sector. Prior to the establishment of the CNCA, credits to peasants were handled by the Regional Development Units, the CFDT and, to a lesser extent, the National Development Bank (BND). The CFDT specialises exclusively in financing cotton farming. The credits take three forms:

> Short-term credits: these are used for fertiliser and pesticide purchases. Such credits, geared to the farming season, are to be paid back as soon as harvests are marketed.
> Medium-term credits: these are intended for the purchase of agricultural equipment and draught animals. Ploughs and sprayers are the kinds of equipment commonly used.
> Long-term credits: these, rather infrequently accorded, are available for up to ten years. Peasants obtaining such loans may use them to purchase machines like Fiat 850 or Boyer tractors, valued at proximately 4,500,000 F CFA in 1982.

Interest rates vary from year to year, as well as according to type of loan. Medium-term loans drew 11 percent interest for the 1982-83 farming season whereas long-term loan interest rates rose from 5.5 percent to 9 percent in 1980.

As a rule, these loans go mainly to village associations, especially in the cotton-producing regions. Peasants are still obliged to go through the Regional Development Units to obtain such loans. Thus, the Black Volta and Hauts-Bassins Regional Development Units which have the largest number of village associations, also get the bulk of available credits. Of total credits extended to the 11 Regional Development Units, these two units obtained 21 percent in 1981-82, 30 percent in 1982-83, and 36 percent in 1983-84. Forecasts for the next three farming seasons indicate that these current trends will continue, with credits accorded staying around 36-38 percent.

One possible explanation for this lopsided credit policy might be that the state lacks the necessary funds for financing agricultural activities, and that it is therefore the funding agencies which determine policy options.

State aid to agriculture fits into the framework of the financial policy followed by the country during the period here under study. As pointed out in the preceding section on the role of the state, in Burkina Faso the public sector, in addition to the central government and the local communities, includes various state enterprises. At the material time there were no less than 120 such enterprises. They derived their incomes from fees and commissions charged for services rendered. This was the situation of financial institutions such as the CNDI, the Banque Nationale de Développement (BND) and even the Regional Development Units. That, at any rate, is the theory. In practice, these corporations are often in the red, so that the state has to pump subsidies into them to bail them out.

It has become increasingly difficult for the government to keep supporting these corporations, especially since the state depends on foreign aid and tax revenue for its survival. An analysis of the country's trade balance shows how weak the national economy is. Apart from 1960 (Independence Year) and 1968, export earnings have consistently fallen below 50 percent of import expenditure.

For a long time the country's balance of payments showed a surplus. This was because of remittances sent back by Burkinabe migrants in the Ivory Coast, and subsidies accorded by France. Deficits began to appear in 1975. The country's financial policy history may be divided into three periods:

1960-66: a period of consistent deficits;

1967-74: the second period was a time of orthodoxy, the type of policy the then Minister of Finance, Tiémoko Marc Garango, advocated.

1975-82: this last period saw a series of new regimes. It was also a time in which governments reverted to deficit financing. The consequence of all the various changes was that the state invested very little of its own resources. The percentage allocated was seldom higher than 7 percent. The bulk of government contributions went into salaries for civil servants in the Ministry of Agriculture.

Agricultural investment is mainly dependent on foreign financing. In the country's various development plans, almost all investment

TABLE 10.6

Percentage of Imports Covered by Export Earnings; & Percentage of Government Expenditure Covered by Revenue (1960-82).

Year	% imports covered by export earnings	% expenditure covered by revenue
1960	52.5	125
1961		115
1962		74
1963		107
1964		88
1965	40.1	96
1966		103
1967		104
1968	58.0	106
1969	44.6	107
1970		109
1971	36.0	107
1972	41.5	103
1973	36.9	116
1974	38.0	79
1975	34.0	99
1976	49.5	120
1977	43.0	98
1978	42.0	98
1979	42.5	87
1980	43.5	92
1981	40.5	93
1982	43.8	96

destined for the rural areas comes from foreign aid. In the Second Plan period such aid totalled 6,600 million F CFA annually. From 1977 to 1980 the annual total was 14,000 millions F CFA. It seems most probable that this situation of dependence will be aggravated. Indeed, the programme designed by the United Nations for 1981-90 forecasts a total of 810,000 million F CFA in foreign aid, out of which 268,200 million will be allocated to rural development.

Any development process which remains dominated by foreign funding faces a number of problems:

> Funds get channelled into those sectors which the government considers really important. In the case of Burkina Faso, agricultural aid has long been focused on cotton farming and market gardening venture. As for cereals, their cultivation

has been judged unprofitable from the viewpoint of the world capitalist market.

It is no easy task to coordinate operations of the various funding agencies.

Recurrent expenditure has to be financed.

The recipient country eventually faces the mountain of principal and debt service payments. A country like Burkina Faso has a tough enough time paying off debt service charges alone.

TABLE 10.7	
Debt Service Charges as a Percentage of Gross Domestic Product	
Year	**% GDP**
1975	11
1976	15
1977	20.3
1978	23
1979	27.2
1980	30.1
1981	32.2
1982	29.3
1983	27

In a nutshell, then, even though there have been noticeable changes in the pattern of agricultural investment, the problems mentioned above still play a crucial role in the development of agriculture.

Government fixing of agricultural commodity prices: Price-fixing is one of the economic areas in which the state makes its weight felt. The situation is quite diversified, since changes in producer prices determined by the government do not seem to indicate any effective correlation between price levels and quantities produced—actual or potential.

As far as cash crops are concerned, their purchase is often handled by state agencies with monopsonic control over particular crop sales. SOFITEX, the state textile corporation, for example, controls all cotton sales. In such a situation, even though there are autonomously managed markets constituting a kind of cotton producers' organisation, individual cotton growers are in fact caught in a monopsonistic situation which is made all the tighter by the fact that prices are fixed beforehand by the state, and there is practically no way to change them.

However, between the quantities of cotton produced and the kilogramme price, one can observe a positive spearman correlation in which R=0.67. And even when the Qt=f (Pt-1) ratio is posed, the correlation is still rather weak; r=0.79.

TABLE 10.8

Comparison of Cotton Production and Price Levels

Farm Season	Price[1] (F CFA/kg)	Quantity (tonnes)
1962-63	32	6.6
1963-64	32	8
1964-65	32	8.8
1965-66	32	7.5
1966-67	32	16.3
1967-68	32	17.3
1968-69	30	32
1969-70	30	32.2
1970-71	30	23.5
1971-72	30	28.1
1972-73	30	32.6
1973-74	31.5	26.7
1974-75	36.5	30.6
1975-76	36.5	50.5
1976-77	36.5	55.3
1977-78	40	38.0
1978-79	40	60
1979-80	40	75.1
1980-81	40	59.4
1981-82	53.5	55
1982-83	53.5	71

1. Average price of 1st and 2nd grade cotton. Grades are determined by SOFITEX buyers.

In the free market economy with free supply and demand, there is a positive correlation between prices paid for a commodity and quantities available. Normally if both land and manpower are available, an increase in the prices of farm products should be accompanied by an increase in production. Otherwise productivity should be played upon. A test of a possible correlation between grain production and official price shows that the spearman coefficient R=0.07 and if one poses the ratio Qt=f (Pt-1), r=0.062. Yet a comparison of relative price changes between cotton and grains shows a much higher ratio in the case of grains. Real producer prices which might be lower than official prices could explain this lack of correlation. Furthermore, grain production depends on a whole lot of factors such as population, agricultural investment, rainfall, prices etc.

Although grain prices are also set by government the grain market is free of any monopsonistic situation. Indeed on the market, besides OFNACER buying grain, there are also traders and individual consumers.

TABLE 10.9

Comparison of Grain Production—Millet, Sorghum, Corn—and Price Levels

Farm Season	Price (F CFA/kg)	Quantity (tonnes)
1962-63	12	847
1963-64	15	885
1964-65	13	1,165
1965-66	11	990
1966-67	13	1,014
1967-68	14	1,028
1968-69	12	1,035
1969-70	12	989
1970-71	12	996
1971-72	12	1,063
1972-73	14	993
1973-74	18	823
1974-75	22	1,065
1975-76	18	1,090
1976-77	21	1,213
1977-78	32	1,050
1978-79	40	1,126
1979-80	40	1,146
1980-81	45	927
1981-82	50	1,125
1982-83	62.5	1,048

Now let us turn our attention to the urban market for local cereals, mainly millet, sorghum and maize. Here we notice that traders do not market all their stocks. They hold back a portion so as to keep prices at previous high levels. Granted, in the last three months of the year—the harvest period—prices charged by traders fall slightly. Overall, however, they remain higher than the official consumer price, and actual prices charged by traders in the urban areas give a comparative idea of the imbalances involved.

Clearly, then, state intervention turns out to be a futile exercise from the point of view of producers and consumers, none of whom gain therefrom. The only group to benefit is the traders.

As we have seen, production levels of grains are almost totally independent of the price structure. Such being the case, we do not suppose that the various increases in agricultural production observed from time to time are stimulated by price incentives.

TABLE 10.10
Discrepancies Between Official Consumer Prices
& those Charged by Traders in Urban Areas

Product	Ratio of price discrepancies
Rice	1 to 5
Maize	1 to 5
Millet	1 to 4
White sorghum	1 to 5
Red sorghum	1 to 5

In practical terms, statements to the effect that producer prices have been doubled or tripled have only rhetorical value, as we have seen. The only official producer prices that can be taken seriously are cotton prices. In our hypothesis, it is possible to compare the price structure for cotton and cereals in their evolution, in order to judge whether it would be more profitable for peasants to continue growing cotton or cereals.

There is no doubt that cereal prices compare favourably with cotton prices. But given the way the market is currently structured, this price advantage does not benefit cereal producers. Better prices for cotton would have been more beneficial to the peasants, because the official price for cotton is at least guaranteed. Under the circumstances, the fact that the price per kg remained low from 1970 to 1980 even though raw material prices on the international market increased, simply indicates that in this instance also, government policy has favoured the CFDT, thereby shortchanging peasant producers.

Furthermore, it needs to be pointed out that the relative improvement of cereal prices was practically forced on the government by the major drought that began in 1973. There was a foreseeable production drop down to the lowest level registered throughout the period of own analysis: 823,000 tonnes. Faced with this drop, the government really had no choice but to raise cereal prices from 14 F/kg to 18 F/kg—a 28.58 percent rise.

TABLE 10.11

**Comparative Listing
of Changing Cotton and Cereal Prices**

Farm season	Cotton % Millet, Sorghum, Maize	Cotton % Rice
1962-63	2.67	1.78
1963-64	2.13	1.78
1964-65	2.46	1.78
1965-66	5.37	1.78
1966-67	2.46	1.78
1967-68	2.29	1.68
1968-69	2.67	1.68
1969-70	2.50	1.58
1970-71	2.50	1.58
1971-72	2.50	1.58
1972-73	2.14	1.58
1973-74	1.75	1.05
1974-75	1.66	1.04
1975-76	2.03	1.04
1976-77	1.74	1.04
1977-78	1.25	0.63
1978-79	1.00	0.63
1979-80	0.00	0.63
1980-81	1.89	0.63
1981-82	0.07	0.81
1982-83	0.86	0.79

THE IMPACT OF STATE POLICY

From previous developments the impact of state intervention in agricultural in terms of extension, equipment and production can be better assessed now.

Evolution

Over the 23 years under discussion, agriculture in Burkina Faso as a whole has been stagnant, but this stagnation has been most serious in the area of cereal production. The country's agricultural sector was structurally modified by state intervention, through the establishment of Regional Development Units and the control of cash crops by a number of transnational corporations. That is why we have differentiated the agricultural sector into two branches: the traditional branch and the modern (cash crop) branch.

This is the perpective within which the agricultural sector may be said to have evolved. In this evolutionary process, the more active branch has been the branch producing industrial cash crops. For example, as we have already observed, cotton production has increased 29.17 times; similarly, sugar cane production, practically nonexistent on the eve of independence, is growing faster and faster. In 1961 the average cotton yield was 111 kg per hectare. By the 1981-82 farming season the yield had risen to 979 kg per hectare—an increase of 8.82 times. These cash crops register good yields. This is not solely because of the quality of the soil. Other factors contribute to their high yields: extension services, farming methods and fertiliser use, as well as other production factors.

TABLE 10.12
Extension Staffing Index per Farm
and per Regional Development Unit

Rural Development Unit	Number of farms	Number of extension staff	Farms per extension officer
Central	83,018	111	748
East-Central	47,792	3	111
West-Central	80,668	84	960
North-Central	68,280	130	325
East	43,857	153	287
Hauts-Bassins	49,657	167	297
Yakuga	53,754	72	747
Sahel	38,600	45	858
Bougouriba	39,000	107	364
Black Volta	60,954	112	616
Comoé	19,300	36	5,361
		1,062	

Thus it appears that it is only in such areas as the AVV and the Bouki plains, among others, that peasants tried to form cooperatives. However, only a few out of the total number of peasants joined such village groupings. Extension staff for peasant farms is quite insufficient.

An aggravating factor is that extension officers have to put up with terrible working conditions. They are recruited from among the "rejects" of the orthodox school system. After that it depends on the various Regional Development Unit projects whether they get paid salaries or not. Given such conditions, the staffing system is extremely weak, except in the cotton-growing areas.

On the whole, peasant society has been poorly structured. Extension staff has remained inadequate. We do not know whether, in the so-called liberal economy as it has operated in Burkina, peasants have tried to procure equipment on their own as individuals, or have relied on the established state structures.

Rate of Equipment Acquisition among Peasants

We are not in a position to present a yearly breakdown of data. However, statistical evidence indicates that production equipment increased from 12,900 units in 1966 to 141,000 units in 1982.

TABLE 10.13

Equipment Acquisition Rates among Peasants

Type of equipment	YEAR 1966 Number	% of total	1982 Number	% of total
Ploughs	2,400	18.6	45,600	32
"Manga" brand hoes	9,000	69.76	48,800	346

It should be pointed out, however, that in spite of this increase in equipment most of the modern tools are used exclusively in cotton-producing areas.

TABLE 10.14

Percentage Distribution of Agricultural Equipment According to Agricultural Production Options

Equipment	Cotton-growing regions 1980-81	1981-82	Other regions 1980-81	1981-82
Plows (all types)	39	21	21	79
Carts (all types)	39	32	61	68
"Manga" brand hoes	0	0	100	100
Harrows	0	0	100	100
Seeders	98	60	2	40
Furrowers	0	0	100	100
"Como" brand sprayers	62	4	38	96
"ULV" brand sprayers	80	38	20	62

It should be pointed out that in the two cotton-growing regions, the donkey-drawn "Manga" brand hoe is not used. With this table it becomes easy to see that a great deal of agricultural equipment is concentrated in the cotton sector. It is also in the cotton-growing areas that the SOFITEX corporation has exerted its ingenuity in the development of agricultural mechanisation.

Peasants also use such instruments as harrows, seeders, furrowers and sprayers. The National Advanced Training Centre for Rural Artisans (CNPAR), which trained about 1,219 crafts-people (including 440 smiths) between 1970 and 1982, has been partly responsible for the rapid development of modern production tools. At the moment this Centre has six regional branches. Under current conditions, the 440 trained smiths are capable of turning out a yearly average of 22,000 ploughs. Still, these figures are extremely low : as late as 1982 the ratio of farms to ploughs was 4 to 1.

As for fertiliser, from 1963 to 1982, imports rose from 287 tonnes to 27,900 tonnes. If, then, this quantity of fertiliser were to be divided up among the country's farms, each farm would get 48.4 kg. Unfortunately, this fertiliser goes mainly into cotton farming. Leaving out the natural phosphates of Kodjari, we note that other types of fertiliser made up only 5 percent of fertiliser used by the Regional Development Units in 1980-81, while in 1981-82 the figure was 8 percent. Of this quantity the two cotton-producing Regional Development Units (Black Volta and Hauts-Bassins) consumed 76 percent in 1980-81, and in 1981-82 they consumed 80 percent.

Quite obviously, this means that in the cotton sector a great deal of effort has gone not only into providing incentives to peasants, but also into supplying them with the material resources indispensable for improved cotton farming.

Interestingly, peasants have been growing other cash crops (groundnuts and sesame) using the old traditional methods. Result: yields have remained stagnant. As for the areas devoted to these crops, they increase and decrease in inverse proportion as the areas sown with food crops decrease or increase.

In sum, the improvement of productivity and the introduction of modern production which could have sparked sustained growth never happened, for the most part. And in the few cases where such developments did take place, their impact was restricted to the cotton-growing zones.

There have also been changes in the social system. The splitting of the agricultural sector into two main branches has resulted in a greater differentiation between peasants based on their purchasing power and on new social relations created by the production process.

As far as such social production relations are concerned, it is difficult to conduct the kind of class analysis designed to reveal class conflicts. The closest one gets to class struggle is the situation of the newly proletarianised Comoé peasants on the one hand and "the industrial set". Possibly, the emergence of agricultural workers and the birth of an agro-industrial system may signal the beginning of changes in agricultural structures. However, even so the state would first have to impart a class character to the process.

The evolutionary changes, in effect, have sparked relatively important social transformations. Under such circumstances it is no longer valid to talk in simplistic terms of "the peasantry" the way it was possible just before 1960. The changes referred to are still of a local nature, depending on the extent to which capitalism has penetrated the country's agricultural sector. They may bring upheavals in their wake, but that is no licence for thinking there has been an agricultural revolution in Burkina Faso. Revolutions imply profound changes. Of such changes there are so far no clear indications.

The Policy of State Intervention: a Failure
The policy of state intervention has been based on no coherent strategy whatsoever. Certainly, in various development plans the intention of giving priority to agriculture is expressed and affirmed. Actual priorities followed in practice, however, have been quite different, whether we look at structural reforms, at production funding or at the organisation of the extension system. Generally speaking, the sector or branch of the economy expected to serve as the driving force in the development process, on which maximum effort ought to be concentrated so that in its growth it can pull along the other branches and sectors, has not been identified. Furthermore, in a country with regions unequally endowed with natural resources, the issue of regional specialisation based on the theory of comparative costs has not even been broached. Quite the contrary: as we pointed out in the study on agro-industry and the multinational corporation,[5] the state has been content to let international capitalist groups find a solution to this issue. The various Regional Development Units have become private happy hunting-grounds for various agencies: USAID in the Eastern Zone, the World Bank in the Western Regions, the Netherlands in the North-Central Region, and so forth.

The third feature of state intervention in Burkina Faso has to do with the distribution of national wealth. To put it more clearly, the issue at stake is the impact of various sectors of the country's economy on the development process. We have not treated this issue in our analyses. However, it is important to point out that the industrial sector, thanks to the Investment Code, enjoys a whole series of advantages, ranging from tax exemptions to the repatriation of profits and salaries of expatriate experts. By contrast, because peasants are not sufficiently organised to demand better remuneration for the work they do, agriculture has consistently been shortchanged. As for the banking and other financial institutions, they have remained closed shops solely devoted to siphoning off savings at low rates for speculative investment on the major international money markets. Meanwhile the national economy, starved of ready cash, is obliged to borrow money at exorbitant long-term interest rates on the capital market.

To put the matter in a nutshell, in all its 23 years at the political helm of the state, the Upper Volta bourgeoisie was unable to develop the productive forces indispensable to its own survival. It remained a backward bourgeoisie managing a state whose institutions operated in total contradiction with the reproduction of the bourgeois system at a higher level, and were incompatible with the aspirations of the country's emergent youth, especially the literate urban youth. Even within a straightforward capitalist system, whenever productive forces get stalled in such a manner, social revolutions are likely to break out. In the case of Burkina Faso this was doubly true. For, with its meagre funds gleaned mainly from taxes, the state still had to continue importing cereals to make up the shortfall in agricultural production.

To put it bluntly, then, the agricultural and food production policies of the Burkina state have been ambiguous. And it is this ambiguity that has led the country into economic failure.

NOTES

1. See Thiombiano, T., "L'Enclave Industrielle: La Société Sucrière de Haute-Volta", CODESRIA, 1983.
2. The survey was on "Nutritional Systems in the Urban Areas: The case of Ouagadougou". We might assume consumption patterns in Bobo-Dioulasso to be identical, even though the main consumer staples in that Western Zone town are rice and maize, not millet and sorghum.
3. Sinkame, Nana, "Les marchés des produits de base: perspectives et implications pour l'Afrique de l'Quest", paper presented at the BCEAO-IMF seminar on "The Adjustrment Programme and Economic Growth", Dakar, April 1984.
4. Ouli, Kamadini, "La dépendance en héritage au Burkina Faso", August, 1985, unpublished article.
5. Thiombiano, op. cit.

APPENDIX 10.1
Per Capita Cereal Production
& Rate of Satisfaction of Per Capita Food Needs
in the Rural Areas

Years	Cereal production ('000)	Total population ('000)	production	Seeds and shoots	Per capita availability (kg)	Self-sufficiency index as % of 215kg standard ((6/3)
(1)	(2)	(3)	(4)	(5)	(6)	(7)
1962	701	4.3	163	20	143	67
1963	876	4.4	199	20	179	83
1964	901	4.4	205	20	185	66
1965	1,187	4.5	265	20	245	114
1966	1,012	4.6	220	20	200	93
1967	1,036	4.7	220	23	200	93
1968	1,057	4.7	225	23	202	94
1969	1,061	4,8	221	23	201	93
1970	1,011	4,9	206	25	181	84
1971	1,018	5.0	204	25	179	83
1972	1,051	5.0	204	25	179	83
1973	991	5.1	194	25	169	79
1974	812	5.2	156	25	131	61
1975	1,160	5.3	219	27	192	89
1976	1,088	5.4	201	27	174	81
1977	1,196	5.5	217	27	190	88
1978	1,025	5.6	183	30	153	71
1979	1,025	5.7	210	29	172	80
1980	1,147	5.8	190	26	162	75
1981	991	5.9	168	25	143	66
1982	1,200	6.0	200	30	170	79
1983		6.1				

The resulting indices are disappointing. From 1962 to the present, cereals in the country have never been sufficient to meet the needs of the rural population, except in a single year, 1965. This index, a decisive one, gives sufficient grounds for calling the state's food policy from 1960 to 1983 a complete failure.

	Appendix 10.2 **Agricultural Investment**			
Year	**Total investment in million CFA**	**Investment as a percentage of value of primary sector production**	**Index of investment volume (1970 = 100)**	**Agricultural equipment index (1970 = 100)**
1960	900	3.0	44	
1961	1,300	4.3	62	
1962	1,700	5.2	79	
1963	1,800	5.4	82	
1964	1,500	4.2	67	
1965	2,200	6.0	98	
1966	1,200	3.1	53	56
1967	1,630	4.4	69	75
1968	1,550	4.3	63	
1969	(2,000)	5.0	80	
1970	(2,500)	6.5	100	100
1971	(3,000)	7.3	120	113
1972	4,000	9.6	151	135
1973	4,500	11.8	150	
1974	5,000	10.7	154	137
1975	6,000	12.2	172	
1976	7,500	15.5	200	
1977	5,240	8.8	119	259
1978	7,820	12.4	174	
1979	11,200	16.0	224	
1980	19,000	23.0	330	
1981	15,700	17.0	209	
1982				610

Source: Lecaillon, J. and Morrisson, C., *Politiques Macroéconomiques et Performances agricoles: le cas de la Haute-Volta (1960-1985)*, OECD, February 1984.

11. THE STATE AND AGRICULTURE IN ZAMBIA: A REVIEW OF THE EVOLUTION AND CONSEQUENCES OF FOOD AND AGRICULTURAL POLICIES IN A MINING ECONOMY

Adrian P. Wood & E.C.W. Shula *

INTRODUCTION

Despite the relatively high level of urbanisation in Zambia and the dominance of the mining in the economy, agriculture has been an important influence upon the pattern and nature of modern economic development in the country. It is the source of livelihood for the majority of the 3.5 million people who live in the rural areas, and it provides a growing number of urban jobs in related activities. More importantly in recent years, with the impact of the world recession upon copper mining and the far-reaching consequences of this in the Zambian economy, agriculture has been recognised as one of the few sectors with good prospects for growth and as potentially a major source of new employment, government revenue and foreign exchange.[1]

However, the achievement of rapid expansion within the agricultural sector is not easy, as many environmental, logistical and economic constraints are faced.[2] The sector is also heavily influenced by a range of government policies and legislation, which are now being reviewed and, in some cases, reformed as part of major economic adjustment policies. Many of these agricultural policies and regulations have long histories, and so it is necessary to take an evolu-

* This paper was written by Dr Wood from conference papers presented by Mr Shula and himself at the CODESRIA meeting in Addis Ababa in 1984. This work was undertaken when Dr Wood held a British Council Fellowship at the University of Birmingham. The support of these two organisations is gratefully acknowledged.

tionary perspective in order to understand their origins, and the way in which they have become part of the framework within which agriculture operates.

This chapter reviews the development of state intervention in the agricultural sector from the early colonial period until the present day. While the focus is upon agricultural and food policies, other economic and political developments which have had an impact upon agriculture are also considered. The colonial period is discussed to identify the origins of many policies and ideas which have become recurrent themes in the debate concerning agricultural policy in the country. For the period since independence, greater attention is paid to both the varying views among government and party officials concerning the nature of agricultural change desired, and the actual performance of the agricultural sector in response to government policies and relative to national needs. The concluding section considers the crisis which occurred in Zambian agriculture in the late 1970s, analyses the varying policy responses which this elicited, and considers the prospects for creating a more dynamic agricultural sector.

THE COUNTRY AND ECONOMY

Zambia is an extensive, but sparsely settled, landlocked country in central-southern Africa. With a population of only 5.7 m in 1980 occupying the country's 750,000 km², and with over 40 percent of that population living in towns, rural population densities are low, averaging only 4.3 persons per km². [3] There are, however, major variations in rural population densities which range from less than one person/km² in areas infested with the tsetse fly (which are often National Parks), to over 200 persons/km² of arable land, in small parts of the south and east where colonial land alienation combined with viable small-scale commercial farming has created considerable population pressure.[4]

Conditions for agriculture vary considerably. The country has a savanna climate with a single rainy season, but rainfall ranges from over 1,200 mm per annum in the north to less than 700 mm in the south and south-west. While harvests are relatively reliable over the northern half of the country, increasing rainfall variability combined with lower rainfall totals often causes crop failures especially in the Gwembe Valley bordering Lake Kariba and in the south-west. Soils are generally poor and only 58 per cent of the country is suitable for cultivation.[5] Extensive areas in the west have sandy soils of limited agricultural use, while in the north the higher rainfall leads to soil acidity if permanent cultivation is practised without liming.[6]

In precolonial times there were several different rural economies practised in the country. In the north-west hunting and gathering were the major economic activities, while in parts of the north, the east and the west raiding and the extraction of tribute supplemented agricultural production.[7] The major traditional crops are finger millet which is most common in the high rainfall areas, and sorghum and bulrush millet which are the staples of the drier regions. Cassava has become more widely cultivated during the last 300 years, having been introduced into the country from the West; while maize, introduced from the East, had been adopted in parts of the east and centre of the country before the advent of colonial rule. Beans, groundnuts and cucubrits are the major traditional subsidiary crops, with tobacco grown in some areas.[8] The keeping of cattle is restricted by the presence of the tsetse fly in almost half of the country. In precolonial times cattle played only a limited role in agriculture through the provision of manure; cultivation in that period being either by digging stick or hoe. The majority of traditional farming systems involved extended bush fallowing in order to maintain soil fertility with either soil selection methods used for the choice of gardens, or *chitemene* practices used to concentrate wood ash in a locality to create adequate soil conditions.[9] Agriculture on the whole was well adjusted to environmental conditions and in most areas produced an adequate subsistence although with different degrees of security. However, this adaptation to local conditions produced considerable diversity in the traditional agricultural systems which in turn has given farmers in different parts of the country varying potentials for agricultural development as circumstances have changed.

Colonial rule brought into one administrative unit a great number of rural communities which had often previously had little or no contact. More importantly, the economies of these communities were subordinated to the needs of the new mining economy. Since the late 1920s copper has dominated the economy and the country's fortunes have fluctuated considerably with the movements in the world price for this commodity. With the boom in copper prices shortly after independence in 1964, the country's dependence on copper increased, with 94 percent of the nation's foreign exchange earned by copper and 47 percent of government revenue provided by mineral taxes.[10] The mining industry has also been responsible for the relatively high level of urbanisation in the country with seven of the ten largest urban settlements being mining towns which contain half of the urban population. The Copperbelt is also the focus of the country's infrastructure with the major routes of colonial penetration by road and rail traversing the country in a north-easterly direction from Livingstone on the southern border to the mines on the Zairean border.

THE COLONIAL INHERITANCE

The colonial period introduced major changes into the agricultural situation in the then Northern Rhodesia. Despite 20 years of independence this heritage still lies heavily upon the country. This is especially true in the structure of agricultural production and in the pattern of market-oriented agricultural production, as the colonial period introduced and formalised an uneven pattern of change in the rural areas. The colonial period also gave sanctity to a rural-urban dualism in development which has also had a major impact upon the rural areas down to the present day. But perhaps one of the most important inheritances from the colonial period is a variety of policies, ideas, and attitudes towards agriculture which, despite much new thinking and many new policies since independence, still influence Zambian agriculture today.

Colonial penetration began in earnest in the last decade of the 19th century through a series of treaties between the British South Africa Company and local chiefs. By 1900 the whole country had been included in a crown charter granted to the company which administered the territory until 1924 when the British government took over.[11]

The British South Africa (BSA) Company sought minerals whose exploitation it hoped would provide profits for its shareholders. However, during Company rule the territory proved disappointing in this respect and the main value of Northern Rhodesia was increasingly seen as a source of labour for mines in other parts of the region where the Company had interests. Partly to "encourage" this labour migration and partly to recoup some of the administrative costs it incurred, the Company imposed a hut tax throughout the territory. With few local opportunities for employment at that time, and with limited local markets for agricultural produce, the African population had little choice but to send their menfolk out in search of employment mainly in the mining and urban areas. In this way an important relationship between the rural and urban areas was established which is crucial to an understanding of Zambian rural development. The most important aspect of this is the attitudes engendered among rural dwellers who have come to regard towns as the source of all that is new and good, and the place to go to if one wants to "get on" in life and obtain money. The result has been increasing rural-urban migration (until the late 1970s) by young adults seeking urban employment.[12] With longer periods of urban residence since the late 1940s this loss of youthful labour from the countryside has had a damaging impact upon traditional agriculture, reducing output, disrupting traditional farming systems and making labour shortage a major constraint in agricultural development.[13] Despite the removal of hut taxes on independence, the movement to the cities accelerated during the 1960s and early 1970s reflecting in part the attitudes outlined above which colonial policies had established, and also the urban bias in national development which again owed much to ideas and policies inherited from the colonial period.

European settler farmers and a new land tenure system to permit their settlement are two other important inheritances from state intervention in the colonial period. The BSA Company's attitude to European farmers was originally rather unfavourable because of its wish to maintain all land rights in its own hands while mineral prospecting occurred.[14] However, as it became clear that mineral deposits were not widespread, the Company began to encourage settler farmers who it saw as capable of generating revenue for the territory through the export of food to the Katanga mines.[15] The Company's main justifications of settler farmers lay in its perception of African farming as primitive and incapable of producing sufficient surplus for the mining and urban populations, and in its view that a plentiful supply of cheap food was essential for the development of a competitive mining industry. Certainly the mines did create a major demand for food, and their development turned the territory from being a food exporter to being a food importer.[16] However, with the rapid increase in marketed production which African farmers achieved in the 1930s, it is doubtful whether settler farmers were really needed, although their more stable marketable surpluses were an important consideration.[17]

Land was alienated for European farmers by the Company using the treaties it had made with the local chiefs.[18] The majority of this alienation was in the east around the Company's headquarters at Fort Jameson (Chipata) and more importantly along the ''line of rail'' from Livingstone to the Copperbelt on the Congo/Katanga border (Fig. 11.1). By 1921 there were over 700 Europeans engaged in agriculture in the country.[19] Their position was strengthened in the late 1920s when the Colonial Office introduced a new land tenure system which effectively favoured increased European settlement. Land up to 20 miles either aide of the line of rail was designated as alienated Crown Land for settlement by European farmers and was cleared of Africans, while the majority of the country was designated as unalienated Crown land where European settlement could take place and so lead to the removal of the African population in the future.[20] This land provision for European settlement proved vastly in excess of the level of settler immigration and in 1947 much of the land reserved for future European settlement was returned to African jurisdiction as Native Trust Land. However, 2.5 percent of the country remained alienated

TABLE 11.1
Land Tenure Categories(%)

	1937	1950
Native Reserves	18.7	18.8
Barotseland (Protectorate)	19.8	19.9
European Farms (Crown Land)	1.4	2.2
Company Concession	3.3	-
Other Land Alienated to Europeans	-	0.5
Unalienated Crown Land	54.0	-
Native Trust Land	-	55.4
Forest and Game Reserves	2.7	0.6
Towns	0.1	0.1

Source: Calculated from Hellen, J.A., *Rural Economic Development in Zambia, 1890-1964*, Munich, Weltforum Verlag (Ifo-Institute für Wirtschaftsforschung Munchen. Afrika-Studien 32), 1968, pp. 79-80.

to Europeans, the majority of this being for farming (see Table 11.1).

One of the new categories of land tenure introduced in the late 1920s was Native Reserves. These were established in areas where competition for land existed between the African and European populations and adjoining areas where Africans were removed from alienated Crown Land. The government's justification of the Reserves was to ensure that sufficient land was left for the African population in these areas. However, both the quality and the quantity of the land "reserved" for Africans were generally far from satisfactory. For example, in the Batoka area of Southern Province, the 1932 Agricultural Survey Commission thought that "any land that had poor soils, inadequate water supplies, low nutrition grasses unsuitable for European cattle or [was] overgrown with impenetrable bush, was not suitable for Europeans and instead should be allocated to Africans".[21] Even after the addition of the Trust Land in the 1940s only 34 percent of the land available to Africans was suitable for cultivation and only 7 percent was high-quality arable land.[22]

Land provision through the Reserves was often inadequate because those allocating the land failed to understand both the nature of the traditional farming systems with their needs for bush-fallow land, and the prospects of African population growth and increased cultivation in response to market opportunities. The result was considerable overcrowding in many of the Reserves, especially in the Eastern Province, which led to soil erosion and ecological deterioration which are major problems in most former Reserves today. Thus the Reserves policy not only excluded Africans from the best loca-

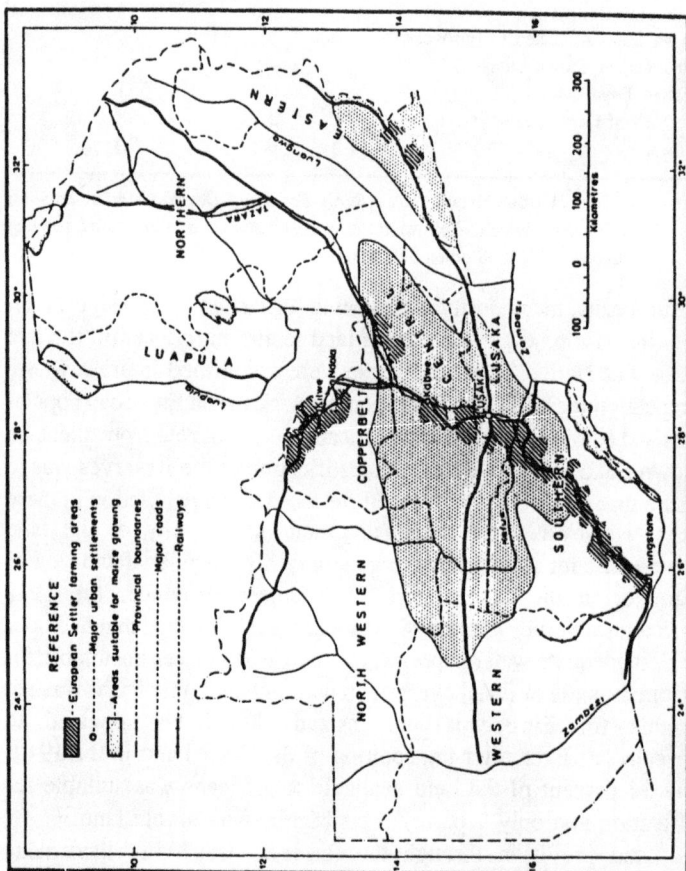

Fig.1 EUROPEAN SETTLER FARMING AREAS, URBAN MARKETS, AND MAIZE GROWING AREAS IN ZAMBIA

tions for market-oriented agriculture by allocating these to the settlers, but also hindered African agricultural development by restricting access to good land.

In contrast, the European farmers had extensive land holdings, their farms varying in size from a few hundred to several thousand hectares. These farms were generally situated on the better soils although often less than a third of the land would be well suited for arable farming.[23] However, the development of commercial farms progressed slowly and by the late 1950s it was estimated that only 5 percent of European farmland was tilled.[24] This "underuse" of alienated land in contrast to the intensive use or "overuse" of neighbouring Reserves has been a major issue in the country's agricultural development since the Reserves were established, and remains so today.[25]

State intervention in agricultural marketing in Zambia dates back to the 1930s. Settler farmers were not very successful in the early decades and often seemed to be struggling to survive.[26] Their maize yields were low and they needed good prices in order to achieve what they regarded as a satisfactory standard of living. Despite the negative implications of the Reserves policy, African farmers in the more accessible areas had begun by 1930 to orientate their farming towards producing surpluses for sale. Between 1930 and 1935 African maize sales increased more than threefold while sales from European farmers increased by only 25 percent (Fig. 11.2).[27] This dynamism in the African farming sector was not initially a problem as the development of the Northern Rhodesian and Katangan copper mines provided adequate markets. However, in the early 1930s the world depression led to the closure of all but two of the mines in Northern Rhodesia and a similar decline in Katanga which greatly reduced the demand for foodstuffs, especially maize. Settler farmers became increasingly concerned about how they could dispose of their produce profitably and saw the rising production by Africans as a major threat which could flood the market and force them out of production. As a result, the 1935 report of the Agricultural Advisory Board recommended the establishment of a Maize Control Board to regulate the marketing of this crop,[28] and this was duly established in 1936.

The Maize Control Board created a number of precedents which have been followed to varying degrees in subsequent agricultural policy. Most importantly, it confirmed that maize was the dominant starch staple for sale. This began a period of almost 50 years in which the state, through marketing arrangements, has encouraged the production of this crop rather than other staples, and so has helped change urban dwellers' food preferences away from their traditional crops, such as sorghum, millet or cassava, towards maize. This preference

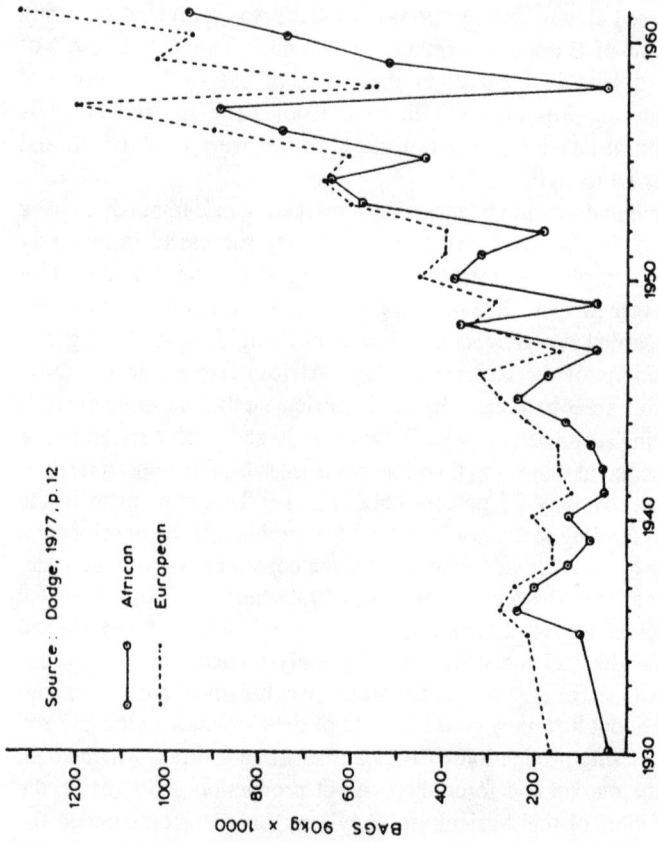

Fig. 2 OFFICIAL PURCHASES OF MAIZE BY PRODUCER GROUP

for maize was partly the result of the agricultural experience the settler farmers brought with them and partly a result of the farming conditions of the areas in which they settled and the crops grown there traditionally. It was also a result of the storage advantages of maize over other crops and of the acceptability of maize to the urban consumers who often found it more palatable than traditional crops.

The Maize Control Board (MCB) operated by dividing the market into two pools, internal and export, and by allocating European producers three-quarters of the more lucrative internal market, while African producers only had one-quarter. Surplus production over internal needs was sold on the external market at lower prices. This system, together with differential pricing, which gave African producers on average only 70 percent of the price received by Europeans, considerably penalised African farmers. Fortunately, the full potential of these measures was never realised as local demand improved in 1938 and there was no local maize surplus until 1953.[29] Indeed, it has been argued that the MCB by guaranteeing a market through its export of surplus maize actually encouraged increased production by African farmers in the areas it served.[30]

The MCB also established the principle of the state as a monopoly buyer, offering guaranteed prices and accepting the responsibility to buy all the maize offered to it. However, the area within which this service was provided was only eight line-of-rail districts for much of the colonial period, and the original depots where maize could be delivered were only along the railway in areas of alienated Crown Land.[31] The method of fixing producer prices was based on a formula which considered the cost of production plus a fair profit, and this has been maintained ever since.[32] This has effectively meant that the home market and producers have been protected from cheap imports, and that producer prices have not responded to import or export parity prices.[33] Control over the maize producer price also extended to the selling price and hence could be used to influence local food prices and the cost of living. The BSA Company's idea of keeping urban wages low in order to help local industry, primarily mining, to be competitive seems to have been influential during much of the life of the MCB, because when imports were necessary to meet local demand, as they were from 1942 until 1954, consumer prices were not raised to cover the costs and the state was responsible for a major subsidy, amounting to over £7 million in this period.[34]

The establishment of the MCB, while representing a political victory for the settler farmers, also marked an economic victory for the farming community *per se* which had effectively forced the state to accept the responsibility for creating a stable internal market for major categories of agricultural produce. This idea was confirmed by Clay in his report concerning development planning after the Second World Ward,[35] and has been an important aspect of post-independence thinking with respect to agriculture. However, it must not be forgotten that in the colonial period this stable internal market was not equally accessible to all producers, as Europeans had preferential treatment, while the majority of Africans who lived outside the MCB area had no guaranteed market.

The operations of the colonial state in favour of European producers included a number of other measures besides the MCB and land policy. Cattle marketing was subject to a Cattle Control Board from 1936. This placed restrictions on the import of high-grade beef which the settlers produced, but allowed an influx of low-grade beef imports similar to that which African farmers produced. As a result the price of high-grade beef rose by 460 percent between 1937 and 1956 while that for low-grade beef rose by only 200 percent. This protected local European producers and increased their share of local sales from 33 to 56 percent between 1936 and 1960. It also helped keep down miners' wages as they consumed much of the low-grade beef.[36]

The provision by the state of agricultural credit and extension advice was also biased in favour of the European farmers. Following settler pressure the government established a Land Board in 1947 which became the Agricultural Land Bank in 1953. While partly a means of attracting European immigrants who had limited funds for farming, the Land Bank was also a source of seasonal credit and so supplemented the commercial banks which had previously been the settlers' major source of credit. African farmers, on the other hand, had no legal title to their land as they were not on Crown Land, and so had no access to official credit until 1960 when the Peasant Farming Loan Fund was established. Even then, only a small proportion of the African farm population was served, these being mainly members of particular government schemes (see below). Extension advice from the Department of Agriculture was primarily directed towards the settler farmers throughout the colonial period. Little or no extension service was provided for Africans before the 1940s because of the fear that their increased production would force Europeans off the land.[37] Despite the establishment of staff posts for African agricultural extension, and the creation of a training school for African agricultural assistants in the 1950s, expenditure on supporting African farming was only a fraction of that received by the settlers; the figures in the late 1950s showing a ratio of 3:1 between services for settlers and those for Africans.[38]

While it is clear that the colonial state greatly supported European farmers, its attitudes and policies towards the African producers were not static. In the period up to 1940 the dominant view of the state was that African agriculture should be maintained in its traditional condition, with sufficient land and other resources to ensure production would meet subsistence needs.[39] This "status quo" view was based on the political situation which saw African agriculture as a threat to the settler farmers, but was also "justified" by the argument that increased market production by African producers would lead to further serious soil erosion and ecological damage.

Policy began to change in the late 1930s as the Colonial Office expressed the view that improved subsistence production should be sought, and as the Northern Rhodesia Department of Agriculture began to realise that the resettlement and conservation measures required in the Reserves should be the first stage of more widespread development measures for those African producers.[40] The Department's view was that through the provision of opportunities for market agriculture African farmers would be encouraged to adopt "better" agricultural practices. However, this was viewed as a very long-term process which would take place within the existing dualistic structure of the rural economy with its differing circumstances for European and African producers. The Clay Report of 1945 developed this idea further by suggesting that European and African agriculture should be seen as complementary. Clay suggested that Europeans should be encouraged to concentrate more on specialist crops and so allow African farmers to have a larger share of the maize market. He also proposed the development of rural depots by the MCB and the introduction in the MCB areas of uniform prices at the depots—despite the variations in transport costs for MCB, in an attempt to reduce the concentration of crop cultivation near depots which was thought to be causing over-cultivation and soil erosion.[41] These measures were adopted in 1951 in the MCB areas along the line-of-rail, while in 1949 the state began to buy maize in the southern part of Eastern Province and established an Eastern Province Agricultural Board in 1952. This acted like the MCB although producer prices were much lower as the state did not wish to encourage maize production in this area relatively far from the market, and was not prepared to subsidise the costs of transport to the market.[42] A change in the state's attitude towards African agriculture occurred in 1957 under the Federation government when the price differential between African and European maize producers was removed, and the range of crops purchased by the state marketing agencies was widened to include sorghum, bulrush millet, finger millet and beans, as well as maize and groundnuts.[43] However with the state still only providing a guaranteed market in the most accessible parts of Eastern Province and the line-of-rail area, such changes in policy had a limited impact on the African farm population.

Somewhat more dramatic changes in African farming were sought during the post-war period primarily in response to Colonial Office pressure. In 1946 the African Farm Improvement Scheme was begun in Southern Province; it was extended into Central Province in 1952, while in 1948 the Peasant Farming Programme was established in Eastern Province, and an intensive Rural Development Programme began in Northern and Luapula provinces in 1956. These schemes all had the aim of transforming a small number of "progressive" African farmers into modern market-oriented producers by imposing on them a series of cultivation practices which were often rather Eurocentric in nature and of doubtful economic viability. Although pursued until independence, they involved less than two percent of the African farm population and produced less than 10 percent of African marketed production despite the major concentration of resources upon these favoured few.[44]

By the end of the colonial period the policies which the state had pursued had led to considerable differentiation in the agricultural sector. On the one hand there was the state-supported settler farming community of about 1,200 households who operated freehold or leasehold Crown Land farms of between 200 and 5,000 hectares. Using mechanised means of production, chemical fertilisers and a considerable African labour force totalling some 45,000, they produced over 75 percent of the country's marketed farm production.[45] In contrast the one-third of a million African farm households, for the most part labouring under the constraints of disciminatory colonial policies, remained primarily subsistence producers in most parts of the country, farming only a few hectares and practising their traditional farming systems as best they could with the labour shortages which mining development and taxation had caused. Only in the more accessible areas, especially along the line-of-rail and in the southern part of Eastern Province, had small-scale commercial production occurred.[46] These producers were responsible for approximately one-third of the country's maize and beef sales at independence, although they also produced almost all of the groundnuts, sorghum, millets and Turkish and Burley tobacco which were marketed.[47]

The distribution of African agricultural development reflected partly the pattern of land alienation and settler farming, and also the distribution of precolonial rural economies growing maize and groundnuts (which had become commercial crops) and keeping cattle (which became a source of draught power for ploughing).[48] The pattern of agricultural change in the African sector also reflected the dominance in government policy of economic considerations, concerned with minimising the cost of food production, over social ideas concerned with spreading opportunities for marketed agriculture more widely

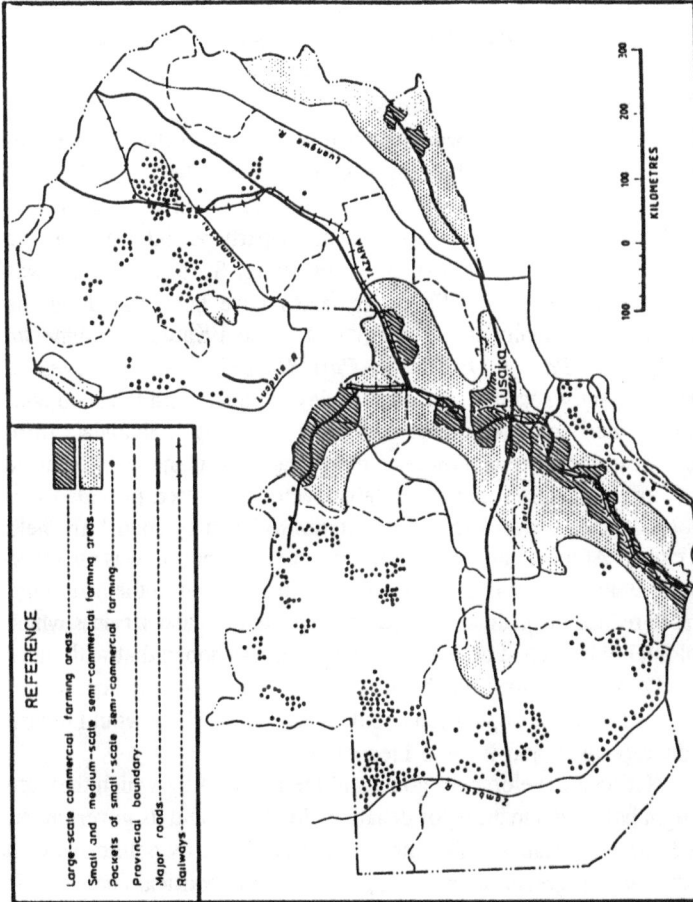

REFERENCE

Large-scale commercial farming areas

Small and medium-scale semi-commercial farming areas

Pockets of small-scale semi-commercial farming

Provincial boundary

Major roads

Railways

FIG. 3 : DISTRIBUTION OF AGRICULTURAL DEVELOPMENT IN ZAMBIA, c. 1975.

in the country. Even in the latter years of the colonial regime, when political considerations made government committees consider social aspects of agricultural policy, the view prevailed that commercialised African agriculture should only be supported and encouraged in areas it had already developed and was most economic.[49] Hence in spatial as well as structural terms Zambia inherited a differentiated agricultural sector.

THE FIRST 15 YEARS OF INDEPENDENCE

The Nature of the Zambian State

Independence in 1964 produced major changes in the character of the state and in its development goals and policies. The United National Independence Party (UNIP) became the dominant political party, forming the governments during the multi-party period and being the sole political organisation since the one-party Second Republic was declared in 1972. UNIP's programmes have been based on the humanist philosophy and ideology of President Kaunda (*Humanism in Zambia, Part I, 1967 and Part II, 1974*), which reveal a predominantly socialist approach to development with the state controlling the "commanding heights of the economy". Kaunda's form of African socialism is, however, more pragmatic than that of Nyerere in Tanzania, with a major capitalist sector permitted provided that its tendencies towards the "exploitation of man by man" are held in check. At independence increased state intervention was also seen as necessary to ensure greater African participation in the economy, and to reduce inequalities in access to opportunities and services which colonial policies had developed. Overall it was hoped that with more positive action by the state the provision of gainful employment opportunities for all the population could be achieved, which would in turn reduce poverty and income inequalities.

National development goals and the methods by which they are sought have been matters for debate within UNIP and its government. In a one-party state a wide range of views are held by politicians with some placing greater emphasis upon social and political goals, while others see economic considerations to be uppermost. Party policy, which is set by the Central Committee, is interpreted by cabinet ministers, and criticised by members of parliament, while it is implemented by civil servants and ministry field staff who may have different views from the politicians and decision-makers. As a result there is often tension between groups with different views at the various decision-making levels, and between different levels in the decision-making and implementation chain. This situation is important in explaining both some of the diversity which has developed in Zambia's agricultural policy, and some of the implementational problems which have occurred.[50] However, financial and control problems have also been influential in explaining implementational difficulties, while oversimplistic and over-centralised conceptions of the agricultural situation have also been partly responsible for the limited achievements in the agricultural sector since independence.

Objectives and Policies
The various government policy documents provide a range of statements concerning the goals sought in the agricultural sector.[51] While varying emphases are given in these, reflecting the range of views within the party and government, seven major objectives seem to dominate the discussions. These are:

a) to increase the importance of agriculture in an attempt to diversify the economy and reduce the dependence upon cooper;

b) to develop agriculture as a source of raw materials for industry and so stimulate both new industries based on local produce, and new agricultural activities to meet local industrial needs;

c) to achieve higher levels of food self-sufficiency in order to reduce the country's dependence on imports from the "south", whilst at the same time keeping urban food costs low;

d) to increase state involvement in agricultural production in order both to secure production by taking over the farmland of departing Europeans, and to meet the ideological goal of increased state influence over the economy;

e) to increase African involvement in marketed agricultural production, and socially diversify production so that a much wider range of people in socio-economic terms have the opportunity to earn cash incomes through farming, (a policy often summed up by the desire for "production by the masses not mass production");

f) to spatially diversify the opportunities for improved and market-oriented farming to all parts of the country from the accessible high-potential areas already developed, and so improve social equity in the country and in turn reduce rural-urban migration, and finally;

g) to achieve rural development throughout the country and improve the standard of living, incomes, nutrition, and access to services of rural dwellers.

These goals show a clear shift from those of the colonial period with increased attention to more balanced national development, both in terms of the position of agriculture and in terms of the wider involvement in development of the rural population, irrespective of location or socio-economic status. This change, of course, reflected the new political realities in the country which meant that the rural electorate, accounting for 80 percent of the population, could no longer be neglected.

In its development of policies in order to achieve these goals the government appears to have had a number of guiding principles. First, it believed that by removing the discriminating colonial policies and by providing African farmers with the same support which Europeans had received in the past, it would be possible for Africans in large numbers to improve their agricultural production and become small- to medium-scale, commercially-oriented producers. At the same time that some policies towards this goal were encouraging farmers to pursue an individualistic and capitalist type of agricultural development, the state also believed in supporting a range of forms of rural production which included parastatal production and cooperative ventures. Finally, in its policies the government tended to stress that rural development was its primary goal with the provision of rural services and social considerations often uppermost in its thinking about the rural sector—a fact reflected by the creation of a new Ministry of Rural Development to include agriculture.

The major revision in state agricultural policies which occurred in the first years after independence produced a number of new nationwide policies as well as interventions and policies to meet specific situations. At the national level the most important policy changes involved the expansion of the network of crop collection and input supply depots to cover almost all of the country and the move towards uniform crop pricing which was achieved in 1974. These measures sought to remove the locational disadvantage which farmers had faced if they lived outside the areas in Eastern, Southern and Central Provinces which had been served by colonial rural marketing arrangements, and to remove the differential pricing of the colonial period which had discouraged production in areas remote from the market—particularly Eastern Province.[52] In making these changes, the new state effectively widened, to cover the whole population, the undertaking which the MCB had first given in 1936 to buy all agricultural produce of specified nature which was offered for sale, and to offer guaranteed prices which reflected the cost of production plus a fair profit. These measures involved a major increase in government subsidies to farmers, especially those in remote locations, and the development of a monolithic nationwide agricultural marketing and input supply organisation, Namboard, which is one of the largest employers in the country.

State control over crop pricing, marketing and retail prices was also extended in several other ways after independence.[53] The government established the Tobacco Board of Zambia to control the marketing of this crop from 1968, while various efforts have been made to assist vegetable growers by providing a guaranteed market through Namboard or Zamhort. Consumer price legislation has been extended as the political significance of urban food prices has grown, while price controls over a large number of consumer items have been introduced.

The Zambian government increased its support for African farmers by expanding the state-funded services which had previously been provided to only a minority of the rural population. Agricultural extension camps were increased to provide national coverage, although this still remains rather incomplete in some of the more sparsely settled parts of the country.[54] Technology transfer to the small-scale producers was seen as a major method by which their production could be increased and consequently further state investments were made in the construction of Farmer Training Centres and in the provision of mechanisation units to provide tractor ploughing services in each district.[55] These subsidised support services were complementary to the improved provision of inputs through the expanded crop collection and input supply network, and to the increased subsidies on the price of chemical fertilisers. To facilitate this technological transition of African farmers the state also radically changed arrangements for the provision of credit by establishing the Credit Organisation of Zambia and by giving it very liberal regulation so that credit became widely available.[56]

Extension, credit, pricing and marketing policies have all been used to encourage agricultural change since independence. In the early years emphasis was also placed on encouraging farmers to adopt new crops, with cotton, sunflowers, rice and kenaf suggested as additions for small-scale production. However, maize has increasingly come to dominate the policies for agricultural change, and the adoption of hybrid varieties has been seen as the main way of progressing from subsistence to semi-commercial levels of production.

State policies and funds have also been used to influence the structure of the agricultural sector directly by creating new forms of production. The major initiatives in this respect have been the encouragement of cooperative farming and the creation of a state farming sector with varying organisational forms. In the first few years after independence, considerable financial help was offered to groups of people who wished to farm cooperatively.[57] This was supported partly for ideological purposes, and also because cooperative organisation was seen as one way in which assistance could be provided in a relatively economical manner to the resource-short rural population scattered over extensive areas at low population densities. Funds were also invested in the development of a state farming sector. In the early years this involved the development of parastatal farming companies using formerly white-owned commercial farms, while from the mid-1970s additional forms of production have included Zambia National Service camps for school leavers, and Rural Reconstruction Centres for rural dwellers without the resources to farm independently. The state has also supported independent capitalist farming through its funding of settlement schemes. Some of these have had varying degrees of central control over production—as in the case of the tobacco production schemes, but others have given the farmers almost total freedom.

Some specific policies have been developed by the state to meet particular problems in the rural sector. One such development has been the reform of land tenure involving the nationalisation of land and the replacement of Crown Land leases and freehold with 99-year leases on State Land.[58] The measure was not introduced until 1975 for fear of increasing the loss of skilled settler farmers who still make an important contribution to the economy. The legislation is also fairly generous towards the settler farmers offering them a possibility of renewing their new 99-year leases.[59]

A final point about agricultural policies is the growing tendency since independence for area-specific rural development projects to be established. This originally began in an informal manner as a result of the clustering of some types of government services in particular areas, generally along the line-of-rail.[60] It became formalised in 1972 with the establishment of the Intensive Development Zones (IDZs) in the more peripheral regions and from these have developed the area-specific Integrated Rural Development Projects (IRDPs). Donor projects, which have increased in recent years, have continued the idea of concentrating support for farmers and rural change in specific localities outside those already more developed.[61] The choice of relatively high-potential areas, but often within the less developed provinces, for these projects reflects the continuing conflict between social and economic goals which goes back to the colonial period and the arguments for and against the policy of reinforcing areas of successful change in agriculture.

Policy Implementation Problems
The experience of the first fifteen years of independence revealed a number of difficulties with the policy measures outlined above. In

some cases their implementation proved difficult, while in other cases they were undermined by changes in the overall circumstances in the rural sector which were the result of other development policies, or neglect in these areas. It also appeared in some cases that policies had been wrongly conceived and that a deeper understanding of the problems was needed prior to policy formulation and implementation.

In general, government crop pricing proved erratic, with producer prices frequently announced late and exhibiting a tendency to lag behind rising production costs.[62] These problems were partly the result of the complex structure of agriculture in Zambia in which farmers' costs of production vary greatly. The lowest-cost producers are the semi-commercial cultivators who use ox-drawn ploughs. They are followed by the large-scale mechanised, State Land, commercial producers. The recently mechanised small- and medium-scale African producers tend to have the highest production costs. With political considerations making it desirable to assist all African producers, but not allow the still predominantly European commercial producers to make excessive profits, crop pricing has been a difficult operation.[63] At one stage, in the late 1970s, maize prices were so low that large-scale producers reduced their acreages, especially in areas where yields were not high,[64] while small-scale farmers using ox-drawn ploughs who lived outside the most ecologically suitable zones also found maize unattractive.[65]

Government support services frequently failed to provide the quality of services which it is increasingly recognised must exist to help farmers make the progression from subsistence to commercial levels of production.[66] Inefficiency and irresponsibility towards farmers were, and still continue to be, frequent complaints about these services, with crop collection depots opening late and not having inputs in stock when they are needed, while farmers often experience considerable delays in receiving payment for their crop.[67] The extension service also failed to support those farmers with the greatest needs, its staff tending to concentrate upon "farmers", those households producing appreciable surpluses for sale, and almost completely ignoring "villagers" or subsistence producers. The research branch for long remained one of the least reformed of the government services, focusing its work upon the needs of the large-scale commercial farmers, and developing crop varieties needing high levels of inputs and good management. Research tended to be directed to raising yields per unit area rather than in relation to labour which is the scarce factor of production in most rural areas.

In some cases the problems experienced with policies involved a mixture of poor technical support, a lack of wholehearted commitment on the part of the officials concerned, and some errors in understanding rural society. These problems all contributed towards the failure of the cooperative farming policy in the 1960s,[68] and were influential in the failure of the Credit Organisation of Zambia (COZ) which collapsed in 1970 with debts of over K22m.[69]

The emphasis upon hybrid maize also showed a serious misunderstanding of the rural situation and its needs. Between 1964 and 1980, as the producer prices for sorghum and millet were allowed to decline drastically in real terms, maize became the sole starch staple purchased by the state (Fig. 11.5). The adoption of hybrid maize was seen as the way subsistence farmers should progress to semi-commercial levels of production. However, this was a relatively difficult and risky route to follow because maize is ecologically suited to less than half of the country (Fig. 11.1) and requires new skills and large labour and capital inputs in comparison to other starch staples. Farmers trying to adopt hybrid maize also became dependent upon the timely allocation of credit, and provision of inputs and hybrid seeds, and upon good extension advice.[70] This myopic attention to maize also encouraged farmers to neglect other crops and produced a major decline in the production of other smallholder crops such as groundnuts and tobacco, which led to dangerous monocropping systems in some areas.[71]

The emphasis upon the technological package for hybrid maize and the increased use of tractor mechanisation to transform traditional agriculture also showed a failure on the part of the government to assess correctly the national capacity to adjust to this radically different production method. As a result mechanisation units have become increasingly poorly serviced, while the requirements for fertiliser have become a serious foreign exchange burden for the country.

A further conceptual error on the part of the policy-makers immediately after independence was their over-simplistic view of the causes of stagnation in African agriculture. Their view that the removal of discriminatory colonial policies and the expansion of services originally provided in support of European producers would be sufficient to allow African farmers rapidly to enter commercial production, failed to recognise the complexity and variety of the problems which were faced and the extent of the damage colonial policies had inflicted upon African rural society and its productive capacity.

But perhaps the most serious misconception which affected the success of post-independence agricultural policies, was the confusion among policy-makers between rural development and agricultural production. The two were generally seen as synonymous, and no difference was recognised in the target groups concerned or in the allocation procedures required. As a result the moral obligation to include the whole of the rural population in the processes of agricultural change and rural development was the dominant influence upon policy, while the distribution of investment was often made on the basis of political patronage rather than economic efficiency. Further, the emphasis which was placed upon rural development stressed consumption rather than production, and led to attempts to control the rural sector and redistribute resources, rather than provide the stimuli and incentives for investment and increased production.[72] The implications of this confusion were numerous and are seen in the failure of both the cooperative farming schemes and COZ. The latter resulted in a severe policy backlash with the successor organisation, the Agricultural Finance Corporation (AFC), having strict regulations concerning the qualifications of its borrowers, which excluded over half of the rural population.[73]

It should also be noted that the poor performance of policies in the first 15 years after independence was to some extent a result of the circumstances in which they had to operate. The most important of these was the overall deterioration in rural-urban terms of trade which between 1963 and 1980 had led to a 65 percent fall in the price of agricultural goods relative to urban goods.[74] This came about through the influence which the urban population, and especially mining unions, were able to exert over the government, which in turn was a reflection of the nature of the dualistic economy inherited from the colonial period.[75] This situation made farming increasingly less attractive as a way of earning a cash income[76] and led to an increase in rural-urban migration which intensified the rural labour supply problems faced by farmers.[77]

The circumstances within which agricultural policies had to operate also deteriorated as a result of a number of other external and internal changes. The Unilateral Declaration of Independence in Rhodesia and the international transport difficulties which this created, the oil crisis and the world recession were some of the more important external events. Internally a lack of experienced manpower proved a major problem especially at policy-making and policy-implementing levels. These difficulties were further exacerbated by the lack of government commitment to its rhetoric concerning agriculture which was exhibited all too clearly during this period in the underspending in agriculture and the absence of strong leadership in this sector of the economy.[78]

Achievements

The new agricultural policies implemented after independence produced only limited progress towards the government's goals. Agriculture did not become the dynamic sector which was envisaged. It failed to develop strong links with industry, and rural wealth did not grow rapidly to create new demands for industrial products. Diversification of the economy also proved impossible, although agriculture did manage to maintain its share of the country's gross domestic product at around 14 percent.[79] Maize production proved erratic with exports in excess of 50,000 tons in 1966, 1967, 1973 and 1976, and imports in excess of this in 1971 and 1972 (Fig. 11.4). In general food production became increasingly unsatisfactory during the first decade of independence with imports growing and exports declining. Overall net food imports increased threefold in real terms between 1965 and 1975.[80] In this rather unfavourable situation urban food prices were kept down and rural services were expended only by increasing subsidies to the agricultural sector.

Progress towards the goals of socially diversifying the market-oriented farming community, increasing the state's role in production, and spatially dispersing market-oriented agriculture was slow, and generally unsatisfactory.** Between 1969 and 1980 the proportion of the farm population who were subsistence producers fell by only 13 percent from 75 to 62 percent, and the number of people living in such households declined by only 68,000 due to the rising rural population (Table 11.2a). Small- and medium-scale semi-commercial producers did increase in importance, from 23 to 36 percent of the rural households, and they increased their share of marketed maize production from around 40 percent to over 60 percent (MAWD, personal communications). The state has become an important producer of beef and milk through its parastatal enterprises but it has not been as dynamic as the large-scale commercial producers and its share of total marketed production probably does not exceed 5 percent by value. to all other provinces, Copperbelt, Luapula, Northern, North-western and Western. These are the peripheral and least agriculturally developed areas.

The changes occurring in agriculture, which included the adoption of sunflower and cotton as important crops for small-scale farmers, were not evenly distributed across the country. Farmers in the already more agriculturally advanced Central, Eastern and

---

** The data used in this section is drawn from a background paper for the Food Strategy Study (Harteveld, K., op. cit.) In this an attempt was made to model the changes in both the structure of the farming population and the *expected* sales of each category of farmer, with assumptions that "normal" rainfall, input supply and credit conditions existed in the years considered. The data used in these calculations was obtained from Namboard, Provincial Agricultural Offices, and the Crop Forecasting Section of the Ministry of Agriculture and Water Development. It is clear that this data was fragmentary and that several unspecified assumptions and interpolations have been employed in developing the model.

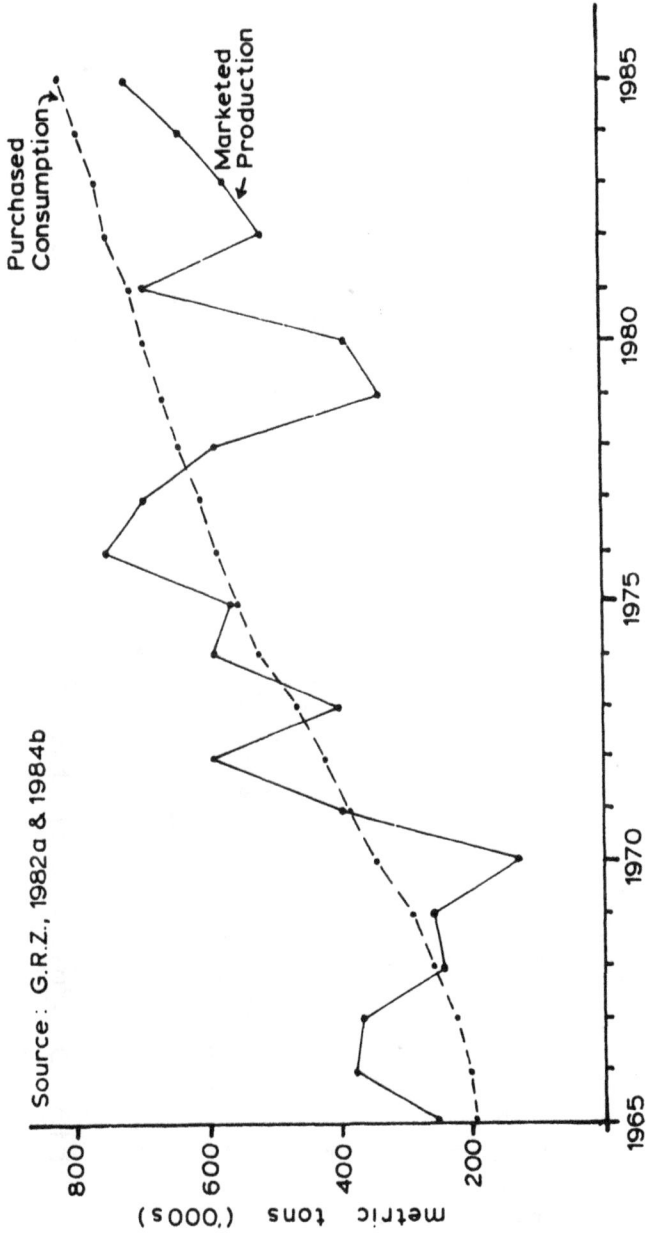

Fig. 4 MAIZE PRODUCTION AND CONSUMPTION

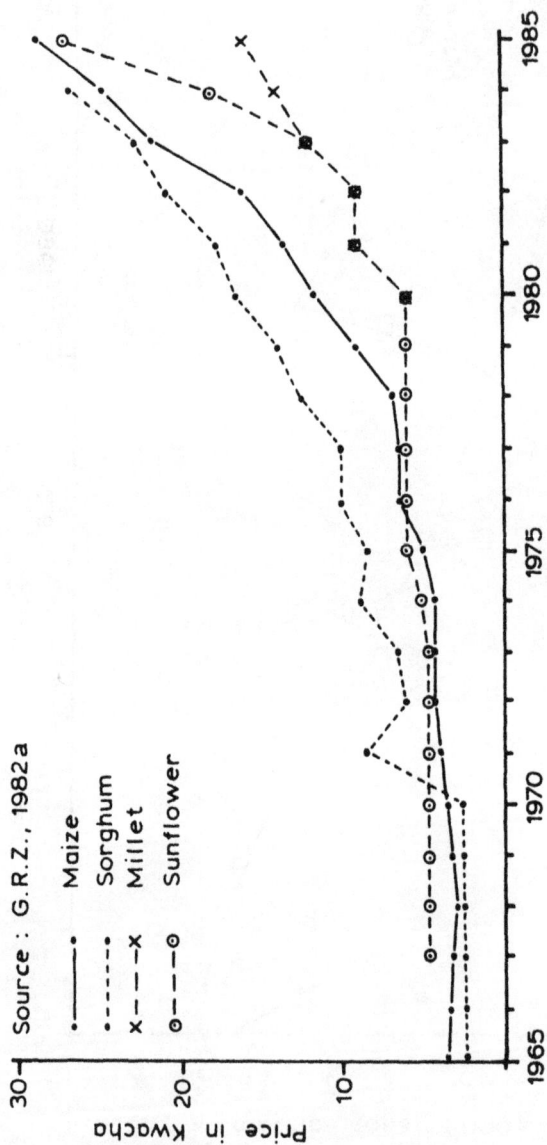

Fig. 5 PRODUCER PRICES OF MAJOR CROPS

TABLE 11.2
Post-Independence Agricultural Change, 1969-80

a) Structure of the Farming Population
CATEGORY OF FARMER[1]
(Percentage of Farming Population)

	One	Two	Three	Four	Total Population (m)
1969	1.7	4.5	18.7	75.0	1.865
1980	1.2	6.3	30.1	62.3	3.239

b) Regional Distribution of Farming Population by Farmer Category

C.E.S. Region[2]					
1969	3.8	9.7	34.8	51.7	1.402
1980	2.4	12.7	49.9	35.0	1.543
Non-C.E.S. Region[3]					
1969	0	0.3	5.2	94.5	1.563
1980	0	0.4	8.8	90.8	1.696

Source: Calculated from Harteveld, K., *Methodology. Supply and Demand Models 1969-1980 Food Strategy Study*, Amsterdam, Royal Tropical Institute (mimeo.), 1982, Tables 3 and 23a.

Notes: 1. Categories of farmers:
One—large-scale commercial (+40 ha)
Two—medium-scale commercial (10-40ha)
Three—small-scale semi-commercial (1-10 ha)
Four—traditional/subsistence (1-2.5 ha)

2. C.E.S. Region refers to Central, Eastern and Southern provinces, including Lusaka province which was excised from Central province. This includes the most agriculturally advanced areas.

3. Non-C.E.S. Region refers to all other provinces, Copperbelt, Luapula, Northern, North-western and Western. These are the peripheral and least agriculturally developed areas.

Southern provinces (the C.E.S. region) found it easiest to benefit from the new policies on Independence, while those elsewhere were affected little (Table 11.2b; Fig. 11.3). Subsistence production households declined from 52 to 35 percent of the rural population in the C.E.S. region between 1969 and 1980, while elsewhere the change was from 95 to 91 percent of the population. Conversely, the growth of small- and medium-scale producers has been concentrated in the C.E.S. region where the population in their households increased by almost

400,000 or 18 percent (580,000 to 960,000; 45 percent to 63 percent), while in the non-C.E.S. region the increase was only 70,000 or 3 percent (86,000 to 156,000; 6 percent to 9 percent). Measured in terms of the proportion of marketed maize produced in the two regions the progress has been slightly more dispersed, the C.E.S. region's share of total output falling from 96 to 92 percent between 1969 and 1980. However, 87 percent of the total national increase in maize production came from the C.E.S. region during this period.

Besides the generally poor production results of agriculture during the first 15 years of independence there was also little evidence of the establishment of a sound basis for widespread rural development. By the mid-1970s increasing evidence was available of the differentiation of rural society with only some households benefiting from the support services provided, while the position of most rural households changed little or deteriorated especially in terms of nutrition.[81] Patronage, personal contacts, and existing wealth increasingly explained who ''got on'',[82] both in the Chiefs' Land farming areas (formerly Reserves or Trust Land), and on the State Land commercial farms where an African farming elite began to develop.[83]

From a policy point of view, the first 15 years after independence was a period of experimentation in which a variety of policies were pursued in an attempt to support a range of goals and types of agricultural production. Government funds, generated in the mining sector, were relatively plentiful until the mid-1970s and, even though agriculture did not obtain the share allocated to it, the sector was able increasingly to call upon subsidies to support policies which in both economic and social terms proved of limited success.

AGRICULTURAL CRISIS AND
THE REFORM OF STATE POLICIES

The Origins of the Crisis

The problems in the agricultural sector began to grow in the late 1970s as the impact of the world recession upon the Zambian economy intensified. The country had made little progress towards diversifying its economy away from mining so that when the copper price was halved in seven months in 1974-75, and continued to fall in real terms for a further eight years, there were serious repercussions throughout the economy. The drastic decline in the profitability of the copper mines, which had provided almost half of government revenue in the early 1970s, led to a major cutback in government expenditure and to a series of government cashflow difficulties. Similarly, the reduction in overseas earnings from copper, which had provided 94 percent of the country's exports, let to a persistent and devastating shortage of foreign exchange.[84]

This situation impinged upon agriculture in a number of ways. The reduced government budget let to a decline in the quality of support services to farmers. Extension staff became less mobile, and were seen less often in the field, while Farmer Training Centres operated for only a few months a year. Veterinary services deteriorated and the tsetse fly control measures almost collapsed. Government cashflow problems meant that late payment of farmers for their produce became more frequent, while producer prices were not increased in line with production costs partly because the government lacked the funds to raise subsidies on urban food.[85] Agricultural credit also suffered from the squeeze on government expenditure and did not grow as fast as the demands and needs for it.[86] The timely availability of fertiliser was hit by the foreign exchange shortage which not only led to the late release of letters of credit for the importation of fertiliser, but also meant that vehicles and tyres required for distributing these inputs were increasingly scarce. A shortage of machinery spare parts for large-scale commercial producers also disrupted production, while peasant farmers were reportedly discouraged from producing surpluses for sale when foreign exchange shortage and price controls meant that there were no commodities in village stores on which to spend their earnings from farming.

These deteriorating conditions led to a progressive decline in marketed maize production from 1976 until 1980 (Fig. 11.4). This saw the country move from a food surplus situation to one where imports equal to half the nation's needs were required. These imports required increased subsidies to maintain the low urban food prices which were approximately half the economic cost (even before importation costs were considered).[87] As a result, in 1980 the country found itself spending 19 percent of the government revenue on agricultural subsidies, and 10 percent of its foreign exchange on food imports.[88] This was an intolerable situation given the scarcity of foreign exchange for the importation of items not available in Zambia, and given the cutback in government services and investments which the growing food subsidies required. Further pressures came from the rising level of urban unemployment with its political dangers, and from the recognition that many of the mines had only a very limited life remaining even if prices improved considerably.[89]

Redefining the Goals for the Agricultural Sector
By the late 1970s there were strong pressures upon the government to review its agricultural policies and create a more dynamic farming sector. Besides the internal economic and political influences, foreign

organisations, including the IMF, the World Bank and Western donor countries, became increasingly forceful in their demands for policy reform. The resulting debate led to a number of different views about the future role of agriculture and about state policies in this sector. From the various speeches and documents, four major goals can be identified. These are:

a) to increase the dynamism of the agricultural sector,
b) to improve food self-sufficiency and security,
c) to reduce the costs of agriculture to the nation, and
d) to re-adjust rural-urban terms of trade, including urban food prices.

A more dynamic agricultural sector is seen as vital to the country's economy in view of the poor prospects for mining. It is suggested that agricultural expansion should be able to generate increased tax revenue, employment and, in the long term, foreign exchange which will help replace the contribution copper used to make in these areas. With foreign exchange increasingly scarce, food self-sufficiency is vital to prevent the unnecessary competition between food and other vital imports, while Zambia's position as a frontline state, combined with its restricted foreign exchange earnings and its large urban population, make food self-sufficiency a matter of national security. The level of agricultural subsidies is seen by many as intolerable for the country given the deteriorating government budget situation, while some policy-makers, and especially the IMF and international donors, argue that the distortions caused by subsidies have led to misallocation of resources and inefficiencies in the national food production system. Finally, it is increasingly recognised that rural-urban terms of trade must be moved in favour of the farmer to encourage his increased production and to discourage further growth of the urban population for whom jobs cannot be found. However, there is also widespread concern, especially among the more ideologically and socially minded members of the party and government, that these last two goals should not produce such major increases in urban food costs that serious hardships result creating political unrest.

Policy Measures
These four redefined goals have different degrees of support across the political perspective within the party and government and they have elicited a variety of policy responses which reflect these positions.

i) Operation Food Production: A major ideological response to the agricultural crisis came from the party in 1980 in the form of a

Presidential speech outlining "Operation Food Production".[90] This provided guidelines, as well as some new policies, concerning how the agricultural crisis should be solved. The speech emphasised the ideological purity of its proposals, stressed the importance of social goals, and made it clear that a capitalist solution to the crisis should be avoided.

The major objectives stated in the President's speech were the achievement of food self-sufficiency, increased agricultural exports, and the creation of employment both in agriculture and through the development of agro-industries.[91] These goals were to be sought primarily through an expansion of state involvement in direct production, with two 20,000 ha state farms established in each province, and the Zambia National Service farms and Rural Reconstruction Centres expanded despite the heavy criticism of their inefficiency which had been made the previous year.[92] Producer cooperatives were also seen as a major production form, the idea receiving its first official "airing" since its demise in disgrace in the early 1970s.[93] Peasant farmers came fifth in order of priority, with a comment that they should cooperate and regroup themselves around government-provided services. Commercial farmers were last on the list and noted primarily for the technical knowledge they could impart to peasant producers. The programme did recognise the problems of inefficiency which have recurred in Zambia's agricultural development and stressed the need for economic viability in the new state ventures and in the improvement of services to farmers.[94] The use of incentives to stimulate production from private producers was also recognised but with the imperative of avoiding the "disastrous consequences of arbitrary high prices of staple food for our masses". Consequently some reduction of production costs was hoped for and increased fertiliser subsidies were announced.[95]

ii) **Pricing and incentives**: An economic or market-oriented response to the crisis began to appear in the Ministry of Agriculture from 1979. In an attempt to achieve a rapid turnaround in agricultural production, major changes in crop pricing policies and the taxation of agricultural income were introduced. Producer prices were considerably increased in 1979 and since then they have risen faster than inflation and in most years faster than the rate of increase in production costs (Fig. 11.5). In 1980, income tax on farm incomes was given a separate flat rate of 20 percent in contrast to the former progressive rate which reached 75 percent of income of over K15,000.*** Since

---

*** One Kwacha (K1) was worth £0.60 in 1980. It fell to £0.40 in early 1985, and to around £0.10 after the Kwacha was floated in the autumn of 1985.

then, the flat rate has been reduced to 15 percent. Foreign exchange payments were also introduced for major non-Zambian producers of maize, soyabeans, and wheat as a further encouragement to increased production, and were still in force in 1986.[96]

These measures were all primarily addressed to the established medium- and large-scale commercial farmers while some, especially the foreign exchange payments and tax benefits, were directed to the large-scale commercial farmers alone. The logic behind these measures was that the quickest way of obtaining increased production was to create favourable conditions for the farmers who already had the resources and skills to produce large marketable surpluses. These were identified as the large commercial producers and to a lesser extent the established medium-scale producers. They were thought to be sensitive to crop prices and taxation, and so these were adjusted to encourage their production.

The discussion which led to these decisions can hardly have been easy in view of the aversion in the party to commercial farming because of its capitalist nature, the historical role it has played in the country, and the fact that most of the largest producers are still Europeans who have not taken out Zambian nationality. Further, the pricing policy of the 1970s appeared to some observers to have tried to reduce the country's reliance upon commercial farmers for maize and allow small- and medium-scale farmers on Chiefs' Land to feed the nation. If this was the case, the change in policy in 1979 and 1980 represented a major reversal of the party and government's attitude to commercial farming.

iii) Crop purchasing policy: A more widely acceptable policy reform began in 1982 when the government announced new crop prices for the long-neglected traditional staples of sorghum and millet, and for the first time created an official market for cassava. The 1982 prices of these crops were still well below the free-market prices but they have been increased more rapidly than the prices of many other crops, and sorghum was priced at 95 percent of maize in the 1985 purchasing season (Fig. 11.5).

These developments appear to herald a major change in crop purchasing policy away from maize as the exclusive staple crop for official purchases and urban consumption. This is justified partly on social grounds, as it is felt that once a secure and competitive market for the traditional staples exists, poor households should be able to progress to semi-commercial levels of production more easily than by adopting hybrid maize with its requirements for new management skills, and major labour and capital inputs. However, it is probable that the dominant rationale for the change in policy is the desire to increase food production as quickly as possible from all sections of the farming community and so achieve self-sufficiency. It is argued that if traditional staples can replace maize in a number of uses, especially for the brewing of local beer and in the production of stockfeed, then the nation will be able to achieve self-sufficiency in maize production for food consumption. A further advantage of this policy is that by encouraging the cultivation of crops which are ecologically suited to the different parts of the country, harvest fluctuations will be reduced and a more regular and secure marketed food supply will be obtained. Further potential advantages of this change in crop purchasing policy are seen in the reduced fertiliser, and hence foreign exchange, requirements of traditional staples, and in the possibility of increasing starch surpluses near to urban centres in outlying regions where maize is not suited, thereby achieving regional food self-sufficiency and reducing crop transport costs.

iv) **Crop marketing and input supply**: In the late 1970s Namboard, the monolithic parastatal organisation responsible for crop collection and input supply in most parts of the country, was increasingly criticised for inefficiency and the poor quality of its services to farmers. Three cooperative unions existed alongside Namboard at that time and on the whole they appeared to provide a better service to farmers. Consequently, as reports increasingly suggested that low production was in part the result of poor services to the farmers, the government decided to reorganise crop collection and marketing. Namboard was stripped of its rural functions and left only with national imports and exports, interprovincial movement of crops and inputs, and national and provincial storage. Its rural functions were handed over to provincial cooperative marketing unions and where these did not exist they were set up.

The ideas behind this policy were partly political and partly economic. In the first instance, it was felt that decentralisation of crop marketing and input supply should take place at the same time that other government functions, especially administration, were being decentralised. The creation of provincial cooperative unions was also part of a renewed political drive to establish cooperative organisations throughout the country and so stress the advantages of this movement. The Zambian Cooperative Federation had recently become affiliated to the party as a mass movement, and it appeared in the early 1980s that the party felt that ZCF and the provincial unions had an important ideological role to play. From the economic point of view decentralisation of Namboard's functions was felt to be desirable in order to improve efficiency, small organisations being seen as easier for efficient management than the monolithic parastatal. It was also felt that by having local representatives from primary cooperative societies on their boards and by being provincially based, unions would become sensitive to local needs and more responsive to farmers' wishes. In this way it was hoped that production would no longer be disrupted by the late distribution of inputs or late payment of farmers. Finally, it was thought that in the long run with primary cooperative societies, developed by the unions, taking over their functions of running depots, crop collection and input provision costs could be reduced as these tasks would utilise the societies' facilities and manpower rather than requiring separate ones.

This policy initiative was drastically reversed in 1985 when problems of corruption and inefficiency in several provincial cooperative unions led to Namboard resuming responsibility for all stages in crop collection and input supply. Early in 1986, changes away from state control over marketing and input supply were announced with the removal of Namboard's monopoly on behalf of the state in these activities.[97] As a result private trading in input provision and crop collection will be allowed in the future alongside Namboard, the hope presumably being to encourage increased efficiency and improved services to farmers.

v) Agricultural research and extension: While many of the above policy reforms have been relatively sudden and are clearly responses to the crisis in Zambia's food supply, a number of changes in agricultural research and extension have occurred more slowly during the last ten years. These should also be considered as they mark major changes in the state's policies towards these aspects of agriculture, and could have considerable consequences depending on how other reforms are pursued. Overall they mark a rather belated penetration of the socialist ideals of the party into these areas of agricultural policy, and show an increased concern with the smallest-scale semi-commercial producers and those trying to progress from subsistence production.

The earliest initiative in this direction began in 1977 with the "lima" programme. This introduced simple extension advice aimed at farmers cultivating only a "lima" (0.25 ha) of commercial crops. Fertiliser and seeds were for the first time packaged in units suitable for such areas, and a national campaign was established to encourage subsistence farmers to become "lima" farmers.[98] Linked with this programme have been attempts to reform the extension service away from its concentration upon "farmers" so that all rural households have the opportunity to benefit from extension advice. Thus redirection of extension advice has involved the adoption of the "training and visit" approach, but with increasing cutbacks in extension funds and resources for small-scale producers little has been achieved.

On the research side the establishment in 1978 of the Adaptive Research Planning Team (ARPT) has marked a major redirection towards the small-scale producers. ARPT concentrates on identifying key constraints to increased agricultural production and market orientation in the farming systems on Chiefs' Land. In conjunction with "villagers" and "farmers" ARPT identifies strategies by which these constraints may be overcome and then undertakes on-farm trials to test these solutions as well as ones produced from the research stations.[99] More recently the research stations have begun to recognise that breeding should consider the reality of the village situation which includes poor soil and shortages of labour and inputs. Hence, in the late 1970s the Research Branch began to place emphasis upon the traditional starch staples apart from maize, and at the same time began breeding programmes with village field trials which sought to produce new varieties suitable for sub-optimal or "lousy" growing conditions.[100]

vi) **Price decontrol, subsidies and future crop pricing**: The questions of crop pricing and agricultural subsidies have produced much heated debate as the conflicts between social objectives and economic considerations in this area have been explored. The first area in which a change was agreed was in consumer price controls. These had been greatly expanded in the years after independence in an attempt to reduce "the exploitation of man by man" through trading. During the inflationary 1970s price controls had been slow to adjust and had led to unrealistically low profit levels both for producers and especially for rural traders in remote areas given the principle of nationwide uniform pricing. This forced companies to cut back production of their less profitable products, and parastatal organisations began to make large losses and require increased subsidies. Traders found it increasingly difficult to make a living and as the recession led to shortages, the absence of consumer goods in many rural areas was reported to be discouraging farmers from producing crops for sale. Faced with the growing financial problems of the parastatal industries, the need to reduce shortages to maintain urban peace, and the desire to stimulate agricultural production, the government decided in December 1982 greatly to reduce price controls and introduce economic pricing for most commodities. However, to date maize meal and a few other essential items have remained subject to price controls, although their prices have been adjusted more rapidly to changing economic conditions in order to keep to a minimum subsidies to parastatal industries.

The expansion of the principle of economic pricing into agricultural producer prices has occurred very slowly. Price controls on wheat were removed in 1985 in an attempt to stimulate production, and a general movement in this direction can be expected in the future as Zambia becomes increasingly dependent upon the IMF. Discussion of regional pricing of maize began in 1984 with the idea of trying to match supply and demand within different parts of the country more closely and so reduce the level of transport subsidies required for moving maize from remote surplus-producing areas, notably Eastern Province, to the major markets of Lusaka and the Copperbelt. Wider application of this policy, it was hoped, would help re-establish the comparative advantage of different regions for particular crops and so lead to a more efficient use of farm resources. Regional pricing naurally met considerable opposition from Eastern Province politicians who saw the proposed price changes as discriminating against their region, while opposition also came from "nationalists" within the party who saw the proposal as divisive and against the national motto of "One Zambia: One Nation". More recently, in early 1986, ministerial statements have suggested that the state will free the producer price of maize and other regulated crops from the 1987 harvest, but it remains to be seen whether this is implemented given doubts concerning the consquences upon production of abolishing the guaranteed producer price and the political consequences of free-market pricing upon urban food prices.[101]

Other developments confirm the trend towards a free-market agricultural system with minimal state intervention. These include a reduction in fertiliser subsidies and discussions of the ways in which the costs of government services to farmers, especially in the veterinary field, can be recouped. Subsidies still exist in the collection and processing of maize meal, but these are being reduced as quickly as the urban population will accept increased food prices. Thus overall, as a result of the deteriorating economic conditions and the growing IMF influence over its policies, the state is attempting greatly to reduce the cost to the exchequer of the country's food production system.

Performance and Prospects

The floating of the Kwacha in late 1985 and the apparent movement thereafter of government policy towards a free-market approach to agriculture make it difficult to predict the outcome of the reform measures outlined above. However, it is possible to review the results achieved so far in response to some of these measures, and to discuss what consequences may result from the apparent dominant directions in policy reform.

The emphasis which the party's "Operation Food Production" (OFP) placed upon state farms as a solution to Zambia's food crisis has not been followed up. Although 18 sites were identified and soil surveys undertaken at these, only two farms had reached pilot production by 1984.[102] The strategy failed to attract the considerable foreign aid funds which it needed and so in early 1986 the farms were handed over to parastatal companies and para-military organisations to subdivide and run as smaller units.[103] However, as government finances have not permitted the infusion of funds into these companies and organisations to allow them to fulfil their roles outlined in OFP the prospects of any increased production from state farms or other state organisations seem very slight. The cooperative movement, although expanding in the early 1980s, is now oriented towards service rather than producer roles.[104] Consequently it has expressed virtually no interest in becoming involved again in the thorny issues of producer cooperatives which OFP allocated to it and which in the early 1970s almost caused the demise of the whole movement.

In contrast, the large-scale and, to a lesser extent, small-scale commercial and semi-commercial farming sectors have experienced a considerable revival since 1979. Although below average rainfall in the 1979-80, 1981-82 and 1983-84 seasons reduced harvests, the area under maize and the marketed production showed a considerable expansion which it is expected will lead to self-sufficiency in 1986 for the first time since 1977 (Fig. 11.4). The changes in crop pricing and taxation, combined with the foreign exchange incentives, led initially to a rapid response by large-scale commercial producers who increased their share of the marketed maize production from around 30 percent in 1979 to nearly 60 percent in some years of the early 1980s. Doubts about government crop pricing policy then arose again in 1983 when producer prices lagged behind increased producing costs for some months. However, the government subsequently sought to reassure farmers that they would always receive fair prices, while foreign exchange incentives have been continued and were confirmed again in early 1986.

Small-scale producers have been slower in expanding their production, but they have shown a vigorous response to the price increase in areas such as Northern Province where government support services have been improved as a result of foreign aid funding.[105] However, elsewhere there has been little or no improvement in the quality of services to farmers following the reforms and in some areas there have been closures of crop collection depots as marketing unions have had subsidies cut.[106] Small-scale farmers are yet to receive any benefits from the ARPT research or the plant breeding for village conditions, although it is hoped that results from these activities will become available within the next few years.

There are also doubts about the impact of the policy favouring the purchase of traditional staples which seemed to offer a more attrac-

tive route to semi-commercial levels of production for many subsistence farmers. The slow progress towards price levels which are
competitive with maize raises doubts about the government's sincerity
with this initiative, while there are still questions as to whether storage
and processing problems can be solved and whether sufficient uses
can be found and demand generated for the volume of these crops
which could be produced.

Probably the most serious threat to the traditional staples policy
and to all the reforms which favour the subsistence, and poorer, rural
households is the threatened attrition of the marketing network in the
peripheral provinces. Crop collection costs in some of these areas
are close to the value of the crops collected. This is a result of both
the low population density and the consequent distances between
depots, and the low volume of marketed production which often result
in partially loaded lorries and under-used depots. Increased production
resulting from the traditional staples policy could result in a more
efficient use of this network, while changes in the network and crop
collection policies might help increase efficiency and reduce the levels
of subsidies required. However, the evidence of reduced restitution
to the provincial unions and pressures upon them to reduce the levels
of subsidies they require suggest that a major reduction in subsidies
and the depot network will be sought. If private crop collection and
input supply does develop it seems likely that this will take the more
profitable areas away from Namboard and so reduce the potential for
that organisation to cross-subsidise from its profitable operations. This
would result in increasingly severe cuts in the loss-making parts of
the depot network in order for Namboard to keep within its subsidy
ceiling. Alternatively, if free-market pricing is introduced, producers
in the peripheral areas will receive very low prices for their produce
to compensate for the high collection costs. Hence most scenarios
suggest that market-oriented production in peripheral areas will
become less attractive in the foreseeable future as a result of the
changes in the state's agricultural policies. Conversely, the dynamism
needed in the agricultural sector and the improved food self-sufficiency
for the nation seem to be sought increasingly from the established
producers in the more accessible regions. The implications of the present policies aimed at reducing the cost of the agricultural sector to
the nation will almost certainly be increased inter-regional variations
in income and welfare, a phenomenon which Zambia has sought to
reduce through many of its policies since independence.[107]

CONCLUSION

During the 80-year history of modern state organisation in Zambia, there have been major changes in the attitudes of the state towards agriculture and in the policies which it has pursued. To some extent a cyclical pattern of change can be seen with the rise and fall of particular policies and areas of state intervention. On the other hand, some ideas and debates have been recurrent features despite the changes in regime which have occurred.

In the colonial period the state increasingly intervened in the agricultural sector in order to support the European settler farmers. A whole range of policies were established to protect this farming community with state-funded services directed primarily in favour of the settlers, and institutional arrangements, such as land tenure, giving them several advantages over African farmers. State intervention in the market was also designed to support European farmers, although this to some extent protected European mining interests by stabilising urban food prices. The colonial state bore some of the costs of its policies towards agriculture, but many of these costs were passed on to the African producer. Colonial policies focussed upon economic considerations, and encouraged market-oriented agriculture only in the more accessible areas and from the European and larger-scale African producers.

The resulting spatially and socially differentiated rural society was a major challenge for the newly independent Zambian state. With its socialist ideals, the new regime sought to provide equal opportunities to farmers whatever their location or their socio-economic status. To achieve this, increased state intervention in the agricultural sector was pursued. Although some policies established in the colonial period, such as the maize preference, state marketing responsibilities, and a two-tier land tenure system were maintained, many policies were drastically changed to provide state services to all farmers, while new policies were initiated which led to state production and the support of new forms of agriculture such as cooperatives. With a democratic electoral system, the state faced increased political pressures which included regional demands for participation in the market-oriented economy, a desire for national unity goals to be seen in agricultural policies, and an increased sensitivity over urban food costs. As a result, the costs of increased state intervention in agriculture had to be borne by the country as a whole out of mining revenue, rather than by one section of the community.

The recession, combined with the costs of the post-independence policies, and external pressures from the IMF and donors, has led to major changes in agricultural policy. These show signs of an increased concern with production and economic criteria rather than with social and political ones, and a trend towards reduced state intervention, especially in agricultural marketing. However, the state at present remains committed to providing some support services for farmers, although their financing is not yet clear, while the state will probably retain a central role in internal food storage and distribution, and a supplementary role in marketing in remote areas.

The state's perceptions of the role of the agricultural sector have also changed over the last 80 years. The BSA Company's originally negative view of settler agriculture became more positive as mineral revenues failed to develop and as it realised the importance of secure and cheap urban food supplies for a competitive mining industry in central Africa. The colonial regime, despite much influence from the settler community, regarded settler agriculture as a minor and primarily service sector of the economy in which mining was dominant. African farming was viewed negatively for much of the colonial period because of its potential for competition with the settlers and with the role of the rural areas as labour reserves. The dominance of economic considerations also meant that African agricultural development could only be supported in a few favoured locations. Since independence the state's view of agriculture has become more positive although the rhetoric of this has not always been matched by funding and implementation. Political and social considerations after independence led to rural development being raised at least in speeches to a priority position, ''a matter of life and death'' as President Kaunda once stated. However, in practice agriculture was neglected in terms of investments, while urban political pressures dominated over rural ones leading to a major deterioration in rural-urban terms of trade. The situation appears to be changing now that the limited economic life of the mines is recognised and the consequences of the recession upon the urban economy are becoming increasingly clear. A growing black Zambian commercial farming lobby, combined with IMF pressures and dangerously high levels of food imports have forced the government to pay increased attention to the agricultural sector and see it as the ''economic salvation'' of the nation.

This increased concern with agriculture has come, however, at a time when the state's ability to intervene and develop new policies is seriously constrained by the economic circumstances and by its creditors, especially the IMF. The present reformulation of agricultural policies is very much a response to IMF and donor pressures, so that the evolving policies are more a result of international influences than of the present nature of the Zambian state. Thus the prospective consequences of these new agricultural policies will almost certainly not be in line with the national philosophy of Humanism which President Kaunda, the party, and the government have publicly espoused for the last 20 years. The resolution of this emerging conflict will have important implications for agriculture and will add another stage in the evolution of the state's policies towards the agricultural sector.

NOTES

1. IBRD (International Bank for Reconstruction and Development), *Zambia: Issues and Options for (Economic) Diversification*, Nairobi, World Bank; GRZ (Government of the Republic of Zambia) *Restructuring in the Midst of Crisis*, Government Printer, Lusaka, 1984.

2. Roberts, R.A.J. and Elliot, C., "Constraints in Agriculture", pp. 269-98 in Elliot, C. (ed.), *Constraints on the Economic Development of Zambia*, Oxford University Press, Nairobi, 1971.

3. Wood, A.P., "The Demographic Situation in Zambia and its Implications for Socio-economic Planning", pp. 81-129 in Ncube, D.D. (ed.), *Agricultural Baseline Data for Planning*, Lusaka, National Commission for Development/University of Zambia, 1983.

4. Beaumont, A.N., "Agricultural Land Use in Southern Province and its Effects on Natural Resources", pp. 19-24 in Johnson, D.S. and Roder, W. (eds.), *Proceedings of National Seminar on Environment and Development*, Zambia Geographical Association Lusaka (Occasional Paper No. 10), 1979; Priestley, J.S.W. and Greening, P., *Ngori Land Utilisation Survey, 1954-1955*, Lusaka, Government Printer 1956; Wood, A.P., *An Interpretation of Quantitative Data Concerning Population Pressure in Southern Province*, unpublished manuscript report to the Southern Province Land Commission, 1981.

5. Schultz, J., *Explanatory Study to the Land Use Map of Zambia*, Ministry of Rural Development, Lusaka, 1974.

6. Ibid.; Hennemann, R., "The Soils and Vegetation of Western province", in Wood, A.P. (ed.), *Handbook to Western Province*, Zambia Geographical Association, in press; McPhillips, J.K., "The Development of Fertilizer Recommendations for Maize with Particular Reference to the High Rainfall Areas of Northern Zambia", pp. 219-29, in Svads, H. (ed.), *Proceedings of the Seminar on Soil Productivity in the High Rainfall Area of Zambia, Lusaka, 8-10 February, 1983*, Oslo International Development Programmes, Agricultural University of Norway (Zambian SPRP Studies, Occasional Paper No. 6), 1983.

7. Ranger, T.O., *The Agricultural History of Zambia*, History Association of Zambia (Paper No 1), Lusaka 1971; Roberts, A., *A History of Zambia*, Heinemann, London, 1976.

8. Trapnell, C.G., *The Soils, Vegetation and Agriculture of North-Eastern Rhodesia*, Government Printer, Lusaka 1953; Trapnell, C.G. and Clothier, J.N., *The Soils, Vegetation and Agriculture of North-Western Rhodesia*, Government Printer, Lusaka, 1957.

9. Allan, W., "How Much Land does a Man Require?", *Rhodes—Livingstone Papers*, No 15, 1949, pp. 1-23; Allan, W., *The African Husbandman*, Oliver and Boyd, Edinburgh, 1965.

10. Wood, A.P., "The Demographic Situation in Zambia and its Implications for Socio-economic Planning", in Ncube, op. cit.

11. Roberts, *History of Zambia*, op. cit., pp. 162-72.

12. Heisler, H., *Urbanization and the Government of Migration. The Inter-relations of Urban and Rural Life in Zambia*, Hurst and Co., London, 1974; Wood, A.P., "Population Trends in Zambia; a Review of the 1980 Census", pp. 102-25 in Findlay, A.M., (ed.), *Recent National Population Change*, Durham, Institute of British Geographers, Population Geography Study Group, 1982.

13. Kay, G., 1969, "Agricultural Progress in Zambia", pp. 495-524 in Thomas, M.F. and Whittington, G.W., *Environment and Land Use in Africa*, London, Methuen, 1969, pp. 503-4; Marter, A. and Honeybone, D., *The Economic Resources of Rural Households and the Distribution of Agricultural Development*, Rural Development Studies Bureau, University of Zambia, Lusaka, 1976.

14. Roberts, *History of Zambia*, op. cit., p. 182.

15. Gann, L.H., *The Birth of a Plural Society*, Manchester, Manchester University Press, 1958, pp. 137-8.

16. Baldwin, R.E., *Economic Development and Export Growth*, Berkeley, University of California Press, 1966, p. 141.

17. Klepper, R., "The State and Peasantry in Zambia", pp. 120-49 in Finch, R. and Markakis, J. (eds.), *The Evolving Structure of Zambian Society*, Edinburgh, Centre of African Studies, 1980.

18. Dodge, D.J., *Agricultural Policy and Performance in Zambia: History, Prospects and Proposals for Change*, Berkeley, University of California Institute of International Studies, 1977, p. 6.

19. NRG (Northern Rhodesia Government), *Report on the Census of Population, 1951, Northern Rhodesia*, Lusaka, Government Printer, 1956, p. 21.

20. Mvunga, M.P., *Land Law and Policy in Zambia*, Lusaka, Institute for African Studies, University of Zambia (Zambian Papers No. 17), 1982.

21. NRG, *The Agricultural Survey Commission Report, 1930-1932*, Livingstone, Government Printer, 1933.

22. Baldwin, op. cit., p. 143.

23. Lombard, C.S. and Tweedie, A.H.C., *Agriculture in Zambia since Independence*, Lusaka, NecZam, 1972, p. 57.

24. Barber, W.J., *The Economy of British Central Africa: A Case of Economic Development in a Dualistic Society*, London Oxford University Press, 1961, p. 137.

25. Williams, S., *Central Africa: the economics of inequality*, London Fabian Commonwealth Bureau, 1960, p. 10; Watson, W., *A New Deal in Central Africa*, London, Heinemann, 1960, p. 135; SPLC (Southern Province Land Commission), *Term of Reference*, Lusaka, Government Printer, 1981.

26. Siddle, D.J., "Achievement, Motivation and Economic Development: Farming Behaviour in the Zambian Railway Belt", *Third World Planning Review*, Vol. 3, No. 3, 1981, pp. 259-73.

27. Baldwin, op. cit., p. 150.

28. NRG, *Report of the Agricultural Advisory Board*, Lusaka, Government Printer, 1935.

29. Dodge, op. cit., p. 13.

30. Baldwin, op cit., p. 154.

31. Dodge, op. cit., pp. 24-5.

32. NRG, *Report of the Agricultural Advisory Board*, op. cit., as discussed in Makings, S.M., "Agricultural Change in Northern Rhodesia/Zambia: 1945-65", *Food Research Institute Studies*, Vol. 1, No. 2, p. 200.

33. Dodge, op. cit., p. 16.
34. Ibid., p. 13.
35. Clay, G.F., *Memorandum on Post War Development Planning in Northern Rhodesia*, Lusaka, Government Printer, 1945, pp. 212-30.
36. Baldwin, op. cit., pp. 153-4, 158-9.
37. Ibid., p. 151.
38. Barber, op. cit., p. 54.
39. Kay, op. cit., pp. 507-9.
40. Ibid., pp. 510-11.
41. Clay, op. cit., p. 211.
42. Dodge, op. cit., p. 33.
43. Ibid., pp. 34-35.
44. Baldwin, op. cit., 166; Morgan-Rees, A.M., *An Economic Survey of the Plateau Tonga Improved Farmers*, Lusaka, Government Printer, 1957 (Agricultural Bulletin, No. 14, Department of Agriculture).
45. Baldwin, op. cit., p. 143.
46. Dodge, op. cit., p. 46; see also Figure 11.3.
47. Baldwin, op. cit., p. 143; Dodge, op. cit., pp. 46 and 48.
48. Kay, op. cit., pp. 501-2.
49. Ibid., pp. 514-16; NRG., *Report of the Rural Economic Development Working Party*, Lusaka, Government Printer, 1961, p. 10.
50. Elliot, C., "Equity and Growth: an Unresolved Conflict in Zambian Rural Development Policy", in Ghai, D. and Radwan, S. (eds.), *Agrarian Policies and Rural Poverty in Africa*, Geneva, International Labour Organisation, 1983, pp. 155-89.
51. GRZ, *First National Development Plan*, Lusaka, Office of National Development Planning, 1966; GRZ, *Second National Development Plan*, Lusaka, Ministry of Development and National Guidance, 1971; GRZ, *Third National Development Plan*, Lusaka, National Commission for Development Planning, 1979.
52. Ocran, M.T., "Towards a Jurisprudence of African Economic Development: A Case Study of the Evolution of the Structure and Operations of Zambia's Food Crop and Cotton Marketing Boards from 1936 to 1970", Ph. D. dissertation, University of Wisconsin, 1971, p. 154; Dodge, op. cit., pp. 82, 94-105.
53. Lombard, C.S. and Tweedie, A.H.C., op. cit., pp. 8-19.
54. Klepper, R., "The State and Peasantry in Zambia", op. cit., p. 132.
55. Honeybone, D. and Marter, A., *An Evaluation Study of Zambia's Farm Institutes and Farmer Training Centres*, Lusaka, Rural Development Studies Bureau, University of Zambia, 1975; Siddle, D.J., "Rural Development Schemes", in Davies, D.H., *Zambia in Maps*, London, Hodder and Stoughton, 1971, pp. 70-1.
56. Lombard, C.S. and Tweedie, A.H.C., op. cit., p. 78.
57. Lombard, C.S., *The Growth of Co-operatives in Zambia, 1974-1981*, Lusaka, University of Zambia, Institute for African Studies (Zambian Papers, No. 6), 1971; Siddle, "Cooperatives", in Davies, op. cit., pp. 64-5; Quick, S.A., *Humanism or Technocracy? Zambia's Farming Co-operatives, 1965-1972*, Lusaka, Institute for African Studies, University of Zambia (Zambian Papers, No. 12), 1978.
58. GRZ, Land (Conversion of Titles) Act No. 20 of 1975.
59. Klepper, "The State and Peasantry in Zambia", op. cit., p. 134.
60. Siddle, "Rural Development Schemes", op. cit., pp. 70-1.
61. Smith, W. and Wood, A.P., "Patterns of Agricultural Development and Foreign Aid to Zambia", *in Development and Change*, Vol. 15, 1984, pp. 405-34; Wood, A.P., "The Growth and Character of Foreign Assistance to the Zambian Rural Development Programme", in Williams, G.J. and Wood, A.P. (eds.),

Geographical Contribution to Development in Southern Africa, Townsville, James Cook University Press, in press.

62. Roberts, R.A.J and Elliot, C. op. cit., p. 278; Dodge, op. cit., pp. 94-105.
63. Dodge, op. cit., p. 92; Wood, D., "Humanism—the Consequences of Political Objectives upon Marketed Output in Zambia, 1964-1975", in Honeybone and Marter (eds.), *Poverty and Wealth in Rural Zambia*, op. cit., pp. 7-22.
64. Wood, "Population Trends in Zambia; A Review of the 1980 Census", op. cit., pp. 102-25.
65. Marter, A., *Cassava or Maize? A Comparative Study of the Economics of Production and Market Potential of Cassava and Maize in Zambia*, Lusaka, Rural Development Studies Bureau, 1978, p.8; Klepper, The State and Peasantry in Zambia", op. cit., p.133.
66. IRDP (Integrated Rural Development Programme) Serenje, Mpika, Chinsali, "The Dynamics of Cropping Patterns and Maize Production in Serenje, Mpika and Chinsali Districts", in Svads, op. cit., pp. 269-88.
67. Roberts, R.A.J. and Elliot, C., op. cit., p. 281; Dodge, op. cit., pp. 84-9; Stollen, K.A., "Socio-economic Constraints on Agricultural Production in the Northern Province of Zambia", p. 350 in Svads, op. cit.; Wood, A.P., "An Economy under Pressure—Some Consequences o f the Recession in Zambia", paper read at Annual Conference, Institute of British Geographers, 1984, pp. 10-12.
68. Quick, op. cit.; Lombard, *Growth of Cooperatives*, op. cit.; Siddle, "Co-operatives", op. cit.
69. Lombard and Tweedie, op. cit., p. 78.
70. Marter, op. cit.; Klepper, "The State and Peasantry in Zambia", op. cit.; Elliot, C., "Equity and Growth: an Unresolved Conflict in Zambian Rural Development Policy", in Ghai and Radwan, op. cit., pp 155-89; McPhillips, J.K., "The Development of Fertilizer Recommendations for Maize with Particular Reference to the High Rainfall Areas of Northern Zambia", in Svads op. cit., pp. 219-29; Roberts and Elliot, op. cit.
71. Ministry of Agriculture and Water Development, *Groundnut Production and Marketing in Eastern Province: A Market Analysis*, Lusaka, Planning Division, MAWD (Special Study No 1), 1983.
72. Elliot, "Humanism and the Agricultural Revolution", op. cit.
73. GRZ, *Census of Agriculture, 1970-71 (First report)* Lusaka, Central Statistical Office, 1974; Klepper, "The State and Peasantry in Zambia", op. cit., p. 136.
74. ILO/JASPA (International Labour Organisation/Jobs and Skills Programme for Africa), *Zambia: Basic Needs in an Economy Under Pressure*, Addis Ababa, ILO, 1981, p. 7.
75. Shula, E.C.W., "Cooperatives and Rural Development Policies in Zambia: A Diagnosis and Evaluation", M.A. Dissertation, Institute of Social Studies, The Hague, Netherlands, 1979, pp. 39-40.
76. Pottier, J. "Defunct Labour Reserve? Mambwe Villages in the Post Migration Economy", *Africa*, Vol. 53, No. 2, 1983, pp. 2-23.
77. Marter and Honeybone, op. cit.; Wood, A.P., "Population Trends in Zambia", op. cit.
78. Ollawa, P.E., "Rural Development Strategy and Performance in Zambia: An Evaluation of Past Efforts", *African Studies Review*, Vol. 21, No. 2, 1978, pp. 101-24.
79. IBRD, *World Development Report*, Washington, IBRD, 1979, p. 130.
80. Wood, D., "Humanism—the Consequences of Political Objectives upon Marketed Output in Zambia, 1964-75", in Honeybone and Marter, op. cit.,
81. Honeybone, D. and Marter, A., *Poverty and Wealth in Rural Zambia*, Lusaka,

Institute of African Studies Communication 15, 1979; Klepper, R., "Zambian Agricultural Structure and Performance", in Turok, B. (ed.), *Development in Zambia: a Reader*, London, Zed Press, 1979, pp. 137-48; Cliffe, L., "Labour Migration and Peasant Differentiation: Zambian Experiences", in Turok, op. cit., pp. 149-69; Hedlund, H., *Small Farm Development on the Periphery: A Study of Two Chiefs Areas in Eastern Province of Zambia*, Lusaka, Rural Development Studies Bureau, University of Zambia, 1980; NFNC (National Food and Nutrition Commission), *A Review of Zambia's Nutrition Policy*, Lusaka, NFNC (mimeo), n.d.

82. Bratton, M., "Ward Development Committees and Village Productivity Committees in Kasama District", in Honeybone and Marter, *Poverty and Wealth in Rural Zambia*, op. cit., pp. 103-25; Pottier, J., "Reciprocity and the beer pot: some thoughts on the changing pattern of Mambwe food production", paper read at Symposium on "Food Production Systems in Central and Southern Africa", held at the School of Oriental and African Studies, University of London, July 1983.

83. Baylies, C., "The Emergence of Indigeneous Capitalist Agriculture: the Case of Southern Province, Zambia", *Rural Africana*, No. 5, 1979; Baylies, C. and Szeftel, M., "The Rise of a Zambian Capitalist Class in the 1970s", *Journal of Southern African Studies*, Vol. 8, No. 2, 1982, pp. 187-213.

84. Wood, "An Economy under Pressure", op. cit.; ILO/JASPA, op. cit.

85. Wood "An Economy under Pressure", op. cit.

86. Due, J.M., *Agricultural Credit in Zambia by Level of Development*, Lusaka, Rural Development Studies Bureau, Unversity of Zambia, 1978; Due, J.M., *Update on Financing Small Holders in Zimbabwe, Zambia and Tanzania*, Urbana, Illinois, Department of Agricultural Economics, University of Illinois, Staff paper 83, E-261, 1983.

87. GRZ. *Food Stategy Study. Main Report*, Lusaka, Ministry of Agriculture and Water Development, 1983.

88. GRZ, *Economic Report, 1982*, Lusaka, National Commission for Development Planning, 1983; GRZ, *Monthly Digest of Statistics*, Lusaka, Central Statistical Office, 1983; GRZ, *Food Strategy. Main Report*, op. cit.

89. GRZ, *Restructuring in the Midst of Crisis*, op. cit.; GRZ, *Economic Report, 1983*, Lusaka, National Commission for Development Planning, 1984.

90. GRZ, *Operation Food Production*, Lusaka, Government Printer, 1980.

91. Ibid., p. 5.

92. Dumont, R., *Towards Another Development in Rural Zambia*, Lusaka, Government Printer, 1979.

93. Quick, op. cit.

94. Ibid., pp. 6 and 11.

95. Ibid., p. 8.

96. NFZ (News From Zambia), "Uncontrolled Maize", No 375, week ended 6 February 1986, London, ZIMCO, p. 2.

97. NFZ, "Namboard Power Cut", No. 373, week ended 23 January 1986, p. 1.

98. Oppen, H.J.V., Shula, E.C.W. and Alff, U., *Lima Target Group Survey, Kabampo, District*, Kabampo; IRDP, North-Western Province, 1983; Stocking, M., "Agricultural Development for the Small Farmer: An Evaluation of the *Lima* Programme in Zambia", *Zambia Geographical Journal*, in press.

99. Kean, "Farming Systems Research and its Role in Soil Productivity Research", pp. 355-63 in Svads op. cit.

100. GRZ, *Proceeding of the Sorghum and Millets Workshop*, 15-16 April 1982, Lusaka, Research Branch, Department of Agriculture, 1982; GRZ, *Plant Breeding*

for Low Input Conditions, Lusaka, Government Printer (Department of Agriculture, Mount Makulu Central Research Station), 1983.

101. NFZ, "Uncontrolled Maize", op. cit.
102. GRZ, *Economic Report, 1983*, op. cit.
103. NFZ, 1986, "State Farms Cut Up", No. 373, week ended 23 January 1986, p. 2.
104. Wood, A.P., *Primary Co-operative Society Development in Western Province: Progress and Potential in Six Communities*, Lusaka, Rural Development Studies Bureau, University of Zambia, 1984.
105. IRDP (Integrated Rural Development Programme) Serenje, Mpika, Chinsali, "The Dynamics of Cropping Patterns and Maize Production in Serenje, Mpika and Chinsali Districts", in Svads, op. cit., pp. 269-88; Wood, "Food Production and the Changing Structure of Zambia Agriculture", op. cit.,
106. Toivianen, M., "The Practical Agronomic Experience of Farmers in Luapula Province", in Svads, op. cit., pp. 289-94.
107. Fair, I.J.D., *Zambia: the Search for Rural/Urban Balance. A Review*, Pretoria: Africa Institute of South Africa, 1983.

12. STATE POLICIES, AGRICULTURAL DEVELOPMENT AND FOOD PRODUCTION: THE ALGERIAN EXPERIENCE

Slimane Bedrani & Naceur Bourenane

STATE POLICY AND AGRICULTURE IN ALGERIA

In every period of Algeria's recent history, the state has played a manifestly essential role in the various transformations affecting the agricultural sector, and in the overall development of the country's agriculture.

From 1830 to 1962, the colonial state was clearly the prime mover in the development of market relations in favour of the metropole and of dominant groups on the local scene. It was thus the key instrument used in forcing the local population off their land and reducing them to poverty. This pauperisation came about not simply as a result of the takeover of land, but also in the wake of crop-destruction and harvest-confiscation expeditions carried out whenever metropolitan and local settler interests so dictated.

After the attainment of independence, the new Algerian state took over the function of prime mover in all processes of social change, initiating them, strengthening them, or, where necessary, aborting them. In next to no time the state asserted itself, both at home and abroad, as the principal agent to be dealt with, if not the sole agent.

The hegemony of the state instigated a dual process with repercussions in the agricultural sector. The first aspect of this process was a steady strengthening of the state apparatus and its operative social agencies, with a consequent extension of its operational range. As a result, a standardised system of agricultural plantation management was applied to the totality of the former settler sector. The second aspect was a steadily widening gap between existing social forces and the state. Of this gap the most visible consequence is that the rate of actual performance in the achievement of objectives set by the country's political rulers is mediocre, in spite of the allocation of substantial material resources.

CONDITIONS FOR THE ESTABLISHMENT OF AGRARIAN POLICIES: THE RISE OF THE MYTH OF AN AGRICULTURAL BOOM

Just after Algeria regained her independence, the country's social, economic and political condition was characterised by a general slump, the result of a particularly bloody war of national liberation coupled with an almost total breakdown of the productive and administrative systems. Apart from hundreds of thousands of casualties, the war occasioned the concentration of 2,157,000 persons, or 36 percent of the rural population, in camps under military guard. To all intents, those thus relocated had to give up farming.

A further consequence of the war was that toward the end of the 1950s the normal process of equipment replacement on mechanised settler plantations (by far the richest) first slowed down, then halted altogether. These plantations were situated along the coastal plains and in the interior lowlands.

The poorest Algerian peasants were the first affected by the actual drift away from farming. But the process continued even after independence. Not only did it continue; it tended to pick up speed, spreading so far as to involve European plantation owners and farm staff, who, like other settlers, eventually left for France. In the space of just a few months, over 80 percent of the European settler population left the country. Behind them they left an impressive accumulation of property—real and movable. All their houses, workshops, plantations, vehicles etc. had become ownerless, subject to no effective administrative control.

So pervasive was the crisis that it accelerated the drift from the rural to the urban areas, while within the rural areas it speeded up population shifts from poor, marginal zones such as the mountains and the scrubby outlands to the interior lowlands and coastal plains, formerly the settlers' happy hunting grounds. This movement, whereby a whole population shifted into space left by another, quickly turned into a general process of upward social mobility: former permanent farmhands on the settler plantations moved up, at the very least, into the shoes of the old European overseers, while former day labourers became permanent workers at least, and so on down the line.

As far as agriculture was concerned, the first observable result was that on the settler plantations (which, as already pointed out, were by far the best endowed), the level of skill among the work force dropped, while the actual cultivated farm area shrank, because small farmers who had hitherto tilled household plots on marginal, low-yield land now shifted elsewhere.

From the foregoing, it would be logical to expect a drop in production. Paradoxically, no such drop occurred. Because bioclimatic conditions at the time happened to be particularly favourable, the country was actually faced with the pleasant problem of agricultural surpluses for two years running (1963-64). Concurrently, the national demand for food commodities decreased in the wake of the departure, in 1961, of over 1,100,000 European settlers. The result was that the normal trend towards increasing imports of food items was reversed. Thus imports for 1966, the year the first development plan was drafted, cost less than 50 percent of the figure for 1960. All this could only lend credence and support to the myth, already current by the time of independence, according to which the old colonial agricultural system was abundantly rich, and thus capable of supplying the country's food requirements and supporting, with income generated from exports, the process of capital formation in other sectors.

In a nutshell, then, right after independence the salient feature of the situation was a paradox: on the one hand, the productive capacity of the agricultural system had gone down; on the other hand, there was an available surplus of agricultural produce. The situation at that time was characterised by the almost total control exercised by foreign companies over the rapidly growing hydrocarbon sector, a deep slump in other productive sectors, and the freakishly fast growth of the tertiary sector (which accounted for 40 percent of the gross domestic product in 1963).

There were other factors at play: resources actually possessed by the Algerian state were meagre; unemployment and underemployment rates were high (25 percent of the active urban population, 50 percent of the farm population); the country was almost exclusively dependent on the former metropole for export outlets, and its economic balance was consequently shaky. This was a direct result of the former colonial state's policy of complementarity, designed to prevent competition with the metropole. In accordance with this policy, commodities from Algeria were to be kept out of the French market during periods when metropolitan commodities were widely available. In return Algerian commodities were protected from competition with more efficient agricultural producers in the Mediterranean basin.

To sum up the various factors: the agricultural system was generating a surplus even though its productive capacity had decreased; there was substantial productive property now lying unoccupied and ownerless; protective tariffs formerly favouring Algerian commodities on French markets had been loosened or altogether dropped; the state possessed only scant resources, and the rural exodus was growing at an alarming rate: all these factors pushed the state into making active intervention in the agricultural sector the key to its economic policy.

In short order the government, acting as much for historical[1] as for socio-economic and political reasons,[2] was obliged to spell out its major operational guidelines and to give priority, in its activities, to the former settler sector. In the years immediately after independence, the definition and implementation of agricultural policies were so closely interlinked that at times it even seemed as if implementation preceded definition. As from 1967, however, the two aspects began to be clearly differentiated, with the state tending to promote a planned economic policy based on a reinforced state sector, and backed up by a specific vision of agriculture and its various functions in the development process. At that time this process was thought of as the redefinition of Algeria's place on the international scene and the world market. It might be pertinent to point out that Algeria's economic policy was presented as profoundly nationalistic and somewhat state-dominated, but not anti-capitalist.

The first implementation measures in the agricultural sector were the confiscatory laws passed against European farmers and plantation owners who had abandoned their estates in 1962, together with the establishment of a set of structures designed to manage the property thus seized. These structures were primarily intended to function as an instrument of state intervention, an extension of the state, in fact. Thus, as early as 18 March 1963, a National Office of Agrarian Reform was set up under the direction of the Ministry of Agriculture, with the head of government serving as chairman of its governing board. This office was given responsibility for the management of the abandoned estates, together with control over credit facilities, cooperation, agricultural extension work and existing supply circuits.

Simultaneously, laws were passed to regulate the operation of production units in the former settler sector, now known as "autonomously-managed" enterprises. These laws came to be referred to as the "March Decrees". They gave succinct form to the option in favour of state domination.

The National Office of Agrarian Reform was abolished in 1969—a move that reflected intensified state intervention in the agricultural sector and a much more definite reduction of the producer's autonomy in the former settler sector, now the "socialist" or "autonomously-managed" sector.

ECONOMIC POLICY AND AGRICULTURAL POLICIES

At all times state intervention in agriculture has been geared to three closely intertwined concerns: the recovery and exercise of national sovereignty; the rationalisation and modernisation of the productive system; and the righting of social wrongs caused by the colonial process. It is in the name of these principles that various measures have been taken; the nationalisation of settler estates in 1962, and of the property of absentee farm owners in 1972 ; the various land reform exercises; socio-spatial income distribution policies involving free health services, universal education within a state-controlled system, subsidies for staple items, the extension of the state consumer commodity distribution system into the rural areas etc. These issues served as unfailing reference points both before 1965 and after 1979, in the rhetoric as well as the deliberation of Algerian political life. Quite naturally, all the successive development plans from 1967 to 1984 were intended to meet these concerns, and give these aspirations material form.

In this paper we intend to survey these plans as a prelude to assessing relevant performance in terms of investment (in agriculture as well as water resource development and forestry), technological effects, and the eradication of social inequalities. Our analysis will centre on the past two decades. We shall pay special attention to the years 1967-80, because that was the period during which the system of state intervention in its various forms was established and consolidated. Such a focus, however, does not preclude our bringing in current data and interpreting them whenever feasible.

Algeria's development plans were designed within a medium-term framework, their terminal date being 1980. The deadline was chosen in accordance with the calculations of planners, according to which from that year the national economy would have to meet a demand for 100,000 new jobs at least. With the target thus identified, planning became geared to three objectives. First, unemployment and underemployment were to be eradicated in order to bring eventual job demand within the prescribed limits. Second, the groundwork necessary for the creation of the new jobs envisaged would have to be done. Finally and simultaneously, the living standards of the poorest people—the peasant population still caught up in "archaic" social, economic and cultural relationships, those who suffered most during the national liberation war, those intensely affected by the rural exodus, by underemployment and unemployment, but who still dream of tasting the fruits of independence—would have to be raised.

Given such an analysis of the issues of income and employment, it would be logical to conclude that agriculture should be given prime operational emphasis. In practice, however, on the basis of the same analysis, the state decided to emphasise non-agricultural activity as a prerequisite for structural intervention in agriculture. This structural intervention was to be started only after the installation of the necessary resource base for a continuous, lasting process of modernisation in the agricultural sector and, by the same token, in the entire economic system. This, then, is the so-called "industrialist" option. Its basic principle is that it is the function of industry to provide the resources needed to modernise agriculture and to raise the living standards of the underprivileged strata on a lasting basis.

Considerations of the type of steps to be taken aside, such a scheme almost automatically orients the planner towards a two-phase implementational approach. In the first phase, the necessary conditions for "take-off" would be established, and the technological, economic, socio-political and cultural underpinnings of this "take-off" consolidated. In the second phase, the initial investment would begin to yield, to multiply and get diversified, simultaneously providing solutions to the job problem, stemming the rural exodus, and thus bringing about a palpable closing of the gap between urban and rural standards.

The first phase was to comprise two stages covering the periods 1967-69 and 1970-73. During the first stage investment levels were expected to be generally low. The aim at this stage would be to provide the various administrative services with comprehensive training in planning, while rationalising the use of existing productive capacity. The second would signal the actual "take-off" stage : investment volume would be tripled, a substantial portion of national income would be earmarked for investments, while the Gross Domestic Product would register high growth rates averaging 9 percent per year. It was during this period that radical changes in the conditions of production and living standards in the rural areas were set to begin.

After this first phase of "integration and take-off", there was to be a second phase featuring accelerated and diversified development. This phase, like the first, would also comprise two stages. In the first stage, lasting four years, work done in the preceding period would be consolidated and deepened, newly established production units would come on line, and other units would be created to complement the existing system. In the second stage, lasting three years, the industrial machinery, now perfectly run-in, would reach cruising speed, this marking the practically flawless integration of the national economy.

All this boils down, as far as the agricultural sector is concerned, to four objectives:

1. "Improved nutrition for a growing population": in other words, to provide more and better food through increased production and the substitution of national products for imports.
2. To improve the balance between production levels and income levels (without supplementary work) through capital and wage injections and the modernisation of farming systems.
3. To consolidate Algeria's position on the world market by increasing

yields, improving produce quality and exploring fresh markets.
4. To implement a "programme for the renovation of the traditional sector" so as to render it "fit" to assimilate new, improved techniques, by carrying out agrarian reforms in the sector and applying a consistent social policy there.

Time has taken its gradual toll of these objectives, so that against the background of actual performance and programmes implemented, only the aspiration to feed the population remains a live issue. As from 1973, no one expected any more from agriculture than that it satisfy "the country's main nutritional needs".

Let us now look at the results in terms of investment, technology etc.

THE STATE'S INVESTMENT POLICY

The state's investment policy with regard to agriculture will be assessed first in financial terms, then in terms of physical installations.

Financial Investment

Investment forecasts. Stated in current dinars, the absolute amount invested in agriculture out of gross total investment has risen very steeply, from 588 million dinars in the three-year plan period 1967-69 to an annual volume of 3,846 million dinars during the first five-year plan period 1980-84 (see Table 12.1). This represents an increase of 6.5 times. Calculated in constant dinars, however, the annual amount allocated to agriculture has gone up only 2.3 times between the two periods.

The second five-year plan (1985-89) explicitly gives top priority to agriculture and water resource development. During this plan period, agricultural credits are to make up 5.5 percent of total credits as compared to just 4.1 percent in the previous plan period.

This increase in investment indicates that as far as financial inputs go, agriculture has not suffered unduly from neglect. However, this conclusion needs some qualification. For the fact is that in relative terms, the proportion of investment resources allocated to agriculture actually nosedived from 18.6 percent in the 1967-69 period to 4.1 percent in the 1980-85 period, with a slight upturn in the 1985-89 period.

One reason for this is the imperative need to endow the country, which was woefully lacking in industrial and other types of infrastructure (communications, transport, educational, health, housing, admini

TABLE 12.1
Disbursement Credits Allocated in Accordance with Financial Statutes
(in millions of current dinars)
Proportions of total investment

	Public works budget investment allocation			Loans for company investments			Total			Yearly average		
											Agricultural credits	
	Total credits	Agricultural credits	%	Total credits	Agricultural credits	%	Total credits	Agricultural credits	%	Total credits	In current terms	In current terms
1967-69	5,378	718	13.4	4,107	1,046	25.5	9,485	1,764	18.6	3,162	588	588
1970-73	12,912	1,493	11.6	21,400	2,589	12.1	34,312	4,082	11.9	8,578	1,020	880
1974-77	35,770	2,996	8.4	70,628	5,011	7.1	106,398	8,007	7.5	26,600	2,002	1,371
1978-79	30,560	1,890	6.2	90,279	6,986	7.7	120,839	8,876	5.3	60,420	4,438	2,213
1980-84	191,591	5,448	2.8	278,866	13,780	4.9	470,457	19,228	4.1	94,091	3,846	1,326
1985-89*	228,878	5,500	2.4	321,130	24,500	7.6	550,000	30,000	5.5	183,333	6,008	-

* Projected expenditure.

strative, military etc.), with infrastructural installations and equipment on a large scale.

A second reason is that as economic activity has picked up volume and speed, investment needs outside the agricultural sector have increased just as fast or even faster than agricultural investment demands. One factor that aggravates this relative disadvantage of the agricultural sector is the fact that investment needs in other sectors cannot be alleviated as easily by increasing imports—the expedient adopted for meeting the country's food demands.

From the first four-year plan period (1970-73) the rhythm of investment was very fast. In terms of current dinars, annual investment in agriculture doubled in this period as compared with the previous period. The increase was sparked by the steep rise in hydrocarbon prices in 1973. Up to the end of the 1970s this rhythm held more or less steady, but during the first five-year plan period it slowed down (in constant terms) as a result of the worsening world economic crisis and the fall in petroleum prices.

With regard to investments from government budget allocations, the absolute amount allocated to agriculture, calculated in constant terms (see Table 12.2) increased until the 1974-77 period; it then decreased until the first five-year plan period (1980-84). Throughout the same overall period, i.e. from 1967 to 1984, total public investment, by contrast, increased steadily and substantially. The pace of agricultural investment, then, was relatively sluggish. But there is no reason to suppose that this was because the agricultural sector was saturated with infrastructural investment or any of the various agricultural support resources. Quite clearly the cause lies in the implementation problems associated with investment in the rural areas—an issue we shall be discussing later in this chapter.

The drop in government budget allocations for infrastructural investment in agriculture beginning in 1978 was offset by a very appreciable rise in resources invested by the agricultural enterprises themselves in 1978-79. Although this rhythm was not maintained during the first five-year plan, the absolute amount invested in agriculture was still substantial. From 1977 to 1980 the internal infrastructure of the plantations and farms underwent remarkable improvement, thus partially making up for the disinvestment that probably took place in the 1960s. In this connection, it is pertinent to note that the second five-year plan makes provision for very substantial increases in investment in agricultural enterprises.

It should be pointed out, however, that compared to the fantastic growth of investment in non-agricultural enterprises, agricultural investment has been lethargic indeed. As we shall soon see, there is in fact an inverse relation between the lethargy of the agricultural

TABLE 12.2
Proportion of Average Annual Investment Allocated to Agriculture
(in millions of constant dinars)

Year	Price Index	Investment allocation from Government budget				Loans for investments by enterprises						Total	
		Total annual credits		Annual agricultural credits		Total annual credits		Total agricultural credits		Overall disbursment		Overall disbursment for agriculture	
		Amount	Index	Amount	Index	Amount	Index	Amount	Index	Amount	Index	Amount	Index
1967-69	100	1,743	100	1,369	100	349	100	3,162	100	588	100	588	100
1970-73	116	2,783	155	322	135	4,612	337	558	160	7,395	234	880	150
1974-77	146	6,725	375	513	215	12,094	883	858	246	18,819	595	1,371	233
1978-79	202	7,346	410	454	190	21,702	1,585	1,679	481	29,048	919	2,133	263
1980-84	290	13,213	737	376	157	19,232	1,405	950	272	32,445	1,026	1,326	226
1985-89*	330	13,871	796	333	139	19,462	1,421	1,485	425	33,333	1,054	1,818	309

* Estimate.

investment process and the dynamism of the non-agricultural investment process.

Financial performance. Even though planned investment allocations earmarked for agriculture were lower than allocations to other sectors of the economy, the amounts made available remained relatively steady. In the area of practical investment implementation, however, performance in the agricultural sector was extremely poor. For the truth, as Table 12.3 shows, is that only a relatively small proportion of funds made available in accordance with the various financial statutes was ever used.

TABLE 12.3
Rate of Investment Performance in the Agricultural Sector (1974-80)

Year	% Rate
1974	55
1975	85
1976	72
1977	74
1978	71
1979	41
1980	47

In principle, the first five-year plan (1980-84) put an emphasis on agriculture. In practice, however, the actual rate of performance was by no means better than before, except for the year 1983.

To understand this mediocre performance in the area of agricultural investment, we have to look at several factors. First, ever since the late 1960s the rate of investment growth in Algeria, in financial terms, has been quite high. In all areas the projects under way have been so vast in scope that the country, under prevailing conditions, has simply been unable to cope with them on any level— economic, social, institutional or administrative. To achieve what economists call improved "investment absorption", it would have been necessary to make profound changes not only in the projects themselves but also in their implementational modalities.

In a situation where investments are largely dependent on imports, while their implementation is controlled by a super-centralised government which leaves no initiative to any of its decentralised levels, and is even less interested in tapping the much-touted "creative initiative of the masses" while still taking care to pay lip service to it, the rapid implementation of investment projects is difficult—unless the system is heir to a long bureaucratic tradition.

Second, credit disbursements from year to year are notoriously slow. Very often cash payments are only made available to the various contractors, businessmen and officials involved well into the financial year. The processes of project preparation, assessment and decision-making are equally slow.

Third, the fact is that under economic circumstances in which skilled workers and capital goods are hard to find, it is simply much harder to implement investment projects in the rural areas than in the urban areas. For example, it is easier to find a contractor to build offices in town than to find one to put up sheepfolds in the countryside.

The fourth factor, linked to the third, is that the insufficient resources available are practically monopolised by the non-agricultural sectors. In these sectors workers receive higher wages, contractors of all sorts are more readily available, supply circuits function more efficiently etc.

TABLE 12.4

Investment Performance Rates in the Agricultural Sector during the Five-Year Plan Period 1980-84 %

	1980	1981	1982	1983	1984
Investments from government instructural investment budget allocation	46.5	39	58	83.7	-
Investments from loans for company investments	38.2	52.2	45.6	75	-
TOTAL	40.2	47.8	50.2	77.2	-

Investment in Physical Plant

Needless to say, the low rate of financial investment performance means that investment in physical plant is equally low. As Table 12.5 makes clear, from the time the planning option was adopted, there has never been a single instance, in any area, of investment plans being satisfactorily implemented.

However, this failure to achieve planned objectives does not necessarily mean that agricultural conditions have been deteriorating uniformly in all areas throughout the period under review, in either the private or the public agricultural sectors.

TABLE 12.5
Physical Implementation Rate Stated as a Percentage
of Planned Investment

	1967 -69	1970 -73	1974 -77	1980	1981 -82	1983
Tractors	54	74	59	56	66	82
Combine harvesters	82	76	68	94	83	22
Trailers	-	-	-	12	56	54
Motors pumps	-	-	-	21	49	41
Stables	-	-	-	73	27	31
Plant nurseries	-	-	-	-	0	22
Vineyards	104	48	44	34	48	25
Orchard plantations	157	78	117	49	68	23
Milch cow	65	54	26	-	24	24
Superficie	-	34	25	-		

The development of means of production in the state agricultural sector.[3] According to Table 12.6, the total number of tractors in Algeria dropped as from 1967, and it was not until 1977, that the total again equalled the 1967 figure. The same situation applied in the "autonomously-managed" sector. It must be borne in mind, however, that the complement of tractors was always somewhat overaged, as shown in Table 12.7. This relative obsolescence occasioned quite a high breakdown rate, thus raising repair costs on the farms and plantations. In specific terms, from 1967 to 1977 the breakdown rate for tractors of all types was never lower than 16 percent. And even at the end of the period, in 1977, the rate was still 20 percent.[4]

Beginning in 1978, tractors in the state agricultural sector were rapidly renewed. The state farms bought 19,989 tractors between 1978 and 1982—a yearly average of 3,998. In addition, the polytechnic community farm cooperatives, which operate a large tractor fleet for servicing farmers of all categories, also bought new tractors.

It seems that the various farm enterprises have also bought substantial quantities of other agricultural equipment, so that the situation has improved appreciably.

As far as land improvement measures are concerned, however, it seems the situation in the state agricultural sector has kept going from bad to worse. Since 1962, the irrigated land area in the "autonomously-managed" sector has only increased very slightly. In the same "autonomously-managed" sector, the number of active wells and boreholes decreased from 857 in 1972 to 760 in 1977.

TABLE 12.6

Total Tractor Complement in the Agricultural Sector

	1967	1973	1977
Private sector	18,960	14,337	14,355
"Autonomously-managed" sector	18,762	17,251	18,974
"Agrarian Revolution" farms	-	-	1,458
Community tractors	1,919	2,000	2,651
TOTAL	39,641	33,588	37,438

Source: Bedrani, op. cit.

TABLE 12.7

The "Autonomously-managed" Sector:
Percentage of Tractors less than Six Years Old

	1967	1970	1973	1976
Wheeled tractors	23	55	59	43
Caterpillar tractors	14	45	52	49

Source: Bedrani, op. cit.

Drainage and maintenance work on the irrigated perimeters as well as soil conservation and rehabilitation measures have been quite inadequate, especially considering the fact that Algerian soils are delicate and susceptible to brine infiltration.[5]

In the "autonomously-managed" agricultural sector, the productive capacity of the land has not been improved through increased fruit growing. Orchard areas decreased at an average annual rate of 351,000 hectares during the 1966-69 period and by 294,100 hectares in 1982.

The development of production resources in the private agricultural sector. From independence to 1966, the official policy toward the private sector was one of tolerance. From 1966 to 1970 the policy changed into one of vigorous encouragement. In specific terms, of total investment credit made available to the country's agricultural enterprises, the proportion allocated to the private sector was: 51 percent in 1966, 39 percent in 1967, and 32 percent from 1968 to 1970. The proportion of credits actually used was even higher: 98 percent in 1966, 92 percent in 1967, 86 percent in 1968, 57 percent in 1969, and 39 percent in 1970.

As soon as the ''Agrarian Revolution'' got under way, however, the private sector's share dropped to 27 percent in 1971, and from 1972 to 1977 it hovered between 2 percent and 6 percent.

This steep drop was partly offset by the serious efforts made to equip the new agency set up by the ''Agrarian Revolution'', a multipurpose service cooperative which makes its equipment available to users in the private as well as the public sector.

In 1978, the private agricultural sector began once again to receive relatively large allocations of investment credit, a trend strongly confirmed in the first five-year plan period (1980-84), as demonstrated in Table 12.8.

TABLE 12.8

Medium- and Long-term Credits Allocated to the Private Agricultural Sector (millions of dinars)

	Yearly average 1966-69	1976	1977	1978	1979	1980
Credits allocated						
Amount	122	100	216	351	651	974
Index	100	82	177	288	534	798
Credits used						
Amount	114	100	217	414	269	743
Index	100	88	190	363	236	652

Equipment purchases in the private agricultural sector partially reflect the credit availability pattern. During the first four-year plan period tractor purchases decreased steeply; in the second four-year plan period the drop was even more drastic. But since 1978-79 there has again been a strong upward surge (see Table 12.9).

TABLE 12.9

Tractor Purchases in the Private Aricultural Sector (units)

PERIOD	Number of tractors	Average yearly purchases	INDEX
1967-69	5,001	1,667	100
1970-73	939	235	14
1974-77	295	74	1
1978-79	2,856	1,428	86
1980-82	10,712	3,571	214

The main tractor buyers are large and medium farm enterprises.[6] Since the abolition of the "works" department of the CAPCS, followed by the sale of its equipment, small farmers have been at the mercy of equipment-renting companies which charge much higher fees than those formerly demanded by the CAPCS.

Developments with regard to other agricultural equipment (for harrowing, processing, harvesting and transport) have been the same: compared with acquisitions in the 1970-77 period, there was a very large increase in purchases from 1978 to 1983.

Private sector investment in land improvement was substantially higher than investment by the state agricultural sector. For example, the index for permanently irrigated land areas, using the 1967-69 figure as a base of 100, rose steadily throughout the 1966-79 period, from 86 to 175. With regard to orchard plantations in the private agricultural sector the situation was similar: between the 1966-69 and the 1974-77 periods the planted area increased by 92,000 hectares. To achieve these results the private sector received hefty subventions, except for the first five-year plan period (1980-84).

TABLE 12.10
Agricultural Equipment Purchases in the Private Agricultural Sector (units)

	1967 -69	1970 -73	1974 -77	1978 -79	1982 -82	1983
Combine harvesters	678	61	41	35	63	30
Food harvesters	1,023	159	146	417	6,855	1,073
Harrows	1,907	955	26	398	2,131	1,068
Crops Covers	1,139	667	63	666	3,465	762
Ploughs	2,682	874	126	917	6,049	1,404
Manure spreaders and soil machines	113	48	6	182	270	21
Processors	n.a.	168	169	726	2,833	460
Trailers	472	451	620	363	1,754	210
Trucks	n.a.	n.a.	n.a.	16	24	n.a.

Investment in Water Resource Development, Soil Conservation and Rehabilitation, and Forestry

Algeria, like most African countries, has been suffering from drought for several years. The persistent drought has aggravated the well-known features of the country's climate: average rainfall very unevenly distributed over the year (a planning hazard), frequent windy spells when the hot sirocco blows in springtime, and late frosts in the northern part of the country.

Water and wind erosion are among the main dangers Algeria's

agriculture has to face. Erosion, worsened by such farming methods as dry farming and ploughing on steep slopes, as well as by overgrasing, the destruction of tree cover and frequent forest fires, removes the often delicate topsoil, chokes up the dams with silt, and has gradually been turning the farms and grazing ranges of northern Algeria to desert sand.

If the long-term productivity of Algeria's land is to be improved it is imperative that the following measures be taken: more surface and underground water must be tapped; measures to prevent water and wind erosion must be intensified; active steps must be taken to rehabilitate the country's soils both by improving humus content and by increasing the arable area; greater effort must be put into clearing and draining irrigated land.

In the light of the above, it is obvious that investments in water resource development and forestry are particularly important.

Investment in water resource development. From the colonial era Algeria inherited about 300,000 hectares of land irrigated from wells, boreholes, wire pumps and dams. Large scale water-works based on dams supplied water for irrigating about 104,000 hectares. In effect, then, under the colonial regime small- and medium-scale water resources were only slightly developed. The policy pursued by Algeria after independence has turned out to be practically identical: it is also primarily based on large scale water-works facilities.

TABLE 12.11
Planned and Implemented Investment in Water Resource Development
(Annual volume per period in million of dinars)

		1967 -69	1970 -73	1974 -77	1980 -83	1980 -84	1985 -89
Forecasts	Amount	333	910	3,650	6,006	4,600	8,200
	Index	100	273	1,096	1,804	-	-
Implementation	Amount	164	336	744	3,016	-	-
	Index	100	205	454	1,839	-	-
Implementation × 100		49	37	20	50	:	-

Source: Ministry of Lands and Planning.

In the area of water resource development, as in agriculture itself, the rate at which allocated funds are actually put to use has always been very low, even though from the first three-year plan period (1967-69) to the first five-year plan period (1980-84) as well as during the second five-year plan period (1985-89) the sums allocated have increased very substantially (see Table 12.11).

As far as physical plant construction and the creation of improved facilities on the irrigated perimeters are concerned, implementation rates have been even lower. For instance, from 1967 to 1977, there were plans for the construction of about twenty dams. In the event, however, only two got built, while two old ones had their crests raised. Similarly, it was planned to equip about 160,000 hectares for irrigation, but only 66,000 hectares were actually so equipped. Worse still, out of the 55,000 newly equipped hectares, only 20,000 hectares were in fact irrigated by the end of the period. On the remainder irrigation was delayed either because of a water shortage, or because the necessary equipment had not been fully installed.

Indications are that the rate of physical plant installation remained poor during the 1980-84 five-year plan period. For example, the plan target for the period called for the equipment of 100,000 hectares to the point where they could be brought into agricultural production.

This overall target implies a yearly objectives of 20,000 hectares equipped and ready for farming. In 1980, however, only 5,400 hectares were so equipped, while the 1982 figure for such new hectarage was even lower: 2,000 hectares.

Apparently, between 1970 and 1977 there was no increase in the total irrigated land area. The reason was that water resources became scarcer, and some irrigated land around the urban centres was taken over for activities other than farming.[7]

The scarcity of water in the agricultural sector was caused by the rapid silting-up of dams coupled with an increase in human water consumption.

The analysis of poor performance in the area of financial and material implementation in the agricultural sector holds true for water resource development as well. There is an extra reason here, however: it is necessary to conduct time-consuming surveys and studies prior to the construction of major water works facilities and dams, and yet this is the type of facility whose construction has lately been promoted, even though medium- and small-scale facilities are easier to construct, and yield faster results. In this connection it is interesting to note the government's announced intention to shift the emphasis decisively toward medium- and small-scale water resource facilities, with special attention to small hillside dams, during the second five-year plan period.

Investment in forestry, soil conservation and rehabilitation. Problems in this area are so enormous that financially it could be extremely costly to try solving them by applying orthodox investment policies. The only long-term solution lies in a permanent and efficient mobilisation of the people.

Relatively ample sums have been allocated for reforestation (see Table 12.12), and implementation rates in this area have been satisfactory. The same is true of the soil conservation and rehabilitation effort in the second four-year plan period.

TABLE 12.12

Investment in Forestry, Soil Conservation and Rehabilitation
(average yearly amounts in millions of current dinars)
(Notes : PS and PCD not included)

	1967-69 plan period	1970-73 plan period	1974-77 plan period
1 *Reforestation*			
Plan Investment:			
Amount	18	85	157
Index	100	474	874
Investment implemented:			
Amount	15	82	95
Index	100	547	638
Investment implemented:			
as % of planned investment	83	96	61
2 *Soil conservation and rehabilitation*			
Planned investment:			
Amount	6	33	51
Index	100	550	850
Investment implemented:			
Amount	6	32	20
Index	100	533	333
Investment implemented as % plan investment	100	97	39

With regard to physical performance, indications are that in the eleven years from 1967 to 1978, only 322,000 hectares were reforested. This is 60 percent of the planned total, a paltry performance considering the great damage done each year by forest fires (which consume thousands of hectares of forest) as well as by the undisciplined exploitation of meadows and woodlands. Physical performance in the areas of soil conservation and rehabilitation has also been poor. While it was planned to work over 400,000 hectares during the period here under review, only 196,000 hectares (49 percent) were in fact worked over. Considering that specialists say that 23 million hectares require conservation and rehabilitation work in northern Algeria alone, this is a mere drop of performance in the ocean of need.

The objectives set by the 1980-84 five-year plan were reasonably modest. Yet, by the end of the third plan year, only 45 percent of the reforestation target (calling for 330,000 hectares of fresh plantations and 100,000 hectares of major reforestation unkeep work) had been achieved. As for soil conservation and rehabilitation, of the 170,000 hectares targeted over the five-year plan period, only 16 percent had been worked over. Meanwhile, each year forest fires destroy almost as many hectares as replanted, but the provision of fire-fighting facilities lags behind planned objectives : only 53 percent of planned construction and maintenance work on fire-break trenches has been done, while the figure for forest range tracks is a mere 28 percent.

TECHNOLOGICAL POLICY AND ITS EFFECTS ON PRODUCTION

Ever since independence, Algeria's agricultural policy in this domain has boiled down to following the model of agricultural production in force at the end of the colonial era, with certain "modernising" modifications. It was a perennial article of faith among agricultural policy-makers that this was the immediate term, while preparing to achieve significant production increases later. Unfortunately, the country's economic, technological and socio-political conditions were not such as to favour the consistent application of this model.

Under these circumstances, a number of questions remain to be answered: what precisely is this technological model? Was it coherently applied? And what was the impact on production?

Description of the Technological Model and an Assessement of Coherence in its Application

The model of agricultural production inherited from the colonial settlers gave pride of place to cereals, wine grapes and, to a lesser extent, fruits. Independent Algeria has followed the same pattern. At the close of the colonial period, within the "European" agricultural system the rotation of cereal planting with fallow periods was practised on 59.4 percent of arable land. In the 1974-77 period, under the "autonomously-managed" agricultural systems that replaced the "European" system, the same cereal-fallow rotation took up 60 percent of arable land on the average. The only significant changes have been a decrease in the area devoted to grape vine (from 12.3 percent to 7.8 percent of arable land) and an increase in fodder planting (from 2.3 percent to 7.2 percent of arable land).

The two main rotational systems in vogue during the colonial era still dominate the post-independence scene : the traditional rotation of cereals with open fallow, and the modern method, rotating cereals with dead fallow. This latter method is at the core of what is known as "dry farming", even though it has often proven notoriously dangerous, since it accelerates erosion and reduces the soil's humus content. Here and there, of course, a few changes may be observed. For example, on some farms a more intensive rotational cycle is being tried out, alternating cereals with fodder. In addition, more efforts are being put into coupling animal husbandry with agriculture. On the whole, however, these changes remain extremely limited.[8]

One feature of the colonial production model was that it was highly mechanised. After independence, attempts were made to keep up this high degree of mechanisation, and to maintain the same manpower-machine ratio. Another feature of the model was the relatively abundant use made of selected seeds, fertiliser and chemical inputs. Here the agricultural policy-makers tried to outdo the colonial settlers. From the late 1960s on, the use of fertiliser was heavily boosted in the state agricultural sector. By 1978, the agricultural system was consuming three times more fertiliser than in the 1961-65 period.

There was another novelty in post-independence agricultural policy : the introduction of pedigree cattle for breeding in the "autonomously-managed" sector. The type of cattle thus brought in were famous for their high milk yields, and they were expected to help cut down on Algeria's milk imports.

In sum, then, the agricultural production model adopted for extension by the Algerian policy-makers—a model they still insist on generalising—is highly mechanised and uses large doses of chemical inputs. As such it requires a relatively highly skilled work force, operating according to the sort of work routine typical of the mechanised agro-industrial complexes of the so-called developed countries.

Given such a model, in order to achieve optimum production levels and returns on investment, every single link in the complex production chain has to be in functioning order. In other words, the entire technological package must be available.

This, unfortunately, has rarely ever been the case. The fact is that this model has never been applied with all the necessary coherence. And the main reason for this failure is that its success depends on imported inputs or on items produced locally but with imported technology (and therefore difficult to regulate). Thus, it is not very useful to increase chemical fertiliser inputs if the necessary weeding, mechanical or manual, remains undone because the requisite weed-killers have either not been imported or not been distributed in good time. Similarly, selected seeds produce only poor yields if previous to planting the soil is not properly tilled because equipment breaks down in the absence of imported spare parts.

Within the same perspective, it seems quite unwise to import herds of high-yield pedigree milch cows if investments for the irrigation of hectarage under fodder, for farm machines, for harvesting and storing fodder, and for building the type of hygienic sheds needed for such herds are not coherently implemented.

Examples of breakdowns affecting the application of the production model which Algeria's agricultural policy-makers have been trying to propagate since independence could be cited at great length. The important consideration, however, is that these breakdowns happen because the society lacks adequate skill in the handling of the imported technologies adopted.

The Application of the Technological Model: its Effects on Agriculture

As shown in Table 12.13, which covers the "autonomously-managed" sector where the technological model has been most consistently applied, achievements in terms of agricultural production have been decidedly mediocre. It is clear, for instance, that average production figures for the main crops, cereals, have never been higher than corresponding figures at the close of the colonial era (1954-57). Exceptions exist, however: there were quite spectacular increases in the

TABLE 12.13
Yields in the "Autonomously-Managed" Agricultural Sector (in 100s of kg per hectare)

	Average annual yield per period			
	1967-69	1970-73	1974-77	1978-80
Winter cereal	8.2	8.8	8.3	8.0
Market gardening produce (including potatoes)	71.3	49.7	66.3	44.8
Citrus fruits	100.4	87.1	57.5	90.5
Fruits (seeds and pips)	39.7	33.6	11.0	12.8
Grape vine	27.6	27.8	18.6	17.3

Source: Agricultural Statistics, Series A and B.

production of dry vegetables, potatoes and synthetic fodder between 1967-69 and 1978-80. But compared to the country's needs, these increases have been largely inadequate.

Value added in the entire agricultural sector, calculated in constant prices with 1978 as a base, rose from 5,200 million dinars in 1967 to 5,600 million in 1973 (an annual growth rate of 1 percent), and to 6,100 million dinars in 1977 (an annual growth rate of 2 percent).[9]

In effect, the country's agricultural system has to all intents been in a state of stagnation, which necessarily means a very serious worsening of the country's dependence on foreign food supplies. For instance, average annual imports of the leading staple, cereals, have risen as follows:

1966-69 : 28 percent
1970-73 : 29 percent
1974-77 : 50 percent
1978   : 65.5 percent

TABLE 12.14

Milch Cow Yields in the "Autonomously-Managed" Agricultural sector (in litres per cow per year)

	1970	1971	1972	1973	1974	1975	1976	1977
Yield	2,531	1,334	2,010	3,153	2,489	2,529	2,362	2,443
Index	100	92	103	125	98	100	93	96
Annual Yield per period	2,657				2,456			

Source: Ministry of Agriculture.

As far as yields are concerned, the "improved colonial" model has disappointed expectations, as can be seen in Tables 12.13 and 12.14. These tables show that between 1967-69 and 1978-80, yields either stagnated or decreased in varying degrees. Statistics for yields per hectare provide a much more accurate index than statistics for absolute production quantities, and in the Algerian case they constitute incontrovertible proof that the intensive form of agriculture based on the technological model advocated by the agricultural policy-makers has been a failure.

STATE POLICY ON AGRICULTURAL STRUCTURES

Since 1962, state policy has tended towards increased control over productive structures.

Policy in the "Autonomously-Managed" Sector

Since independence, there was a determination to establish large agricultural units, i.e. to push a process already well under way to its logical conclusion. The President of the Republic, addressing the National Assembly in December 1962, proclaimed that "... to divide up the land, to satisfy the petty bourgeois craving for individual property, is to turn one's back on modernisation". So between 1963 and 1968 the number of production units decreased, while the size of each unit, in terms of land area, increased.

Land nationalisation and reallocation exercises[10]			
	1963	**1964**	**1965**	**1968**
Area (hectares)	2,132,860	2,332,860*	2,332,860	2,301,602**
No. of units	3,000	2,191	2,188	1,623

* An increase due to the expropriation of major Algerian land-owners.
** This decrease is apparently connected with the restoration of expropriated land to a few Algerian landowners.

To justify this process, a number of reasons, mainly of a technical nature, have been put forward. These reasons refer to the low level of available personnel, as well as the need to ensure profitable management and efficient resource allocation.

Since 1980, the state, having trained a relatively large number of agricultural officers, and having realised that the management of giant plantations is a complicated business, has been trying to restructure the state agricultural sector. As of now finishing touches are being applied to the new scheme.

The justification for this restructuring was that it was necessary to "rationalise" the existing potential in order to bring it under greater "technical" and "economic" control. Thus, on the basis of agrotechnical criteria, attempts were made to cut up the large estates once again in such a way as to create individual units, each as homogeneous as possible. In this exercise there was a tendency to overlook the social aspect—the existence of communities with particular social customs and a special approach to the organisation of their relationships with each other and with their environment. The result was the creation of 935 new units partly formed through the absorption by existing "autonomously-managed" units of the cooperative farms established during the "Agrarian Revolution" that affected the private sector between 1972 and 1978.

It might be pointed out that the real objective remained the same: to give the state greater control over the productive system, or at least over that part of the productive system it considered profitable. For the restructuring exercise begun in 1980 caused the abolition of the sector established during the "Agrarian Revolution". Part of the dismantled sector was merged into existing units in the "autonomously-managed" sector, or turned into new "autonomously-managed" units. Most of the remaining land was divided up into individual plots and parcelled out among small peasants.

By 1983, the arable land total in the state sector had already dwindled to 2,833,212 hectares as compared to 3,400,000 hectares before the restructuration exercise—a 17 percent decrease.

Policy in the Private Sector
From late 1971, there was talk of a determination to carry out an "Agrarian Revolution" . The objectives of this process would be to eliminate "all forms of exploitation of man by man" including wage labour, to intensify farm work, to use income generated from the land for the benefit of those working the land, and to promote agricultural cooperation. It was expected that as a result of the scheme's success, private producers would in the long run come voluntarily together to form "economically viable" units.

In short order the activities conducted within this framework also came to look like one more instance of the determination to extend the boundaries of the state's rationalised control. For, by conferring a special status on collective forms of cooperation, by indiscriminately nationalising the land of all types of landowners (big, medium and smallholders such as part-time farmers with paid jobs), by setting up a maze of conditions and constraints around the normal supply and marketing circuits, the state was aiming at a dominant position for itself and its way of conducting business, and ultimately at the imposition of its objectives. In the event, the state has in fact had to revise its operations by making the conditions for providing extension services to the sector more flexible. This change did not result from peasant pressures threatening the state's survival, but from the state's own need to organise the sector more effectively in order to achieve its initial objectives.

Overall, the "Agrarian Reform" extended the state's material base over an area of 1,446,000 hectares, two-thirds of which (1,100,000 hectares) were arable. The reform was supposed to mobilise all the landless together with the poorer, marginal farmers. In practice it mainly involved people in the landless category. That was why, as late as six years after the "Agrarian Revolution" began, large tracts of land had still not been allocated. In 1978, 40 percent of the land had still to be allocated. And of the 60 percent already allocated more than half needed clearing and improvement before cultivation.

The reaction to both the restructuring of the "autonomously-managed" sector and that of the private farm sector was muted, because income from agriculture constitutes only a minor part of peasant income. This fact in turn is linked to the social distribution of the country's oil income, as well as to the employment boom produced by the industrialisation process.

STATE POLICY ON THE ERADICATION OF INEQUALITIES BETWEEN THE URBAN AND RURAL AREAS

At the close of the colonial era, the situation of the overwhelming majority of the rural and farm population was one of extreme poverty. Rural consumption standards, both communal and individual, were very low. It is perfectly true that independence brought better living conditions to the masses. It is equally true, however, that independence or no independence, the urban-rural gap has remained unbridged. In some areas, in fact, the gulf has widened, thus providing continued impetus to the drift from the countryside and from farming, while worsening production conditions in the rural areas.

As will soon become clear, agricultural income levels have risen slowly since independence, with an accelerated spurt in the last couple of years, while variations exist between the different social strata in the rural areas. At the same time, it is plain that public consumption levels in the rural areas are far lower than in the urban areas.

Income Patterns among the Agricultural Population

Studies on income patterns in both the urban and the rural areas are rather scarce. In addition, the few that exist are difficult to interpret because the estimates they present often vary, sometimes widely.

Indications are that from 1962 to 1973, income from agricultural work increased only slightly. This slight increase, incidentally, is reflected by the development of agricultural production in the period. At the same time, income outside the agricultural sector must also have increased only moderately, since the economy as a whole was somewhat sluggish, picking up only with the first four-year plan (1970-73). It has been estimated that from 1967-68 to 1973, consump

TABLE 12.15

The Development of Minimum Wages in the Agricultural and Non-agricultural Sectors

(Wages per 8-hour day, in dinars)

	1961	1963	1964	1970	1972	1974	1976	1978	1979	1980
1. Agricultural wage	7.06	-	7.54	7.54	9.80	12.25	15.30	20.00	28.00	33.00
2. Index	94	-	100	100	130	162	203	265	371	438
3. Non-agricultural wage	9.66	9.66	-	10.88	13.84	16.64	19.20	25.28	33.68	33.68
4. Index	89	92	-	100	127	153	173	232	309	309
Correlation (1/3 x 100)	73	-	-	69	71	74	80	80	83	98

tion expenditure among the agricultural population only increased at
an average annual rate of 1 percent.[11]

According to estimates for 1978 issued by the Ministry of Lands
and Planning, "average per capita rural income is approximately half
the urban average". The estimates also give the per capita rural
average as 2,000 to 2,200 Algerian dinars per year, though it could
occasionally fall as low as 500 Algerian dinars.[12]

Agricultural wages stayed practically stationary from 1962 to
1970, but after that date they increased faster than non-agricultural
wages, while still remaining lower than the latter (see Table 12.15).
At the start of the first five-year plan (1980-84), the minimum wage
in the agricultural sector finally equalled that outside the sector. In
spite of this development, it remains likely that average wages outside
the agricultural sector are higher than average wages within it, for
comparable skill levels.

Agricultural incomes other than wages have risen even faster in
comparison with non-wage income outside the agricultural sector.[13]
According to Ministry of Planning estimates, average annual increases
in gross income earned by entrepreneurs in the agricultural sector
were as follows: 12 percent from 1967 to 1969, 6.8 percent from
1970 to 1973 16.6 percent from 1974 to 1977.

Comparable increases for entrepreneurs outside the agricultural
sector were respectively 2.7 percent, 7.8 percent and 12.9 percent.
Nevertheless, non-wage income in the agricultural sector still seems
markedly lower than non-wage income outside the sector, on the
whole.

To sum up, it seems perfectly clear, as M. Boukhobza has stated,
that "the gap between rural and urban income levels persists, and
may even have widened".[14]

Unequal Access to Public Consumption

In the provision of public and community amenities, progress
accomplished in the rural areas since independence has been
remarkable as compared to what had been done before. However,
there are still considerable inequalities between the urban and the rural
areas. We propose to examine these inequalities in the areas of education, housing and health.

Education in the rural areas. People in both the urban and the rural
areas now identify success with the possession of degrees and certifi-
cates. Possibly, with the steady eradication of illiteracy, this sentiment
is losing its hold. But it has played an important role in shaping rural
attitudes to formal education, and for a lot of people it continues to
do so. Every effort has to be made, so it is felt, to send all or at least

some of one's children to school. Villagers do not hesitate to board out their children with relatives in town or in other villages with schools, if they can afford to do so. If they cannot, they ruthlessly abandon their small farms, pull up stakes and go to settle in the urban areas, taking their chances on the job market there.

School enrollment in the rural areas rose from 32.6 percent in 1966 to 53.1 percent in 1977. During the same period, enrollment in the urban areas rose from 53.7 percent to 88.5 percent. The inequality is not merely quantitative; it is also qualitative. Schoolteachers in the rural areas are more poorly trained than their urban counterparts; their living conditions are much tougher, and they have far fewer teaching aids to work with.

Housing in the rural areas. Strictly speaking, housing is not a public amenity. But the state does play an important role in this field, either by providing direct subventions, or indirectly, through the work of the public construction corporations.

Judging by 1966 and 1977 census data, between the two dates housing conditions in general and those in the rural areas in particular deteriorated. The overall average occupancy figure per housing unit was high: 7.1 persons in 1977, compared to 6.1 in 1966. In the rural areas the figures were higher: 7.2 persons per unit in 1977, compared to 6.3 in 1966. The comparative urban figures were 7 in 1977, as against 5.7 in 1966. With regard to the number of persons per room, the situation was identical. The overall national average for 1977 was 3.2, compared to 2.6 in 1966. But in the rural areas the 1977 figure was 3.3, compared to 2.8 in 1966, while the urban figure for 1977 was 2.9, compared to 2.4 in 1966.

In addition, household amenities and utilities in the rural areas are quite inadequate as compared to urban amenities. For instance, there is a desperate shortage of electricity, tap water, cooking gas, sewage and garbage collection systems in the rural areas (see Table 12.16).

The actual achievement, as illustrated above, was poor even

TABLE 12.16
Utility Supplies to Housing Units from 1966 to 1977 (%)

		Electricity	Tap water	Cooking gas	Sewage
Urban areas	1977	84	82	37	77
Rural areas	1977	2.5	21	-	14
Algeria	1977	49	46	13	40
Algeria	1966	31	39	-	-

though more intense efforts had been put into housing construction in the rural areas than in the urban areas. Specifically, from 1967 to 1979 more housing units (194, 191) were constructed in the rural than in the urban areas (148, 148), mostly because private initiative predominated in the rural areas. Planned housing construction in the rural areas came nearer to achieving set targets (41 percent) than in the urban areas (17 percent) mainly because large numbers of people in the countryside adopted the do-it-yourself approach to housing construction.

In recent years (1980-84), a major do-it-yourself housing programme has been started in the state agricultural sector. This approach promises satisfactory results even though people trying to build their own houses in the rural areas face a number of problems due mainly to the fact that building materials are hard to obtain, and in the remote areas they are extremely expensive.

Health problems in the rural areas. In the area of health, the inequalities between rural and urban areas are even more disturbing. Here also the situation has improved a great deal since the colonial era. But health conditions are still so poor both in absolute terms and in relation to the population's higher aspirations, that, along with the lack of educational and training facilities, they constitute the main cause of the rural exodus.

In the rural areas, general living conditions, which ultimately determine public health standards, have improved very slightly. As late as 1984, a balanced, adequate diet, healthy housing conditions and a normal water supply system were still, for large numbers of Algeria's rural population, not realities but aspirations.

The country's medical and health policy, in spite of the introduction of free medical services, has put the rural areas at a greater disadvantage as compared to the urban areas, in terms both of available health facilities and of staff allocations.[15]

In the 1980-84 plan period, the medical system was liberalised, even though the free medical service scheme was maintained. This liberalisation was likely to put the rural areas at a further disadvantage.

THE EFFICIENCY OF AGRARIAN POLICIES

The state's agricultural policy has not resulted in a reversal of the colonial tendency to make the country increasingly dependent on food imports while worsening the already shaky balance of its productive base. There are various ways of assessing these tendencies. One such method involves an assessment of the changing role of imports in the

TABLE 12.17
Percentage of National Needs Supplied from Imports

	1967-69	1970-73	1974-77	1979-82
Cereals	28	29	50	65
Dry vegetables	12	8	40	85
Eggs	4	35	47	75

TABLE 12.18
Percentage of Average Food Intake Made up of Imports

	Average food intake, 1984 Ministry of Planning estimates (in kg or litres/ person/year)	Quantity of imports per capita in 1983 Customs figures (in kg or litres)	Estimated percentage of food intake made up of imports (in calories)
Cereals[1]	175-180	137.4	76.3
Potatoes	38	16.7	43.9
Dried vegetables	8.7	4.4	50.5
Fresh vegetables	77	-	-
Fruits	44	1.3[2]	23.0
Sugar[3]	20	34.2	171.0
Red meat	9.5	1.6	16.8
White meat	7.5	-	-
Eggs	4	3.4	85.0
Fish	2.9	0.1	3.4
Milk[4]	90	60.2	66.8
Oil[3]	14.9	15.6	104.6
Butter[3]	2	3.1	155
Total calories	2,896	2,401[5]	82.9

Notes
1. Barley and maize not included.
2. Dried vegetables.
3. Quantities imported exceed quantities consumed, possibly because of smuggling or hoarding. A further reason may be that sugar used for industrial purposes (drinks) and pastry-making is not included in the first column.
4. It is assumed that one third of imported tonnage was in powdered form, and that 1982 and 1984 tonnages were identical.
5. Assuming imports of sugar, oil and butter are higher than Ministry of Planning consumption estimates, the imported intake rises to 2,205 calories per day, i.e. 76.1 percent of total average intake.

satisfaction of the nation's needs. In this respect, a Ministry of Planning report based on the country's foreign trade data plus information from a 1979-80 consumer survey indicates that Algeria imports two-thirds of its calorie intake.

In precise terms, the proportion of imported food needed to meet the nation's needs has increased steadily year by year. In the 1977-79 period, the proportion of average food intake covered by imports was estimated at 60 percent; by 1974 this proportion had most probably soared to 70-80 percent (see Table 12.18). Incidentally, it is also probable that the 1983 import percentages listed in Table 12.19 are underestimated, at least as far as certain items are concerned. After all, it would be logical to assume that 1984 imports would be higher than those for 1983.

In the initial years of the first five-year plan period, the value of imported commodities rose rapidly. From a 1979 base index of 100, food imports rose in value to 171 in 1982. The percentage of total import value accounted for by food imports was as follows: 1979 : 15 percent; 1980 : 19 percent; 1981: 18 percent; 1982: 17 percent.

The value of food imports expressed as a percentage of hydrocarbon export income was as follows: 1979: 13 percent; 1980: 13 percent; 1981: 16 percent; 1982: 17 percent.[16]

In the seven years from 1976 to 1983, quantities of an overwhelming majority of imported items have risen two to ten times (see Table 12.19). Import percentages of such so-called basic commodities as cereals, sugar, fats, edible oil and potatoes are particularly high.[17]

In addition, it must be noted that imports of dairy and animal husbandry products as well as animal feed have increased very substantially. The following list gives the factors by which itemised imports have increased, in order of decreasing magnitude:

Meat	9.5
Fish	7.6
Eggs	7.4
Other food items	7.1
Fruit	6.8
Animal food	6.5
Barley	5.6
Potatoes	4.2
Butter	3.5
Canned vegetables	3.4
Cheese	3.3

TABLE 12.19
Food Import Patterns

	Quantities (Tonnes)			1983	Value (million of dinars)		
	1976	1982	1983	1976	1976	1982	1983
02 Meat	3,566	27,030	33,766	9.47	28.8	210.1	371.8
03	5	2,925	3,795	75.9	0.3	22.1	28.0
04 Milk	97,297	180,599	253,358	2.60	257.1	890.1	1,261.5
04 Butter	18,742	47,752	66,393	3.54	97.8	526.1	602.2
04 Cheese	5,380	15,014	17,605	3.27	20.8	152.2	90.8
04 Eggs	9,758	66,043	72,360	7.41	53	430.1	382.1
07 Vegetables	82,620	203,707	350,629	4.24	79.7	203.1	296.9
07 Dried vegetables	39,484	92,010	?	-	67.6	265.7	-
08 Fruits	4,294	56,923	29,045	6.76	10.3	170.8	96.5
09 Coffee, tea, spices	46,681	64,706	109,156	2.34	383.7	664.9	1,125.0
10 Wheat[1]	1,195	1,934	2,129	1.78	1,094.5	1,876.5	1,796.0
10 Barley	66,685	465,423	373,088	5.59	43.6	387.5	237.3
10 Maize	95,163	372,790	305,290	3.21	55.5	289.9	241.7
10 Rice	8,141	18,083	5,599	0.69	11.9	36.9	7.2
11 Flour	352,208	932,316	666,000	1.89	439.6	1,327.9	860.3
12 Oil seeds	69,928	?	40,821	0.58	85.6	-	110.7
15 Oil	164,848	?	306,121	1.86	324.3	-	743.6
16 Canned meat & fish	1,779	11,247	1,517	0.91	10.1	20.0	25.2
17 Sugar	352,117	512,253	714,436	2.03	650.5	861.1	897.4
18 Cacao	1,276	1,956	1,271	1.00	12.7	29.1	18.1
20 Canned vegetables	12,907	20,083	44,307	3.43	25.7	69.2	176.5
21 Other food products	1,729	8,114	12,258	7.09	8.5	61.3	96.4
22 Alcoholic drinks	390	410	69	0.18	2.9	2.0	0.2
23 Livestock feed	34,426	233,492	222,865	6.47	57.9	368.9	346.0
24 Tobacco	7,078	20,892	18,201	2.57	44.5	236.8	198.4

Notes 1. Thousands of tonnes.   *Source:* Algerian Customs Services.

The point is that imports do not merely supply an increasing quantity of Algeria's food needs; they are also getting steadily spread over an increasingly wider range of food items. Most significantly, these imports tend to promote the perpetuation and even the extension of a European-type consumption model, characterised by a heavy emphasis on livestock products and commodities put out by the agricultural and food industries.

Care must be taken to point out, however, that the import boom cannot be explained solely by a simplistic correlation between decreased food production or stagnation on the one hand and population growth on the other. A further cause is that there has been a clear improvement in dietary standards among all strata of Algerian society.

Another way to evaluate the efficiency of agrarian policies is to assess the development of imported inputs. Here again we are confronted with obvious evidence of the strong foreign orientation of Algerian agriculture. The fact is that the nation's agriculture is in a situation of dependence even in such sensitive areas as seed production, and this dependence is steadily getting extended as well as intensified. And as far as input production is concerned, it seems unlikely that the industrial restructuring already under way or the establishment of plant seed and nursery facilities can, in the short and medium terms, reverse the present trend.

A third aspect needs to be highlighted: income from agriculture has been accounting for a smaller and smaller portion of the total livelihood of groups and persons partly dependent on farm work. So steady has this decrease become, that nowadays agricultural income makes up less than 20 percent of the total resources of such people, even though prices of production factors have remained unchanged since 1974, while producer prices have risen appreciably, from an index of 48 in 1974 to 130 in 1981 (on a base of 100). Algeria happens to be a country with substantial mineral resources. Whatever problems it has faced, there is no denying that the country has been able to adopt a definite social and economic policy. Against such a background, the existence of such serious gaps between set objectives and actual performance in the agricultural sector raises questions about the ability of a state-dominated agricultural system to bring about the hopes for development, or even to initiate growth in agricultural production. In a context of international crisis, these questions become even more acute.

TABLE 12.20
Value of Imports, 1979 to 1982 (in millions of Algerian dinars)

	1979	1980 1	1980 2	1981 1	1981 2	1982 1	1982 2
1. Food and drinks items	5,174	7,782	+ 50	9,103	+ 17	8,872	+ 3
11. Basic commodities	2,169	3,459	+ 32	3,339	+ 3	3,883	+ 15
111. Mainly for industrial use	1,933	2,719	+ 41	2,481	+ 7	2,753	+ 11
112. Mainly for household consumption	687	740	+ 8	855	+ 16	1,080	+ 26
12. Processed items	2,555	4,323	+ 69	5,674	+ 33	5,039	+ 13
121. For industrial use	1,674	2,597	+ 55	3,541	+ 36	2,725	+ 23
122. For household consumption	881	1,726	+ 96	2,224	+ 29	2,314	+ 4
Total imports	32,378	40,519	+ 25	48,637	+ 20	49,383	
2. Exports							
21. Food items	401	431	+ 7	519	+ 20	328	– 37
3. Total exports	35,011	58,953	+ 58	57,384	– 4	52,700	– 8
31. Hydrocarbon exports	37,116	59,021	+ 59	56,223	– 5	51,612	– 8

Column 1 : Value
Column 2 : Increase over previous year
Source : Ministry of Lands and Planning.

TABLE 12.21
Development of Per Capita Consumption (in kg)

I T E M	National consumer survey 1979-80	A A R D E S consumer survey 1966-67	Model diet for Algeria (proposed by M. Autret)
CEREALS	185.33	264.44	180
Wheat	173.92	213.0	187
Barley	9	48.8 (18.9)	8
Rice	0.98	0.64	-
POTATOES	34.40	21.7	30
VEGETABLES	8.26	3.4	6.5
FRESH VEGETABLES (including leguminous items)			
FRESH FRUITS	30.0	21.2	
DRIED FRUITS AND OLEAGINOUS			45
ITEMS	2.79	7.6	
SUGAR AND DERIVATIVES	15.82	14.25	22.5
Sugar	15.17		
MEAT AND OFFAL	15.68	8.67	-
(including AID mutton)			
EGGS	1.06	0.47	3
MILK AND DAIRY PRO-DUCTS) (in milk equivalent)			
FISH	2.2	1.35	4
OIL AND FATS	15.29	8.83	13

1. Extra statistics supplied by us (in brackets) are taken from Autret, M. *Analyse nutritionnelle de l'enquête nationale sur la consommation et les budgets des ménages— Algérié*, UND-FAO, Rome, 1978.
2. Table from the provisional print-out on preliminary findings of the 1979-80 households consumer survey conducted by A. Mokaddem, June 1981 (National Office of Statistics).

NOTES

1. The reasons were mainly related to the intensity of popular remembrance of the way Algerians had been dispossessed of their land, as well as to the political platforms of various nationalist parties active during the colonial period.

2. Including the need to prevent all countervailing centres of power from building up a material base, especially since, beginning in March 1962, various factions within the Algerian National Revolutionary Council had become involved in ferocious infighting.

3. The state agricultural sector includes the whole of the "autonomously-managed" sector (the former European settler estates) and the sector created by the Agrarian Revolution from arable land in the public domain as well as from the property of absentee landowners and major landlords, nationalised as from 1972.

4. See Bedrani, S., *L'Agriculture algérienne depuis 1966: Etatisation ou privatisation*, OPU, Algiers, 1981, p. 49.

5. Cf. Bedrani, op. cit., pp. 26 et seq. In the 1983 *Plan Implementation Report*, the Minister of Agriculture blames the low level of fruit production on the fact that "the citrus groves are extremely old, their replacement rate is slow because available irrigated land is limited, and the drainage network is in a poor state of repair", MARA, *Rapport de synthèse de l'exécution du plan annuel 1983*, April 1984 (mimeo.), p.11.

6. It takes a farm area of at least 50 hectares to make a tractor purchase worthwhile.

7. See Ministry of Agriculture and the Agrarian Revolution, *Evolution de l'Agriculture de 1969 à 1978*, September 1979.

8. See Bedrani, S., Bourenane, N. and Molina, I., *Les politiques agraires en Algérie: vers l'autonomie ou la dépendance*, CREA, Algiers, 1982.

9. Ministry of Lands and Planning, *Synthèse du bilan économique et social de la décennie 1967-1978*, Algiers, May 1980, p. 89.

10. Cf. Bedrani S., Bourenane, N. and Molina, I., op. cit.

11. Bedrani, S., op. cit., p. 167.

12. Ministry of Lands and Planning (MPAT), *Bilan de la décennie 1967-1977*, p. 209.

13. Ibid., p. 220.

14. Boukhobza, M., *Disparités des revenus et pouvoir d'achat en Algérie 1968-1979*, INEAP, February 1981, p.152.

15. See Bedrani, S., *L'Agriculture algérienne depuis 1966*, pp.199-203.

16. Source: Ministry of Lands and Planning.

17. The Ministry of Planning gives a figure of 2 kg of butter. This is an obvious underestimate. After all, by 1979-80 average butter consumption had already reached 2 kg and since then market supplies seem to have grown more plentiful, thanks to imports. This also applies to sugar, of which the average consumption figure had already reached 15 kg (not counting sugar inputs for pastry, cakes and drinks).

13. THE STATE AND FOOD POLICIES IN NIGERIA

S. T. Titilola

INTRODUCTION

In the views expressed on food problems by African policy-makers in the Lagos Plan of Action in 1980 they noted *inter alia* that at the root of the food problem is the fact that member states have not usually accorded the necessary priority to agriculture either in the allocation of resources or in giving sufficient attention to policies for promoting productivity and improving rural life.[1]

Most nations are concerned with agricultural productivity but under various development stages. Some nations have achieved annual growth in food production sufficient for their people while others can barely provide sufficient food to maintain their already poor diet. For the latter set of nations, to which Nigeria belongs, the problem is compounded by rapid population growth and declining productivity. Therefore, there is an urgent need to increase farm output in order to lessen the misery, threat of meagre food supplies, hunger and malnutrition and to aid in national economic development.

Faced with the developmental problems of meeting basic human needs (such as water, health care and sanitation) agriculture has not performed satisfactorily the traditional role expected of it such as: a) increased production of food to meet the need of a growing population and rising industrial production; b) increased production and processing of export crops in order to expand and diversify the country's foreign exchange earnings; and c) the expansion of employment opportunities to absorb the increasing labour force.

There is no doubt that agriculture in Nigeria today is an economic and political issue. Due to their adverse effects on other sectors the problems of the poor performance of this sector cannot be separated from the general health of the economy. These problems are of interest and concern not only to the consumers whose incomes are daily being eroded by high food prices, but also to the government whose role it is to give direction for the smooth working of the economy. The current food and agricultural crisis has had a gestation period during which very little attention was paid to the events that led to the consequences we are now witnessing, which of course is due either to neglect or to a lack of understanding and appreciation of the changes which took place within the agricultural sector, and the interrelationships between such changes and the other sectors of the economy. The problem of agriculture and agricultural policy in Nigeria, therefore, is one of understanding the sector in all its ramifications, especially with respect to different actors involved in the food system (i.e. state, producers and consumers), and, based on this knowledge, to formulate appropriate and workable policies to enhance its growth.

BACKGROUND AND BASIC OBJECTIVES OF STUDY

In spite of growing urbanisation and increased revenue from the oil sector, agriculture is the mainstay of the Nigerian economy. Historical experiences have shown that there are no cases of successful development of a major country in which the rise in agricultural productivity did not precede or accompany (industrial) development. Nigeria is predominantly rural in character with about 75 percent of the population directly or indirectly dependent on agriculture. The land area is very diversified reflecting various ecological conditions. Most of the farmers are subsistence smallholders farming 1-2 hectares with outdated levels of technology; consequently, the productivity level is low. The intensity of exploitation of the existing arable land is estimated at about 30-35 percent.

In the past, the traditional farmers have been able to feed the country but in recent times, their output has been insufficient to satisfy levels of demand generated by the rapidly growing population, increases in real per capita income and rapid urbanisation.

Closer attention to the farm families and the farm sector is now required. Inculcation of modern farming methods to farmers in addition to workable policies must be pursued vigorously because experience has demonstrated that farmers are usually willing to accept and adopt new technologies if they are profitable and socially acceptable and if they are provided with other necessary incentives to facilitate their adoption. For national development, farm families constitute, by their number, a vast potential and while many governments recognise this, they often do very little to utilise them. This may be due in part to limited resources, lack of knowledge of how to reach them and the political implications of such mobilisation. However, it is important to give the utmost attention to this aspect because rural families make up a large portion of the population and their farms represent an important part of the nation's assets.

For agricultural and food production, the country has abundant resources that if properly harnessed are capable of producing a wide variety of food for local consumption. These have however been under-utilised. On this Aribisala noted that:

> Out of the country's total land area of 98.3 million hectares, the cultivable land is about 71.2 million hectares or 70 percent. Out of this, only about 34 million is under cultivation. Thus, about one-third of the country's total land area and about 50 percent of the total cultivable land is at present utilised.[2]

Inspite of the very good potential for increasing agricultural production by intensive methods of cultivation, Nigeria has failed to produce enough food. Indications of this are: high food prices and shortages of domestic supplies. This failure has led to massive food imports. The issue in Nigeria's food problem is not that of domestic production alone, but also that of demand as the pattern of consumption is changing constantly due to rising real income leading to importation of more expensive grains such as wheat and rice. Evidence of food importation into Nigeria is reflected in the 40-fold increase in the food import bill between 1970 and 1981 (Table 13.1).

The oil sector and the attendant oil boom had some impact on the agriculture and food sector. Due to increased revenue from oil, the Nigerian economy has grown at a fast rate and has also increased the demand for food. In addition the oil economy has had an adverse effect on the agricultural sector. This phenomenon has caused many farmers to desert their farms for the urban areas in search of easy and quick profit, thus leaving mainly older people on the farms. Migration has also deprived the rural areas of farm labour. Consequently, the number of urban dwellers dependent on the market for their food has increased, which implies that existing farmers must produce more and proper arrangements need to be made to transport food to the cities. The agricultural sector in Nigeria is a victim of neglect and poor policy formulation. In this regard Aribisala warned that:

> all is not well with our agriculture and if we allow things to continue as they are we shall be worsening our present food problem. We must no longer be complacent with the present one commodity economy, based on availability of fossil fuel which in the past years made our economy buoyant, while our agriculture is not producing enough food and other raw materials.[3]

TABLE 13.1
Food Importation in Nigeria 1962-82

Year	Value (N million)	Annual percentage change	Food inputs as a percentage of total imports
1962	46.98		11.6
1963	43.80	- 6.7	10.6
1964	41.24	- 5.8	8.2
1965	46.07	11.7	8.3
1966	51.57	11.9	10.0
1967	42.56	- 17.5	9.5
1968	28.39	- 33.3	7.4
1969	41.73	46.9	8.4
1970	57.69	38.2	7.6
1971	87.91	52.4	8.1
1972	95.10	8.2	9.6
1973	126.26	32.8	10.3
1974	155.71	23.3	8.9
1975	277.86	78.4	8.0
1976	438.93	57.9	8.5
1977	702.01	59.9	10.4
1978	1,108.66	57.9	12.4
1979	1,105.90	- 2.49	10.3
1980	1,437.50	29.9	11.3
1981	2,115.10	47.1	
1982	950.60	- 55.1	

Source: "Character of Nigerian Agriculture", Awoyemi O., in Bulletin, vol.VI, no. 4, Central Bank of Nigeria.

There are several factors/constraints responsible for the poor rate of growth of Nigerian agriculture. These factors include: a) high cost of land clearing; b) land tenure problems which make land acquisition for large-scale farming difficult and expensive; c) lack of suitable credit facilities; d) lack of infrastructural facilities; e) poor inter- and intra-state food trade; f) political interference in the execution of projects; and g) maintenance of soil fertility and control of pests and diseases.

The problems listed above are those that policy measures are supposed to address and eliminate.

Objectives

The overall objective of this chapter therefore is to look at state policies (programmes and projects) and agriculture/food production in Nigeria and the effects or consequences of these policies, programmes and projects for agricultural and food production and overall national development.

The term *policy* is used in this paper, as defined by Wells,[4] to include the entire range of government activities directed at the agricultural sector. Expanding this definition, Bates defines *food policy* as those choices made by government, which affect prices in the market which determine the real income of farmers : the market for the commodities they produce, the inputs they employ in farming and the goods they purchase from urban manufacturers. *By and large we can conclude that government activities and actions towards the agricultural sector are largely responsible for the success or failure of the sector.*

A REVIEW OF AGRICULTURAL AND FOOD POLICIES

Background and Principal Objectives of Agricultural Policies

The agricultural development concept adopted by the government attempts to replace the past uncoordinated method with an integrated approach to the problem of agricultural development. This approach takes into consideration the realities of the existing situation—social, economic and institutional—which hitherto have inhibited a rapid rate of agricultural development and food production.

Unlike the observation made by Bates[5] that food policy in Africa is a derived policy in the sense that choices made with respect to food production are to a high degree made to serve the interests of groups other than producers, the philosophy of the integrated approach emphasises the organisation of and assistance to smallholder farmers as the centre-piece of development activities. In this regard, therefore, small farmers are mobilised to achieve the aims of increased productivity, area expansion and higher incomes.

The approach involves a number of inter-related actions. First, it identifies the social, economic and institutional needs of the agricultural sector. Second, the formulation of policies is based on the realities so identified and the design of strategies by which policy goals can be achieved. Last, the strategies are translated into programmes of action and projects which are executed and monitored. The principles guiding the programme are (1) that of food self-sufficiency, (2) realisation that people are the essence of development and (3) that small-scale producers are the bedrock of development programmes. In addition, provision of basic needs and infrastructures is essential for improved productivity and higher standards of living.

The objectives of the programme can therefore be summarised as: (1) self-sufficiency in food and fibre production; (2) reduction in the rise of food prices; (3) diversification of the country's sources of foreign earnings by the rejuvenation of agricultural exports; and (4) production of raw materials for local agro-based industries.

AGRICULTURAL POLICIES AND PROGRAMMES IN RETROSPECT

There have been a number of government policies and programmes on agricultural development during the last 15 years. Some of the policies were aimed at the restructuring of the marketing board systems for export crops, the creation of marketing boards for grains and root crops and the use of various price incentives and input policies to promote agricultural production. Some of the programmes referred to above include:

(1) National Accelerated Food Production Programme (NAFPP). The programme started in 1972. The primary objective is to raise farmers' income, accelerate the rate of diffusion of new agricultural technology and serve as a medium for testing and adopting agricultural research findings to on-farm conditions. The cooperating institutes involved include the federal and state governments, selected research institutes, extension service experts and private farmers. The crops emphasised are grains and cassava. In evaluating the programme, Sano noted that:

> the programme met with all the problems a rural development project can encounter: lack of supervision by the responsible authorities, shortage of competent staff, late arrival of inputs, reluctance on the part of peasants to adopt the recommended packages and logistical problems.[6]

(2) Agricultural Development Projects (ADPs). ADPs were very similar to NAFPP projects in their general approach but more ambitious in scope. The ADPs began in 1975 and they are partly being financed by the World Bank. The programme calls for the establishment of a number of agricultural development projects in various parts of the country. The objectives of the projects are to promote integrated rural development by providing facilities for intensive extension services, modern input supplies and distribution system and rural infrastructure especially feeder roads. The projects are expected to function in all the states of the country. There are at present 10 ADPs in 10 states in Nigeria.

A related or an extension of the programme is the Agricultural

Development Area (ADA). It applies the concepts of the ADP but adopts the Local Government Area as the basic development unit. The World Bank, because of its involvement, has fairly close control over the selection of personnel and the administration of the projects.

(3) Livestock Development Projects. The programme began in 1976. The objective was to commercialise beef production by the establishment of large-scale public breeding ranches, small-scale private ranches, development of grazing reserves and the provision of supervised credit for smallholder fattening schemes.

(4) Operation Feed the Nation (OFN). The programme started in 1976—the year of massive food importation—as a nationwide campaign to mobilise the general public and create awareness for agricultural pursuits. The objective of the programme was twofold. According to Sano[7] the first objective was to make the urban working and middle classes aware that something was being done to improve the food situation, and also to restore the respectability of farming to the peasant communities in order to make the rural youth abstain from migration into the towns. The urban dwellers were thus encouraged to take up backyard farming. Second, the Operation aimed at raising production and productivity by distributing fertiliser and seeds. As a rural development strategy the programme was a failure and was abandoned after a few years. Some of the problems that confronted the Operation were: bottlenecks in fertiliser distribution, inadequate extension services and insufficient planning.

(5) Agricultural Credit Guarantee Scheme. The scheme was created under the aegis of the Central Bank in 1977 in an attempt to mobilise private bank capital for rural producers. The scheme guaranteed loans from the private banks to peasants, while the banks were required to establish a certain number of branches in rural areas and commit a minimum of six percent of their loans for agricultural purposes.

(6) Nigerian Agricultural and Cooperative Bank (NACB). The NACB was created in 1973 with a few branches in state capitals and with rather strict procedures for the granting of loans. The bank caters primarily for agro-industrial firms and big farmers who are able to provide collateral against the credit obtained. In order to liberalise the strict lending policy of NACB a smallholder Direct Loan Scheme was introduced early in 1981. In this programme the federal government guarantees the loans to the peasants.

(7) Land Use Decree. The decree seeks to bring the existing systems of land tenure under one common law. The issue of land tenure had been a sensitive issue and the traditional land tenure systems had been a constraint on the effective exploitation of the land resources for agricultural purpose. This point was emphasised by the government in the *Guidelines for the Fourth National Development Plan 1981-85.* It states that:

> The land tenure system has long been a bottleneck in the establishment of large-scale farms by private operators. With the establishment of the recent Land Use Decree... private sector involvement in large-scale agriculture activities should receive a boost during the next plan period... The reform should promote better security of tenure and also encourage consolidation of holdings and large-scale operation. It should be easier to attract foreign entrepreneurs and foreign capital into agriculture.

Thus the decree can be considered as an effort introduced to pave the way for a new food and agricultural policy. Its objectives as pointed out by Sano[8] were to make land transferable and available for expropriation in order that it could be used for the flexible purpose of a dynamic agricultural policy.

(8) Green Revolution. Launched in 1980, the Green Revolution is intended to replace Operation Feed the Nation (OFN) thereby removing the deficiencies in the earlier programme. In essence it was designed to meet the needs of small farmers and spread the benefits of rural development to the local level. Presently most programmes, such as FAPPs and ADPs, operate under the Green Revolution programmes.

In retrospect one can conclude that despite the various programmes introduced the performance of Nigeria's agriculture has not been encouraging, especially since 1970. This discouraging conclusion is based on the following points: a) the share of agriculture in GDP declined from 61 percent in 1964 to 18 percent in 1980; b) the share of export earnings attributed to agriculture declined from about 71 percent in 1964 to less than 5 percent in 1979; c) the share of food and live animals in the total import bill rose from about 8 percent in 1964 to 11.3 percent in 1980; and d) a widening domestic food supply-demand gap brought about a relatively high rate of growth of 3.5 percent per annum in demand as against only 1.5 percent per annum in production thus leading to a high rate of increase in domestic food prices.

The Importance of Agricultural Policy

The importance of agriculture (and general development policy in Africa) needs underscoring because most countries are still basically agricultural. Also, agriculture is the sole activity for subsistence living for millions of farmers and people in the rural areas. People in the urban areas also depend on agriculture for food and employment. If, therefore, agriculture fails, the impact can be catastrophic.

The main reason for the critical and poor contribution of the agricultural sector is the failure of government and government policies to come to grips with the problems of the rural areas. The background to this problem is therefore one of many interdependent problems i.e. social and economic development in the farming communities, the relative emphasis devoted to agriculture and industry and the relationships between rural population and technology.

The element of policy as suggested by Wells[9] involves two related but separable components, namely: a) activities which directly effect the allocation of resources; and b) those which deal with the institutional framework within which resource allocation takes place.

The farmers are associated with the problem of allocating specific resources to and within the agricultural sector, while the latter will include the range of activities usually grouped under the heading of "policy". The latter, however, include items such as land tenure, marketing and price policy. The determination of both allocative and institutional policies requires the full range of activities associated with government policy-making. Goals must be obtained and options defined, evaluated and selected. Having made the decision, it must be implemented via the use of appropriate means.

In this case, the instrument for allocative policy always involves direct budgetary allocation while with the institutional policy variable instruments are more diffused. They may involve laws and administrative rules.

In Nigeria, allocative policy has been known to be predominant especially in the agricultural sector. The index of the importance of agriculture has usually been the absolute amount of allocation made to the sector which has been increasing yearly without explicit setting targets against which the increased allocation will be judged.

Reasons for the primacy of allocative policy have been offered by Wells[10] to include: a tendency to envisage planning as essentially involving the direct allocation of capital resource and the notion that variables involved in allocative policy are quantifiable.

Agricultural Policies and Policy Measures

There are several ways of stimulating agricultural production. These different ways may be categorised into two groups: indirect stimulation, and direct stimulation

In Nigeria, indirect stimulation policies include the following: credit provision; input price subsidisation; farm input supply; mechanisation (including irrigation); agricultural product pricing and marketing; agricultural extension; agricultural processing; and research and development.

Policies for direct stimulation are those related to large-scale capital-intensive production often financed and managed by government. They include: state farms (food production companies and agricultural corporations) and farm settlement programmes.

Indirect Stimulation Policies

Agricultural Credit. Because of the subsistence nature of agriculture, farmers' potential for capital formation is severely constrained and savings from farm sources can only contribute to the expansion of farm capital at a rate far below what is required for raising farm productivity and income. This illustrates the vicious circle phenomenon often present in subsistence agriculture. Therefore farmers must look beyond farm sources for a sizeable portion of the capital needed for development.

According to the Federal Ministry of Agriculture,[11] the need for increased supply of capital is based on the following reasons: capital requirements of farmers are increased due to scale expansion and the adoption of innovation; because of the subsistence nature of most farms, little surplus is generated from which little savings for investment could be made; and because of the seasonal nature of farm production, a need for short-to-medium-term farm loans is necessary for financing production activities.

Attempts at obtaining credits at the informal sector level have not been encouraging due to its limited scope, irregularities and high interest rates. Equally, the commercial banks are not forthcoming in that they restrict the flow of loanable funds to agriculture because of the present low capital productivity in farming, lack of adequate security and perhaps lack of appreciation of the potential in agriculture. The agricultural cooperative banks however grant loans to a small percentage of farmers through cooperative societies but since there are only few cooperative members, only a few farmers are affected. To overcome this problem, a series of actions were taken including the following: that six percent of commercial banks' loanable funds should go into agriculture; the establishment of Nigerian Agricultural and Cooperative Bank (NACB); and the establishment of a Credit Guarantee Scheme.

Despite all these efforts, the flow of capital to agriculture has not increased as significantly as one would expect.

Input supply policies. Increased farm output can only be achieved by the adoption and use of modern farm inputs. However, as a result of structural deficiencies in input procurement, distribution and the pricing system, input prices tend to increase faster than the farm gate output prices thereby disadvantaging the farmers. Also, the rate of adoption has been severely limited by the delivery and pricing mechanism. Government in its efforts to improve the situation has introduced policy measures with the following aims: to encourage large-scale adoption of improved inputs by reducing the cost of inputs through subsidy; to ensure timely procurement and delivery of farm input in adequate quantity and at the appropriate time; and to eliminate intermediaries.

Input subsides. Input subsidies which have been used in the past for the promotion of export crops have now been extended to food crops with the following objectives: to introduce farmers to modern farming technology; to increase farmers' profit by lowering input cost; and to serve as an incentive for inducing new entrants into farming.

Lack of proper coordination in the past with regard to the kinds of inputs to subsidise and the level of subsidy has led to significant differentials in prices of farm inputs and has also led to substantial movements of farm inputs across state borders to neighbouring countries.

Mechanisation policy. Farm labour now constitutes the most severe constraint in Nigerian agriculture as reflected in the high share of labour in the total cost of production. The genesis of this problem is the neglect of the rural areas leading to rapid rural-urban migration. If one remembers that the predominant source of farm power is supplied by human muscle, any expected increase in output will definitely compel us to substitute an appropriate form of mechanical power for human muscle.

In this effort government has formulated strategies aimed at improved farm tools adapted to different ecological zones.

Also, government in its effort to improve the situation created institutions such as the Ministry of Science and Technology, and Universities of Science and Technology.

Although various ministries of agriculture perform land clearing activities to facilitate tractorisation and expansion of the area under cultivation, it is well known that the resources available to this programme are not sufficient.

Agricultural pricing and marketing policies. In an attempt to close the food supply and demand gap in Nigeria, the price mechanism has been employed essentially to focus the acceleration of output. It is a universally accepted fact that farmers do respond to price incentives by using currently available resources more intensively as well as accepting improved inputs and production practices. Price manipulation is, however, a tricky business in agricultural policy formulation especially in developing countries. The complexities involved in price policy are illustrated by Mellor.[12] For example, if you raise the price of rice you are shifting resources from something else to rice production. What is the employment effect in rice and the commodity that is being shifted away from? What are the income distribution implications? If you raise the price of a commodity with regional specialisation, will resources move to this region? What sort of regional problems does this raise?

The intended positive effects of price dynamics can be easily nullified by the influence of non-price factors. Hence price incentives can only be effective if the necessary farm supplies, equipment and marketing channels are available and adequate. Pricing manipulation has not been the panacea government expected it to be, as shown by the current high food prices (see Table 13.2). Supply response to high food prices has been low.

The point is that any price policy designed to increase supply will only be meaningful provided the necessary conditions for achieving the increased supply are present. Also, higher prices which should normally lead to greater input usage is not happening in Nigeria because demand for such inputs is already constrained by inadequate supply and an ineffective distribution system.

Agricultural research policy. The importance of searching for new ways of improving agricultural productivity cannot be over-emphasised. The high rates of return to agricultural research have been clearly substantiated in many parts of the world. Agricultural research in Nigeria is not new but the various research institutes have been a *qualified success*. In his study Idachaba[13] itemised reasons for ineffective agricultural research in Nigeria to include: inadequate research findings; inadequacy of research staff; research staff instability; lack of research material; and lack of effective system for delivering research results.

To make agricultural research purposeful, the government has established a Ministry of Science and Technology with eighteen research institutes operating under the umbrella of this ministry. It is hoped that with proper coordination useful results will emerge.

TABLE 13.2

Nigeria Consumer Price Index*

All Items and Food 1960-1977 (1960=100)

Year	All items	Food
1960	100.0	100.0
1961	106.4	109.8
1962	112.0	118.0
1963	108.9	106.6
1964	110.1	105.7
1965	114.4	110.5
1966	125.5	133.1
1967	120.3	119.3
1968	120.3	112.6
1969	132.3	133.9
1970	150.6	164.5
1971	174.7	211.4
1972	179.6	216.6
1973	189.3	223.6
1974	214.6	258.9
1975	285.4	367.7
1976	348.1	464.7
1977	425.1	592.2

* For low income urban group.

Source: Sano, *The Political Economy of Food in Nigeria 1960-82.*

As observed by Abalu,[14] the efficiency of agricultural production in any country is a reflection of the level of technology in it. The low rate of production efficiency suggests that the efficiency with which scarce agricultural resources are converted to food has been inadequate.

Extension policy. In Nigeria, agricultural extension is essentially the responsibility of state and local governments. To this extent, there are various standards of extension services depending on the relative emphasis placed on it by respective governments. The federal government is, however, involved in the provision of infrastructural facilities to strengthen the extension services.

In some states, extension is in a state of paralysis or extinction. The ratio of farmers to extension workers is very high. To revamp the extension services certain actions are needed and these include: increasing the number of extension workers by formulating attractive conditions of service; constant training to improve the effectiveness of extension workers; and provision of adequate working materials especially transportation for effective dissemination of new ideas.

Direct Stimulation

The direct stimulation of agriculture in Nigeria has been limited to the establishment of corporate farms, agricultural companies and large-scale plantations (including state farms and farm settlements).

This approach was adopted based on the assumption that large-scale farms are associated with the following attributes:[15] all year-round employment of labour at wage rates comparable with those of the urban sector; considerable scope for the use of modern inputs which will reduce the drudgery of farming and make agriculture attractive to school leavers; higher productivity arising from economies of scale; and limited absorptive capacity of the smallholder system of agriculture to provide sustained growth in output.

In spite of the huge investment in large-scale farming, over 90 percent of the total agricultural output still comes from the smallholder farmers. Recently, a critical review of this approach revealed the failure of the public sector programme of large-scale production because such programmes are beset by bureaucracy, high overhead costs and management problems.

Government has now realised that large-scale production in which the public sector is involved may not be feasible in the present setting in Nigeria for the obvious reason that government has limited resources which must be shared among all sectors. Therefore it is only logical and expedient that government should desist from this approach and assist small farmers in order to maximise their production.

ANALYSIS

The main policy of the government for the agricultural sector, is manifested in at least two programmes being pursued. They are: irrigation (River Basin Development Authorities—(RBDAs) and Agricultural Development Projects (ADPs)

Government has devoted considerable resources and manpower to these programmes, and all other policies are supposed to function under the canopy of these two programmes. We shall now analyse the implications of these two programmes with respect to the following objectives: self-sufficiency in food production; rural development and reduction of inequality (class content/bias of the programme); dependency on international agencies for the achievement of the goals of the programmes; and environmental and long-term implications.

TABLE 13.3
Agricultural and Petroleum Exports 1960-80

Year	Agricultural Exports			Petroleum Exports		
	Quantity ('000 tonnes)	Value (million)	Share of total exports (%)	Quantity ('000 tonnes)	Value (million)	Total exports (%)
1960	1,218.0	251.0	75.8	5.2	8.8	2.7
1965	1,506	289.0	54.9	81.9	136.2	25.9
1970	799.1	253.4	28.6	324.5	570.0	57.6
1971	785.4	232.8	18.0	451.7	953.0	73.7
1972	629.6	155.0	10.8	541.2	1,176.2	82.0
1973	718.5	224.5	9.9	594.1	1,893.5	83.1
1974	494.7	254.1	4.4	607.3	5,365.7	92.6
1975	417.7	216.1	4.4	538.7	4,563.1	92.7
1976	529.8	261.0	3.9	604.2	6,321.7	93.7
1977	391.2	364.9	4.8	741.4	7,072.8	92.7
1978	285.8	408.5	6.7	604.0	5,401.6	89.1
1979	305.5	459.8	4.2	803.4	10,166.8	93.8
1980	240.1	337.7	2.4	729.5	13,523.0	96.1

Source: Sano, *The Political Economy of Food in Nigeria 1960-1982.*

Self-sufficiency

Crops conducive to irrigation are wheat, rice and sugar. These commodities ranked very high in the list of imported items and they have a large import-substitution potential. The two programmes have a potential to enhance the growth of production and self-sufficiency in the medium- to long-term range, if properly managed, but not in the short-term as envisaged by Nigerian planners.

There were during 1982-83 some problems in the operation of the 11 River Basin Development Authorities (RBDAs) which caused a slowdown in their activities. According to the Central Bank *Report*,[16] there were a widespread reports of financial mismanagement. Other problems are the scarcity of essential farm inputs, especially machinery spare parts, and hostility from rural farmers to the acquisition of their land as a result of the alleged inefficient system of paying compensation for such lands. Available information shows (Table 13.4) that the total area developed and irrigated by the RBDAs fell by 34.9 and 47.2 percent respectively. The number of dams built and roads constructed also declined. The effect of the slowdown finally affected the level of agricultural production.

TABLE 13.4

**Operation of the River Basin
Development Authorities 1982-83**

Item	1982	1983	Percentage change (%)
1. Finance (N million)			
a) Capital allowance	429.6	623.2	45.1
b) Actual exportation	287.4	317.6	10.5
2. Land area developed			
('000 hectares)	108.6	70.0	−34.9
a) Under irrigation	14.4	7.0	−47.2
b) Other land	94.2	63.1	−33.0
3. Number of farmers resettled	3,992	8,518	113.4
4. Production			
a) Grains ('000 tonnes)	346	287.7	−16.8
b) Fish (tonnes)	890.2	1,983.1	122.8
c) Poultry (No.)			
i) Layers	62,684	82,722	32.0
ii) Broilers	69,070	70,218	1.73
iii) Turkeys	7,140	1,407	−80.3

Source: Central Bank of Nigeria, *Annual Report and Statement of Accounts*, December 1983.

Rice is a complex commodity in Nigeria. Though local production has increased tremendously, importation has also increased simultaneously. From Table 13.5, importation of rice increased from 65,400 tonnes in 1976 to about 245,000 tonnes in 1980. During the same period local production increased from 534,000 tonnes in 1977 to 850,000 tonnes in 1980.

Wheat may face more problems as present yields and production are very low.

Sugar importation increased from 114,800 tonnes to 500,000 tonnes between 1975 and 1980. However, there are at least two local production plants in Nigeria, viz : Bacita (Kwara State), and Savannah (Gongola State). These projects however have run into some difficulties which were mainly managerial and organisational involving problems in coordinating large projects. The detailed documentation of the problems involved is found elsewhere. However, the prospects of achieving self-sufficiency in wheat and sugar are still small.

The ADPs are specifically addressed to the production of staple foods. The rate of growth of food production has fluctuated widely in recent years. This performance cannot be attributed to the poor performance of the ADPs.

TABLE 13.5

Production and Import of Principal Food Commodities
1970-80 (In '000 Metric Tonnes)

Year	Production	Import	Rice Production	Import	Sugar Production	Import
1970	19	267.1	345.0	1.7	27	94.42
1971	20	405.1	383.0	0.3	25	150.4
1972	20	316.9	447.0	5.8	28	127.4
1973	15	454.7	487.0	1.1	33	136.9
1974	18	325.4	525.0	4.8	33	72.3
1975	18	407.6	515.0	1.0	35	114.8
1976	20	735.5	534.0	65.4	38	178.1
1977	21	769.7	667.0	427.4	40	358.1
1978	21	1,363.3	695.0	563.83		593.5
1979		955.6	850.0	245.0		535.9
1980		1,176.4		387.0		500.0

Notes:
1. Wheat is wheat and flour in wheat equivalent.
   Sugar is in raw and refined in raw equivalent.
2. Rice production figures are from Aribisala, "Nigeria's Green Revolution", op. cit., p. 8.
3. *Sources:* FAO, *Trade Yearbooks;* FAO, *Production Yearbooks.*

All the ADPs stepped up their rural development programmes in 1982-83. These programmes include an increase in bore-holes, the number of farming centres and the volume of agricultural inputs distributed to farmers. Similarly output from all ADPs increased substantially. This was however due largely to an appreciable increase in the total acreage (Table 13.6).

The impact of the ADPs on the production of staple foods especially in the northern states has been encouraging but not without problems. However, in their evaluation the World Bank has claimed production gains of more than 5 percent annually for maize, millet and sorghum.

Inequality and Rural Development

The growing differentiation among the rural people as a result of programmes and projects to increase food production is becoming a major problem. In its original conception, the World Bank ADPs projects emphasised the "smallholder" or the "rural poor" as the principal beneficiaries of the rural development projects but in implementation, it appears that this original intention is being subordinated to that of efficiency and quick production results. In implementing the

TABLE 13.6
Operation of Integrated Agricultural
Development Projects 1982-83

	1982	1983	Percentage change (%)
1. No. operational	9	10	11.1
2. Capital expenditure			
(N Million)	604	38.0	−93.7
a) Federal	183.4	9.6	−94.8
b) States	223.0	11.3	−94.7
c) IBRD	197.6	17.1	−91.3
3. Infrastructure provided			
a) Roads (km)	4,112	1,624	−60.5
b) Dams (No)	33	21	36.4
c) Boreholes (No)	360	718	99.4
d) Farm Services Centres (No)	128	797	522.7
4. Farm inputs supplied			
a) Fertiliser (tonnes)	45,054	68,292	51.6
b) Seeds (tonnes)	2,162	3,331	54.1
c) Tractors (No)	40	-	
5. Production ('000 tonnes)			
a) Rice	103.7	9.4	−90.9
b) Beans	61.3	29.2	−52.4
c) Guinea corn	519.2	798.1	53.8
d) Millet	344	681.0	98.0
e) Maize	201.1	270.6	34.6
f) Yam	1,327.7	250.0	−81.0
g) Cassava	425.9	55.9	−87.0
h) Cotton	38.0	70.6	75.8

Note: 1983 data are provisional.
Source: Central Bank of Nigeria, *Annual Report and Statement of Account*,
December 1983.

programmes, differentiation has been observed when some farmers are selected as the primary target group and accorded preferential treatment. The so-called progressive farmers receive special attention in the distribution of inputs.

The World Bank is aware of this and has tried to explain it in terms of "trickle-down" effects, ability to take risks and innovativeness. Though the overall goal is the production of food it may be necessary to subordinate the class bias part of the programme to the overall goal.

The socio-economic implications of the irrigation projects may be more serious than those of the ADPs. Though no up-to-date reliable information is available, common knowledge however indicates that irrigation projects involve large-scale alteration of the utilisation of land and technology and disruption of the existing systems of land use. In addition, most farmers in some project areas have not been adequately compensated or resettled.

Hence, introduction of irrigation projects has thus led to the creation of new social relationships and a much more radical transformation of the rural society than the ADPs. The socio-economic implications of such projects are crucial for overall development.

The Land Use Decree[17] of 1978 is the current principal instrument of allocating land in the country. Much has been said by various interest groups on the merits and demerits of the decree. The decree vested all right to land in the state and the individuals had only occupancy rights over the land while the state retained ultimate control and formal ownership. This therefore enables the state to control distribution more easily and base it on need.

The objectives of the decree can be identified as: the creation of large-scale farms; the abolition of Communal Land; and the introduction of state control. Hence the decree can be useful if properly applied to all dynamic food and agricultural policies.

The emphasis of the decree was on use rather than ownership. The social and economic implications of this decree are many. However, since the decree came into effect in 1978 it has not been vigorously analysed or appraised with respect to its social, economic and political implications.

However, Okpala,[18] in evaluating the decree, contended that the decree is revolutionary only in scope because state lands have always existed and government has the power of compulsory acquisition of land. Government in fact has exercised this right on many occasions. The power of government to acquire and distribute land was formalised and wider in scope with the promulgation of the decree. Okpala then looked at the degree of fairness, selflessness and responsibility with which public lands were administered before the advent of the decree. He however concluded that public ownership and control of land as embodied in the Land Use Decree does not necessarily (by itself) ensure the achievement of those objectives envisaged by the decree. If anything, previous experiences show that the reverse can be the case. Realisation of the potential of the measure would depend more on the degree of fairness, integrity and equity involved in the administration or implementation of the decree, than on the fact of the decree as such.

Public ownership may not imply land policies which are more productive and egalitarian than the traditional system of land ownership and control. If human tendencies towards inequality can be restrained, the decree's objectives are wide enough that when properly implemented they can help in the achievement of positive social and economic goals.

Dependency

The indication of dependency was apparent during the 1970s when Nigeria was dependent on oil export and food imports. The present food policies might be able to reduce Nigeria's food imports in the near future. The paradox, however, is that they are likely to increase Nigeria's dependency on manufactured inputs and knowhow. Both the ADPs and the irrigation projects have a high import content hence we are likely to move from food dependency to technological dependency and, even if government decides to establish local manufacturing plant, the issue of technological dependency will still be very much around.

Environmental Implications

No doubt, the introduction of large-scale irrigation and new technology has environmental implications. Large-scale farming still requires proper management and continuous monitoring to mitigate the environmental impact. New seeds will increase the vulnerability of the cultivation system because they can be susceptible to disease and pests as they are not adapted to the environment in which they are introduced.

FOREIGN INVESTMENT IN NIGERIAN AGRICULTURE

Because of the interdependency of the world economic system, one country is necessarily affected or influenced by investment from other countries. In the case of Nigeria, especially in the field of agriculture and related activities, foreign influences have a considerable impact. Foreign investments in Nigerian agriculture are diverse but they are mostly in the area of agricultural inputs such as fertiliser plants and agricultural machines and tractors in joint ventures between the government and private overseas firms.

Other types of arrangements exist like government-to-government programmes, for example, the Joint Agricultural Consultative Committee (JACC) between the Nigerian and US governments. This venture is supposed to be wide-ranging in activities including animal feeds, poultry, piggery and soya bean farming.

The British, Americans, Scandinavians and Brazilians are interested in different aspects of Nigerian agriculture. This may not exhaust the list of foreign interests' involvement in Nigerian agriculture, but the list nonetheless shows that an attempt to increase food production may involve a considerable level of imports of other countries' skills and technology thereby involving foreign intervention and investment.

Apart from the government programmes, the multinationals are also involved in input supplies such as chemicals, seeds and other farm equipment. These multinationals are very important in the control and transfer of technology to developing nations.

To promote large-scale agricultural production, the government has provided a package of incentives with the hope of encouraging foreign investors to participate in direct agricultural production. These incentives include: income tax relief for pioneer enterprises; duty-free importation of farm machinery and raw materials for the manufacturing of livestock feed; five years' tax holiday for investment in agriculture and transfer of integrated agricultural production and processing from schedule II to III of the Enterprise Promotion Act, thus enabling aliens to own up to 60 percent of the equity of such enterprises.

Despite the generous incentives, foreign investors have not been involved in large-scale agricultural production as envisaged. Even those foreign companies already in Nigeria have similarly not engaged in food production even though such production can be useful as raw materials for their final products.

Of recent, a few Nigerian private enterpreneurs have taken up large-scale farming for reasons which may not be unconnected with the recent food shortage, but the number of such entrepreneurs is small. It is important to realise that the incentives to lure aliens into such a basic industry as agricultural and food production may be fraught with future potential problems. Examples exist, especially in Central and Latin America, of situations where such companies became too interwoven with national development to the extent that they acquired political leverage and thereby could dictate the direction of national development.

Our choice at present therefore lies in the direction of encouraging Nigerian citizens with resources to engage in large-scale production. In the conclusion of a seminar held at Ahmadu Bello University, the conference warned that it was opposed to the recolonisation of the country by allowing foreign companies and enterpreneurs to engage in direct and indirect agricultural production. It concluded that the country has the necessary manpower and resources to embark on massive food production.[19]

SUMMARY AND CONCLUSION

The objective of this paper was to review and analyse the food policies pursued by government in Nigeria in order to solve principally the current food problems and other allied social problems of equity and provision of basic needs.

The author is of the opinion that Nigeria has in the past, and most especially since the 1970s, devoted considerable manpower and resources to the agricultural and rural sector but the record so far has not been very encouraging. One reason may be that perhaps we have reacted more to the symptoms rather than to the causes of the problems. There are several reasons given as being responsible for poor agricultural performance in Nigeria, ranging from the early neglect of food crops in favour of export crops to the Sahelian drought, the civil war, lack of appropriate technology, and basic social and physical infrastructures in the rural areas. But as earlier mentioned, the motive force behind the present crisis can be traced to the transformation which the country went through in the 1970s manifested in the oil boom: rapid but disproportionate development, higher real per capita income, changing patterns of food demand and consumption, rural-urban migration and population growth. This rapid development reveals the structural nature of the crisis thereby bringing into focus, and perhaps for the first time, the relationships between economic growth and transformation and falling levels of food production.

The prospects for food self-sufficiency, reduction in inequality, dependency on foreign technology and long-term environmental impacts were viewed under the two major programmes currently going on in the country: River Basin Development Authorities (RBDAs) and Agricultural Development Programmes (ADPs). The author opined that these programmes will be able to achieve food self-sufficiency in the longer time period as opposed to the time period of 1985 for food crops as envisaged by Nigerian policy-makers. With respect to the reduction in inequality, the ADPs are better suited to achieve this aim than the RBDAs, but both programmes have high import contents.

Finally, the currently available agricultural projects, programmes and infrastructures must be consolidated, monitored and improved upon. This cannot be achieved unless the country improves on its data-gathering methods and techniques. Good policy formation is dependent on relatively reliable information.

NOTES

1. OAU, *Lagos Plan of Action for the Economic Development of Africa 1980-2000*, 1981, p.11.
2. Aribisala, T.S.B., "Nigeria's Green Revolution Achievement, Problems and Prospects", NISER's Distinguished Lecture series, 1983, p. 6.
3. Ibid., p. 9.
4. Wells, J., *Agricultural Policy and Economic Growth in Nigeria*, Oxford University Press, 1974, p.107.
5. Bates, R.H., "Food Policy in Africa, Political Causes and Social Effects", *Food Policy*, Vol. 6, No. 3, Aug. 1981, p. 147.
6. Sano, H. O., *The Political Economy of Food in Nigeria 1960-82. A Discussion on Peasants, State and World Economy*, Research Report No 65, Scandinavian Institute of African Studies, Uppsala, 1983, p. 27.
7. Ibid., p. 3.
8. Ibid., p. 56.
9. Wells, op.cit., p. 107.
10. Ibid., p. 105.
11. Federal Ministry of Agriculture, *Information Bulletin on Nigerian Agriculture*, Lagos, 1984, p. 23.
12. Meller, J.W., "African Food Policy in a Global Context", paper presented at symposium on "*Food Problems in Africa*". University of Illinois, Urbana-Champaign, 1981, p.181.
13. Idachaba, F.S., "Agricultural Research Policy in Nigeria", Report No. 17, 1980, IFPRI, p. 9.
14. Abalu, G.O.I., "Solving Africa's Food Problems", *Food Policy*, Vol. 7, No. 3, Aug. 1983, p. 253.
15. Federal Ministry of Agriculture, *Information Bulletin on Nigerian Agriculture*, Lagos, 1984, p. 21.
16. Abalu, G.O.I., "Summary of Processing and Recommendations on the First National Seminar on the Green Revolution in Nigeria", *Nigerian Journal of Political Economy*, Vol. 1, No. 1, April 1983, p. 19.
17. Federal Republic of Nigeria, Decree No. 6, Land Use Decree, 1978, supplement to *Official Gazette Extraordinary*, Vol. 65 of 27 March 1978.
18. Okpala, D., "Land Use Decree of 1978. If the past should be Prologue", *Journal of Administration Overseas*, Vol. XVII, No.1, 1979, p.2.
19. Abalu, "Summary of Processing and Recommendations", op. cit., p. 101.

INDEX